Glasses and Glass-Ceramics

Edited by

M. H. Lewis
Centre for Advanced Materials Technology
University of Warwick, Coventry, UK

London New York
CHAPMAN AND HALL

First published in 1989 by Chapman and Hall Ltd
11 New Fetter Lane, London EC4P 4EE
Published in the USA by Chapman and Hall
29 West 35th Street, New York NY 10001

© 1989 Chapman and Hall

Printed in Great Britain at the
University Press, Cambridge

ISBN 0 412 27690 9

All rights reserved. No part of this book may be reprinted, or reproduced or utilized in any form or by any electronic, mechanical or other means, now known or hereafter invented, including photocopying and recording, or in any information storage and retrieval system, without permission in writing from the publisher.

British Library Cataloguing in Publication Data

Glasses and glass-ceramics.
1. Glass 2. Glass ceramics
I. Lewis, M.H., 1938–
666′.1

ISBN 0-412-27690-9

Library of Congress Cataloging in Publication Data

Glasses and glass-ceramics/edited by M.H. Lewis.
p. cm.
Includes bibliographies and index.
ISBN 0-412-27690-9
1. Glass. 2. Glass-ceramics. I. Lewis, M.H.
TL857.G55 1988
666′.1—dc 19

Contents

Contributors ix
Preface xi

1 **MAS NMR: a new spectroscopic technique for structure determination in glasses and ceramics** **1**
R. Dupree and D. Holland
1.1 Introduction 1
1.2 Glasses 7
1.3 Binary glasses 12
1.4 Ternary and mineral glasses 16
1.5 Devitrification 22
1.6 Phase separation 28
1.7 Ceramics 31
1.8 Materials produced by low-temperature processes 34
1.9 Conclusion 36
References 38

2 **X-ray absorption studies of glass structure** **41**
R. F. Pettifer
2.1 Introduction 41
2.2 Basic theory of EXAFS 45
2.3 Glass structure studies by EXAFS 50
2.4 Conclusions 56
References 56

3 **Volume nucleation in silicate glasses** **59**
P. F. James
3.1 Introduction 59
3.2 Summary of classical nucleation theory 60
3.3 Experimental studies in 'simple' one-component systems 63
3.4 Effect of glass composition on nucleation kinetics 80
3.5 Heterogeneous nucleation: experimental results 83
3.6 Non-metallic nucleating agents 87
3.7 Effects of amorphous phase separation on crystal nucleation kinetics 91

WITHDRAWN LIBRARIES
WITHDRAWN FROM STOCK LIBRARIES NI

3.8 Summary and conclusions 100
References 103

4 Oxynitride glasses and their glass-ceramic derivatives 106
G. Leng-Ward and M. H. Lewis
4.1 Introduction 106
4.2 Oxynitride glass formation 107
4.3 Oxynitride glass preparation 116
4.4 Oxynitride glass structure 123
4.5 Properties of oxynitride glasses 129
4.6 Crystallization 137
References 153

5 Optical properties of halide glasses 156
J. M. Parker and P. W. France
5.1 Introduction 156
5.2 Glass formation and structure 156
5.3 Melting techniques 160
5.4 Optical transmission characteristics 164
5.5 Infra-red absorption 167
5.6 Ultraviolet absorption 176
5.7 Intrinsic scattering 177
5.8 Minimum intrinsic losses 180
5.9 Extrinsic losses 182
5.10 Refractive index and dispersion 187
5.11 Other properties 192
5.12 Fabrication 195
5.13 Conclusions 196
References 196

6 Applications of microporous glasses 203
N. Ford and R. Todhunter
6.1 Introduction 203
6.2 Phase separation 203
6.3 The Vycor process 208
6.4 Reverse osmosis 210
6.5 Antireflection coatings and optical waveguides 216
6.6 Resistance thermometers and superconducting materials 219
6.7 Nuclear waste disposal 221
6.8 Refractory foams 221
6.9 Enzyme immobilization and catalyst supports 222
References 223

7 Glass-ceramics in substrate applications **226**
G. Partridge, C. A. Elyard and M. I. Budd
7.1 Introduction 226
7.2 Bulk crystallized glass-ceramics 229
7.3 Bulk glass-ceramics via powder techniques 246
7.4 Glass-ceramic coated metal substrates 253
7.5 Conclusions 270
References 271

8 Glass-ceramics for piezoelectric and pyroelectric devices **272**
A. Halliyal, A. S. Bhalla, R. E. Newnham and L. E. Cross
8.1 Introduction 272
8.2 Ferroelectric and non-ferroelectric materials 273
8.3 Selection of glass compositions 277
8.4 Preparation of glass-ceramics 279
8.5 Compositions of glasses 284
8.6 Heat-treatment and microstructure 284
8.7 Dielectric properties 288
8.8 Pyroelectric properties 292
8.9 Piezoelectric properties 297
8.10 Surface acoustic wave (SAW) properties 303
8.11 Connectivity model for piezoelectric and pyroelectric properties of polar glass-ceramics and tailoring the properties 305
8.12 Summary 313
References 314

9 Interfacial electrochemical aspects of glass in solid state ion-selective electrodes **316**
R. E. Belford and A. E. Owen
9.1 Introduction 316
9.2 The glass–metal interface 321
9.3 The glass–solution interface 328
9.4 Conclusions 334
References 334

10 Fibre reinforced glasses and glass-ceramics **336**
K. M. Prewo
10.1 Introduction 336
10.2 Composite systems 336
10.3 Composite fabrication 338
10.4 Composite properties 343
10.5 Summary 363
References 366

Glass systems index 369
Glass-ceramic phases and other compounds index 371
Subject index 373

Contributors

R. E. Belford Department of Electrical Engineering, University of Edinburgh, UK
A. S. Bhalla Materials Research Laboratory, The Pennsylvania State University, USA
M. I. Budd GEC Research Ltd, Engineering Research Centre, Stafford Laboratory, UK
L. E. Cross Materials Research Laboratory, The Pennsylvania State University, USA
R. Dupree Department of Physics, University of Warwick, UK
C. A. Elyard GEC Research Ltd, Engineering Research Centre, Stafford Laboratory, UK
N. Ford Ceramic Developments (Midlands) Ltd, Corby, Northants, UK
P. W. France British Telecom Research Laboratories, Ipswich, UK
A. Halliyal Materials Research Laboratory, The Pennsylvania State University, USA; present address: Du Pont Electronics, Wilmington, Delaware, USA
D. Holland Department of Physics, University of Warwick, UK
P. F. James University of Sheffield, UK
G. Leng-Ward Centre for Advanced Materials Technology, Department of Physics, University of Warwick, UK
M. H. Lewis Centre for Advanced Materials Technology, Department of Physics, University of Warwick, UK
R. E. Newnham Materials Research Laboratory, The Pennsylvania State University, USA
A. E. Owen Department of Electrical Engineering, University of Edinburgh, UK
J. M. Parker Department of Ceramics, Glasses and Polymers, Sheffield University, UK
G. Partridge GEC Research Ltd, Engineering Research Centre, Stafford Laboratory, UK
R. F. Pettifer Department of Physics, University of Warwick, UK
K. M. Prewo United Technologies Research Center, East Hartford, CT, USA
R. Todhunter Ceramic Developments (Midlands) Ltd, Corby, Northants, UK

Preface

The emergence of synthetic ceramics as a prominent class of materials with a unique combination of properties has been an important part of the materials-science scene over the past 20 years. These 'high-technology' ceramics have varied applications in areas utilizing their exceptional mechanical, thermal, optical, magnetic or electronic properties. A notable development of the 1970s was that of 'Si-based' ceramics (Si_3N_4, SiC and 'Sialons') as high-temperature engineering solids. More recently the zirconia-based ceramics have evolved as a class of material with significant improvements in fracture-toughness. In the 1980s we are on the threshold of development of ceramic-matrix composites with the promise of overcoming major limitations in engineering design with 'brittle' ceramics and the development of novel properties unattainable with monolithic microstructures. Throughout this period there have been significant but less well-publicized developments in the field of glass-ceramics and glasses. It is the purpose of this publication to review selected topics within this important area of materials science.

A key element in understanding the relation between properties and microstructure is a knowledge of atomic arrangement in ceramic phases. Recent developments in NMR and X-ray absorption spectroscopies have had considerable impact on studies of atomic co-ordination in glasses and crystalline ceramic materials and are reviewed in Chapters 1 and 2.

Glass-ceramics are derived from the parent glasses by controlled crystallization and have properties dictated, in part, by the efficiency of crystal nucleation within the glass volume. Current theoretical and experimental understanding of nucleation kinetics is surveyed in Chapter 3.

Although the majority of glasses and glass-ceramics are 'silicate'-based, novel systems have been explored with alternative anions. Significant property modification accompanies the partial substitution of nitrogen for oxygen in glasses and offers the possibility of unusual oxide/oxynitride phase combinations on crystallization. Chapter 4 is a review of oxynitride glass and glass-ceramic systems. Recent research on halide glasses, reviewed in Chapter 5, is motivated by their exceptional infra-red optical properties with potential for fibre-optical communication systems.

Subsequent chapters review current or potential applications for silicate-based glasses and glass-ceramics, utilizing properties such as microporosity induced by phase separation, dielectric or piezoelectric properties coupled with ease of fabrication, thermal expansion or chemical durability. The final chapter is a review of progress in fabricating high-strength composites with glass or glass-ceramic matrices utilizing their comparative ease of fabrication and identifying interfacial characteristics which have an influence on the philosophy for subsequent development of structural composites.

These reviews are not exhaustive but identify growth topics in the field and necessarily reflect the work of the respective authors.

This volume is dedicated to the memory of the late Professor P. W. McMillan in recognition of his contribution to the science and technology of glasses and glass-ceramics. Peter McMillan is remembered especially for his research in 'glass-ceramics' and the publication of a text, representing a landmark in this field, which has been reprinted in several languages since its first appearance in 1964 (*Glass Ceramics*, Academic Press, New York and London). It was written during his period as Head of the glass and ceramics group at the Nelson Research Centre of GEC Power Engineering. Peter McMillan subsequently joined the academic research community at a then newly established University of Warwick and, influenced by remarkable foresight on the part of the founding Head of Physics, Professor John Forty, developed a ceramic materials research group which has become prominent internationally and is the basis for the Centre for Advanced Materials Technology. In addition to his prolific publication of original research papers Peter McMillan was editor of a number of journals: the *Journal of Materials Science*, *Glass-Technology*, and the *Journal of Non-Crystalline Solids*. Many of the authors of this review volume are former associates who have continued to research in the field of glasses and glass-ceramics. This publication is both an appropriate memorial and a timely review of developments within this field.

M. H. Lewis
Warwick 1987

Peter W. McMillan 1928–1984

1

MAS NMR: a new spectroscopic technique for structure determination in glasses and ceramics

R. Dupree and D. Holland

1.1 INTRODUCTION

NMR has long been used to obtain structural information about glasses. Fairly recent reviews of standard wide line NMR have been given by Muller-Warmuth and Eckert (1982), Bray *et al.* (1983) and Bray and Gravina (1985). The information given by these studies has, in the main, been concentrated on quadrupolar nuclei, for example the use of ^{11}B to determine the relative amounts of 3 and 4 fold co-ordinated boron in glasses. Other main themes have been concerned with using dipolar broadened lineshapes to test for the probability of particular local configurations.

Much of the structural information potentially available from NMR experiments is masked in solids by various static anisotropic nuclear interactions which broaden the line and make the small differences in position undetectable. These include the magnetic dipolar interaction, the anisotropic chemical shift interaction and, for nuclei with spin $I > 1/2$, the quadrupolar interaction. The dipolar interaction arises from the magnetic field at one nucleus produced by neighbouring nuclei and varies as the inverse cube of the distance between nuclei. The chemical shift is caused by bonding of the atom to its surroundings so that each crystallographic site will have a particular shift with three components corresponding to the three principal axes. For a site of cubic point group the chemical shift will be isotropic, for sites of lower symmetry the shift will be anisotropic and in microcrystalline and glassy (it is the local symmetry that is important) samples lineshapes such as those shown in Fig. 1.1 will be observed. The three components of the chemical shift are chosen so that $\sigma_{33} \geqslant \sigma_{22} \geqslant \sigma_{11}$ and the asymmetry in the site surroundings is characterized by the asymmetry parameter η where $\eta = (\sigma_{22} - \sigma_{11})/(\sigma_{33} - \sigma)$. Thus for axial symmetry $\sigma_{22} = \sigma_{11}$ and $\eta = 0$. For lower symmetries $0 \leqslant \eta \leqslant 1$. The isotropic shift σ

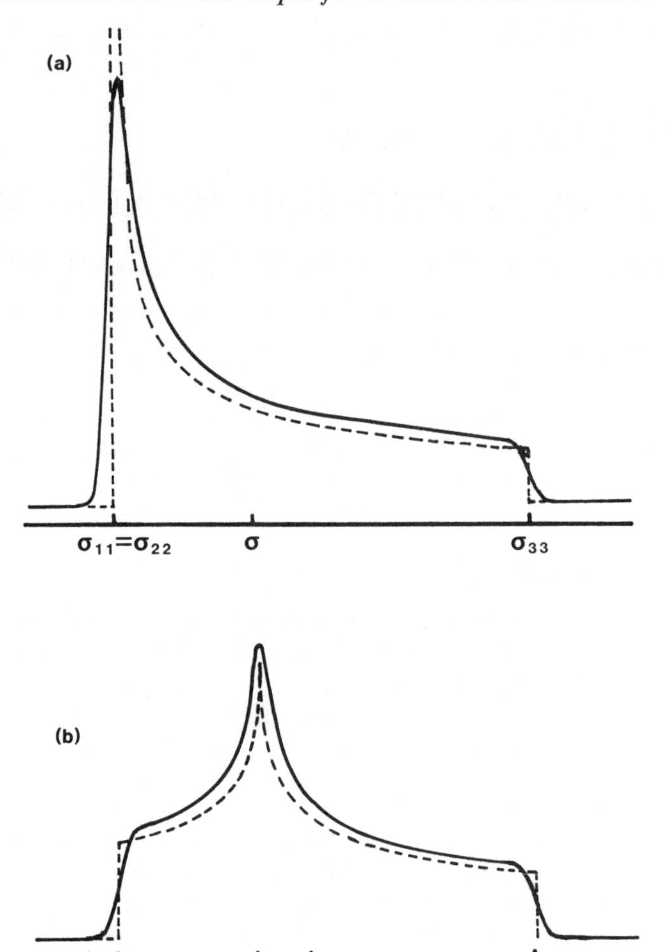

Fig. 1.1 Powder pattern lineshapes for (a) axial site symmetry, (b) lower site symmetry.

is defined as $(\sigma_{11} + \sigma_{22} + \sigma_{33})/3$, thus σ, $\Delta\sigma = \sigma_{33} - \sigma_{11}$, η are sufficient to describe the chemical shift parameters.

For nuclei with spin $I > 1/2$ the interaction of the nuclear quadrupole moment (eQ) with the electric field gradient ($\partial^2 V/\partial Z^2 = eq$) at the nucleus will alter the nuclear energy levels depending upon the relative orientation of the electric field gradient and the applied magnetic field and to first order in the quadrupolar interaction will give the powder pattern shown in Fig. 1.2(a). If e^2qQ/h is sufficiently large only the central ($\frac{1}{2} \rightarrow -\frac{1}{2}$) transition is

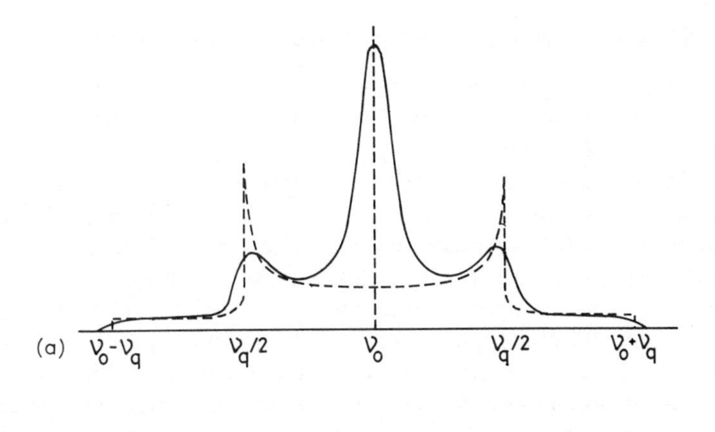

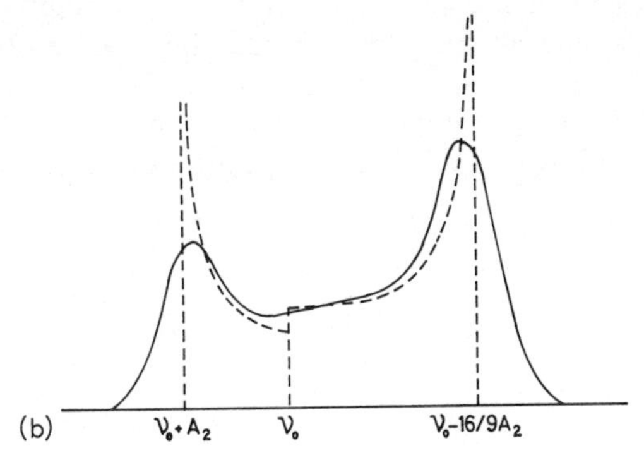

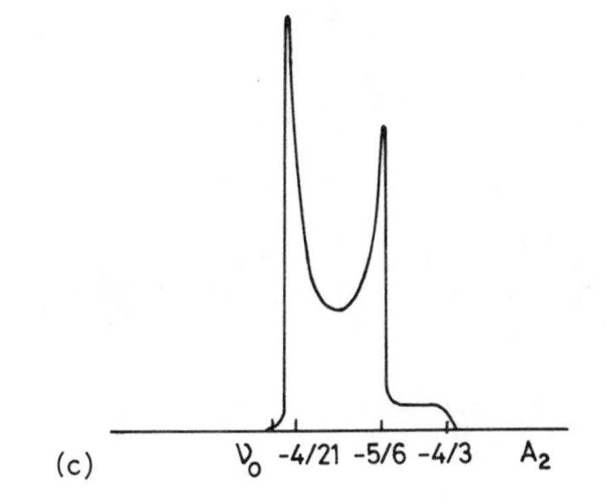

Fig. 1.2 (a) Powder pattern lineshape for $I = 3/2$ nucleus (e.g. ^{23}Na) due to first order quadrupole interaction, i.e. small e^2qQ/h in axially symmetric environment. Dashed line is without dipolar broadening; full line with dipolar broadening. ($\nu_q = \frac{1}{2}e^2qQ/h$.) (b) The central $\frac{1}{2} \rightarrow -\frac{1}{2}$ transition lineshape is observed in an axially symmetric environment for larger e^2qQ/h. ($A_2 = (3/64)\,[(e^2qQ)^2/h]\,(1/\nu_0)$.) (c) The effect of magic angle spinning on this transition. Note that the resonance is displaced from the true chemical shift position.

observable in powders, and it will be broadened and can have a complex shape dependent on the quadrupolar asymmetry parameter; examples are shown in Fig. 1.2(b) and (c).

In a liquid the NMR lines are usually very narrow because all of these broadening mechanisms have, in first order, a $3\cos^2\theta - 1$ angular dependence and this averages to zero under the rapid motion present. Recently various line narrowing techniques, such as magic angle spinning (MAS) where the sample is orientated at 54° 44′ ($\cos\theta = 1/\sqrt{3}$) to the magnetic field and spun rapidly (Andrew, 1981), have been developed such that in favourable crystalline materials a reduction in linewidth of more than two orders of magnitude can be achieved and the underlying structural information present in the spectrum obtained. The improvement is less marked in glasses because of the disorder but the resolution is often sufficient to give much unique structural information.

Figure 1.3 shows, as an example, the static and spinning NMR spectrum of ^{27}Al in a sample of $YAlO_3$ (YAP). Both spectra were acquired for the same

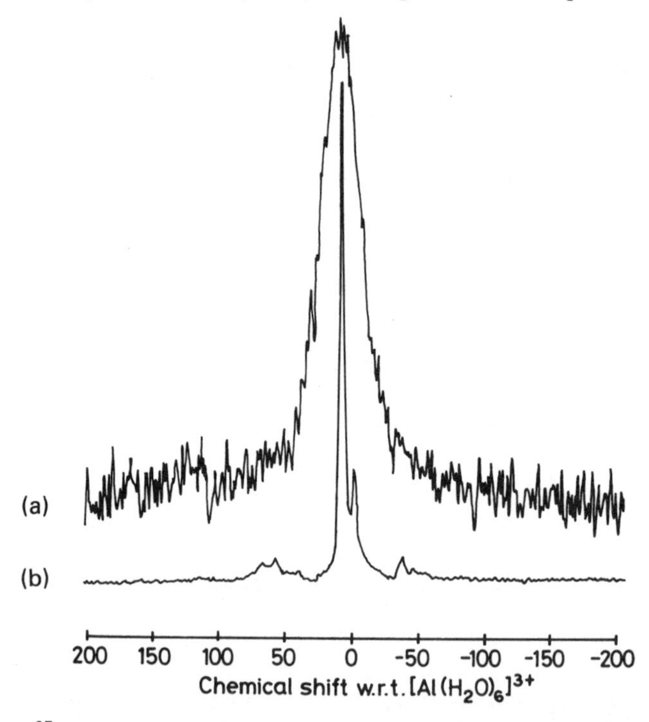

Fig. 1.3 ^{27}Al spectrum of $YAlO_3$: (a) static, (b) MAS. The main peak in (b) corresponds to $[AlO_6]$ in $YAlO_3$ at +9.5 ppm, the small peak at 0.8 ppm and the broad bump centred on ~60 ppm are from $[AlO_6]$ and $[AlO_4]$ units in ~5% $Y_3Al_5O_{12}$ present as second phase. Two small spinning sidebands are also visible at −40 ppm and +60 ppm.

time; not only does the MAS spectrum have a much better signal-to-noise ratio but also it clearly shows the presence of ~5% $Y_3Al_5O_{12}$ (YAG) as a second phase. (The main peak at 9.5 ppm is the octahedral (AlO_6) in YAP, the small peak at 0.8 ppm is the octahedral (AlO_6) in YAG and the broad 'bump' is from tetrahedral (AlO_4) units, shift 72 ppm, in YAG. Also visible are two small, spinning sidebands from the octahedral YAP.)

The value of the chemical shift is very sensitive to the local environment and thus is determined by many variables. Amongst the factors which influence the shift are:

1. The co-ordination number of the atom concerned – the shift range of ^{29}Si is −142 ppm to −221 ppm when six co-ordinated to oxygen compared with ~−60 ppm to −128 ppm when four co-ordinated;
2. The type of neighbouring atom, e.g. the ^{29}Si shift in quartz where silicon is tetrahedrally co-ordinated to four oxygens is −107.1 ppm compared with −48.5 ppm when tetrahedrally co-ordinated to four nitrogens in β-Si_3N_4. The replacement of one nitrogen by oxygen in Si_2N_2O gives a shift of −63 ppm;

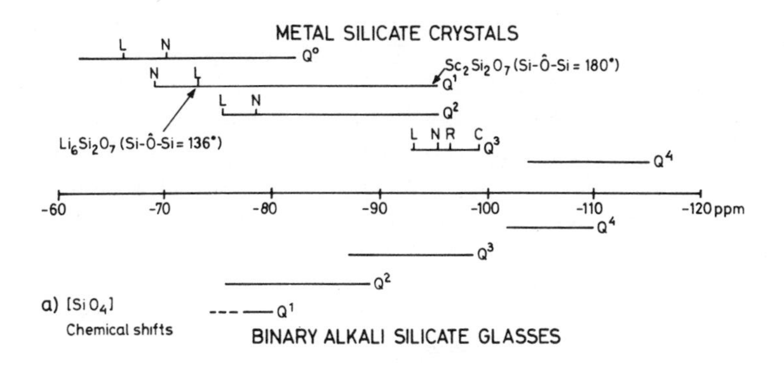

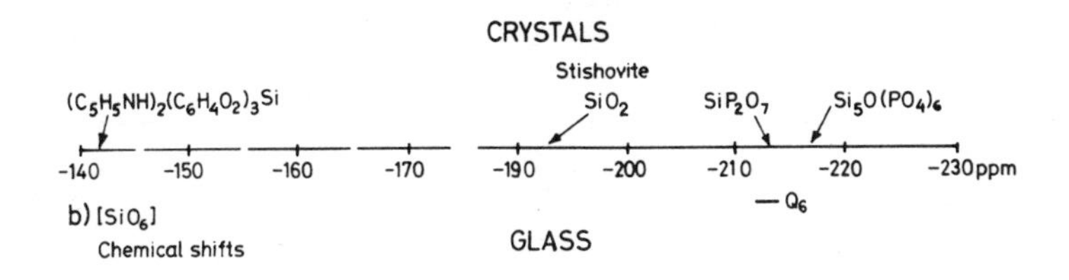

Fig. 1.4 Chemical shift ranges for ^{29}Si in crystalline and glassy oxide environments. (L = Li, N = Na, R = Rb, C = Cs.)

3. Bond angles/lengths – the value of the Si–O–Si bond angle in $Si_2O_7^{4-}$ units has a strong effect on the chemical shift which ranges from ~ -95 ppm for a 180° angle to ~ -72 ppm for a 136° angle. Correlations with Si–O bond length within a particular structure have also been made (Grimmer, 1985); typically a change in shift of ~1 ppm occurs for a bond length change of 0.0001 nm;
4. Connectivity of the structural unit – the connections of SiO_4 tetrahedra in minerals and glasses can be described in terms of Q^n units where $0 \leqslant n \leqslant 4$ is the number of bridging oxygens to other Q units. The shift ranges for ^{29}Si in metal silicate minerals and in alkali silicate glasses is shown in Fig. 1.4. In general the shift becomes less negative as the number of bridging oxygens decreases. There is considerable overlap in the shift range for each Q^n, however, due to effects such as those described in items 3 and 5;
5. Second co-ordination sphere – the replacement of a silicon in the NNN shell by an aluminium will shift the ^{29}Si resonance by +5 ppm;
6. Third co-ordination sphere – the chemical shift of a Q^n silicon connected to others will vary depending upon the Q type of the neighbouring silicon, i.e. Q^4–Q^3 is different from Q^4–Q^4. This is readily observed in glasses.

To date ^{29}Si has the most well-documented set of chemical shifts in minerals and glasses, but similar effects will determine the shifts of other nuclei and Fig. 1.5 gives some data for ^{27}Al. It is with the aid of such information that the identification of sites can be achieved. However, as is clear from Fig. 1.4, a range of chemical shift values is found with each particular structural unit, e.g. Q^n silicon. The significant overlap of these ranges means that additional information, e.g. on stoichiometry or on the environment of other nuclei present is often needed for unambiguous

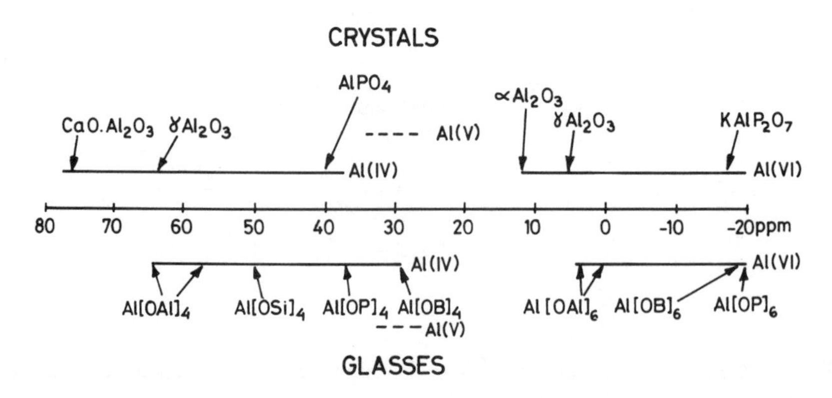

Fig. 1.5 Chemical shift ranges for ^{27}Al in crystalline and glassy oxide environments.

interpretation of the MAS NMR spectrum. Since the intensity of a line is directly proportional to the number of nuclei producing the signal, quantitative information on the types of site present and on the relative order at a given site can often be obtained relatively quickly; some caution in quantitatively interpreting data is necessary, however. For spin $I = 1/2$ nuclei spin lattice relaxation times can sometimes be very long and care must be taken to ensure that relaxation has not affected the results. Nuclei with quadrupole moments ($I > 1/2$), such as ^{27}Al, in a strongly distorted environment may experience a sufficiently large electric field gradient that the resonance from them is broadened beyond detection. It is essential therefore in quantitive work to compare the signal intensity with a known standard, something that is not always done.

In this chapter we will concentrate on describing the new information about glasses and ceramics obtained using MAS NMR. The isotopes most commonly used are ^{29}Si and ^{27}Al partly because of their relatively good NMR properties although many other nuclei are potentially useful. Fortunately, silicate glasses and ceramics are open to investigation. Unfortunately, experiments using either oxygen (^{17}O) or nitrogen (^{15}N) as the probe are not viable without the use of expensive isotopically enriched materials.

1.2 GLASSES

1.2.1 Glass structure

An all-encompassing model for glass structure does not exist and, considering the wide range of elements and bonding types involved in amorphous materials, no simple model can be envisaged. However, we can still consider the possible effects of coulombic and steric contributions on the choice of atomic arrangements in glasses. This should be applicable to all glass systems, although most information is available for silicate-based glasses. The chemical shift of an individual nucleus is dependent on the shielding of the nucleus, i.e. on the local electron density which in turn reflects the distribution of other atoms about the said nucleus. As yet, the calculation of chemical shift for a nucleus in a particular bonding arrangement is not possible and thus deduction of environment from measured chemical shift can be done only by comparison with materials where local structure can be determined by other means. Thus for glasses the chemical shifts for nuclei in known crystalline structures must first be determined. However, this does not give sufficient information since glasses provide a much wider range of environments than observed in crystals because the factors which control structure differ between the two states. In crystals, the dominating factor is preservation of long-range order: bond angles, interatomic distances and site occupancies are dictated by the need to maintain a certain symmetry such that the bulk free energy is minimized. In a glass it is the minimization of local free energy which is important and the local bonding arrangements which achieve this state can be very different from those required to give a long-range-ordered bulk structure. Thus one occasionally observes that the local environment of a nucleus in a glass may be more symmetric than in the corresponding crystalline environment.

The basic concepts of oxide glass structure are that certain oxides can form strong ionic–covalent bonds in a 3D network; this presents a kinetic barrier to bond rearrangement and crystallization. Other modifier oxides can react with the network to break these bonds producing mobile ions and increasing the degrees of freedom of the network such that rearrangement becomes easier and the glasses less kinetically stable. The usual simple example of this involves the soda–silica binary. In pure silica, 4 strong Si–O–Si links exist between $[SiO_4]$ tetrahedral units and a rigid network results. Addition of modifier oxide Na_2O produces $Si–O^- \ Na^+$ units which reduce the rigidity of the network since Na^+ are mobile and $Si–O^-$ make more degrees of freedom available to the network than Si–O–Si. At an Na_2O content of 33.3 mol % there is enough added oxygen to give an average of 1 non-bridging oxygen [nbo] per silicon and 3 bridging oxygen [bo] remaining. At 50 mol % Na_2O the average is 2 [nbo] per silicon which now have only 2 strong links to other silicons, i.e. the 3D network has degenerated into chains. Any further disruption of the linkages results in a rapid decrease in stability and glasses can only be formed with considerable difficulty.

So far we have referred to an average number of [bo] and [nbo] per silicon. The properties of the glass will in fact depend quite critically on the exact way in which the different bond types are distributed, i.e. the quantities of different Q^n species present. There are several possible arrangements:

1. Statistical distribution (unconstrained random) – the distribution of [nbo] is determined only by composition and statistics;
2. Binary distribution (constrained random) – [nbo] repel each other leading to maximum dilution which in turn gives successive formation of lower Q^n types and a homogeneous distribution of no more than $2Q$ types result;
3. Clustered – [nbo] cluster, eventually leading to phase separation.

Figure 1.6 illustrates the possible arrangements of Q units for a glass of composition $Na_2O \cdot 2SiO_2$. The occupancy of the first co-ordination sphere of Si by varying numbers of {O–Si} and $\{O^- \ Na^+\}$ will lead to pronounced changes in the electron density at the nucleus, i.e. a change in chemical shift. Thus MAS NMR is a potential tool for distinguishing different $Si(OSi)_n \ (O^-)_{4-n}$ or Q^n type silicons and identifying the particular arrangement in individual glasses.

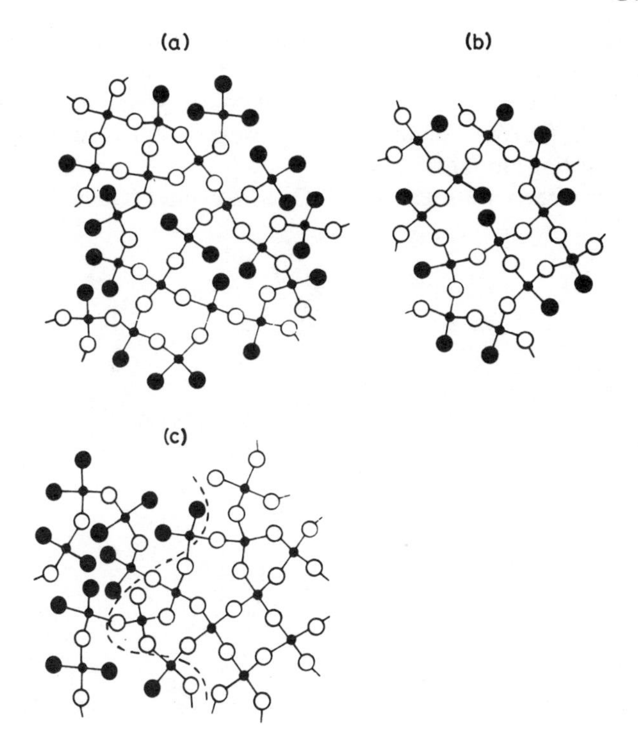

Fig. 1.6 Schematic diagram of types of disorder in a binary alkali disilicate glass: (a) random distribution of Q types; (b) all Q^3; (c) phase separated into Q^4 and Q^2. The basic unit is a silicon atom (small circle) surrounded by four oxygens (large circles). Bridging oxygens are open circles and non-bridging oxygens filled circles (after Gaskell, 1985).

1.2.2 Silicon dioxide

SiO_2 can be regarded as the prototype silicate glass and there has been much interest in determining its structure. Whilst SiO_2 tetrahedra have long been recognized as providing the basic unit in this glass the interconnection of the tetrahedra usually described in terms of the Si–O–Si bond angle is of fundamental importance in the structure. Several elastic scattering experiments have been performed on vitreous SiO_2 but because of the difficulties in going from raw data to the details of the structure considerable uncertainty exists about this interconnection. The chemical shift of various SiO_2 polymorphs (and also several dealuminized zeolites) is found to correlate very well with the secant of the Si–O–Si angle and Dupree and Pettifer (1984) suggested that, with suitable analysis, the Si–O–Si bond angle distribution in vitreous SiO_2 could be deduced from the MAS NMR lineshape. The resonance presented in that paper turns out to be atypical of vitreous SiO_2. In later work SiO_2 samples prepared in different ways have been found to have slightly different chemical shifts and lineshapes. Gladden *et al.* (1986a) found small differences between two samples containing differing amounts of OH.

Much larger effects are seen when SiO_2 is densified (Devine *et al.*, 1987)

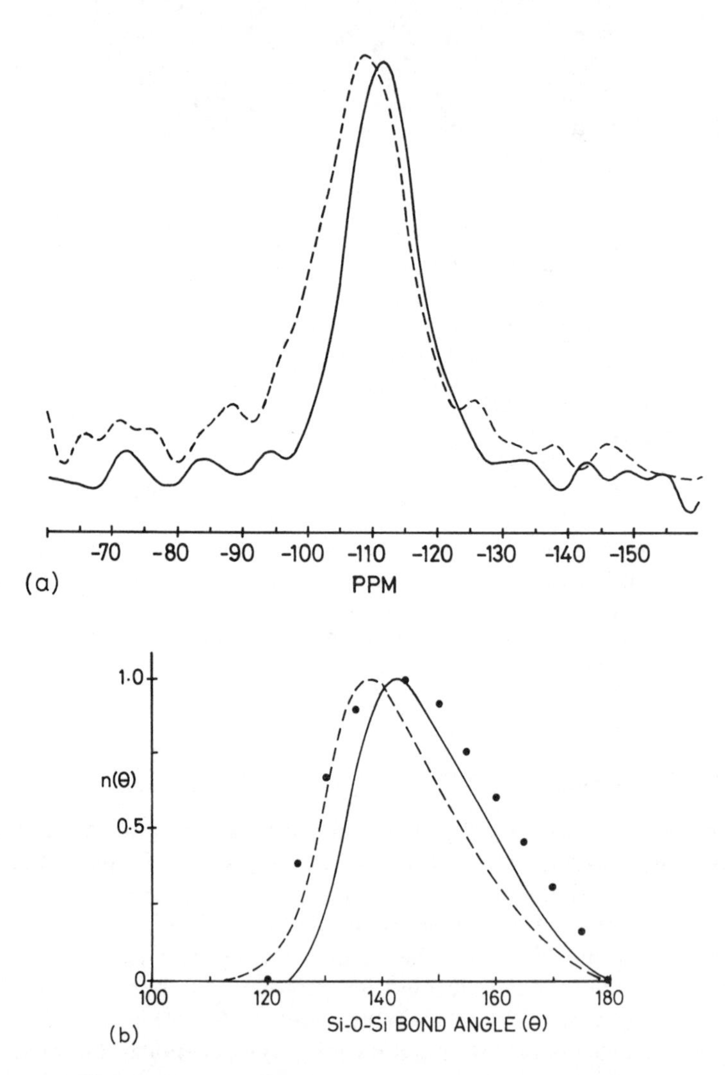

Fig. 1.7 (a) ^{29}Si spectrum of Suprasil I ——— and after densification by ~16% ----. (b) The deduced Si–O–Si bond angles in undensified ——— and densified ---- Suprasil I. Filled circles are the bond angles for undensified silica deduced from X-ray scattering.

(i.e. subjected to a change in density which remains once the pressure is removed). The ^{29}Si resonance of Suprasil I (OH ~ 1200 ppm) in normal form and after densification is shown in Fig. 1.7(a). A shift of + 2.5 ppm and a broadening (FWHM) from 11 ppm to 14 ppm is observed for the densified SiO_2. The corresponding Si–O–Si bond angle distribution is shown in Fig. 1.7(b) where it can be seen that the main consequence of the densification is to move the peak in the distribution by ~ 5° from ~ 143° to ~ 138° perhaps through modification of the SiO_2 ring configurations in the network.

The ^{29}Si spectrum for amorphous SiO_2 produced from $SiCl_4$ by the hydrolysis of water is shown in Fig. 1.8(a). This sample has been dried under vacuum at 200°C overnight to remove surface water but has a considerable amount of OH present and two peaks are distinguishable, one at − 101 ppm, the other at − 108 ppm. The effect of OH can be readily seen from the spectrum shown in Fig. 1.8(b) which uses cross-polarization from the protons to the silicon to emphasize those silicons close to an OH. The peak at − 101 ppm corresponds to a Q^3 with one OH group, that at − 91 ppm to a Q^2 with two OH groups. It should be noted that the Q^4 peak is shifted from the 'normal' vitreous range ~ − 110 to − 112 ppm, indicating a rather different structure with a smaller mean bond angle.

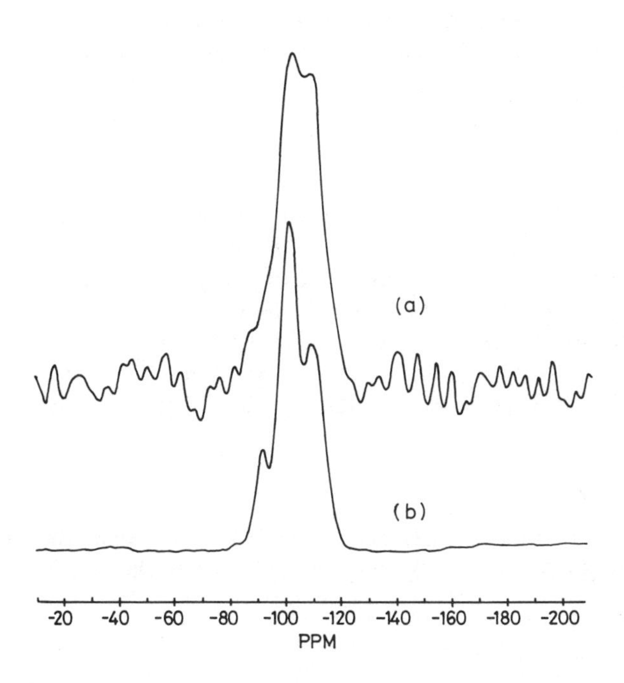

Fig. 1.8 (a) ^{29}Si MAS spectra of amorphous SiO_2 produced by hydrolysis from $SiCl_4$; (b) cross polarization spectrum of same sample contact time = 10 ms.

1.3 BINARY GLASSES

1.3.1 Alkali silicates

The distribution of [nbo] has been studied for all the alkali metal silicate glasses using ^{29}Si. Grimmer and co-workers have looked at the Li_2O, Na_2O and K_2O systems; Schramm *et al.* (1984) have also investigated the Li_2O–SiO_2 glasses; and Dupree *et al.* (1984c, 1986a) the Li_2O, Na_2O, Rb_2O and Cs_2O systems.

The ease of study of each system depends on the chemical shift range of silicons of each Q type. This is found to differ with each alkali metal and thus controls the degree of difficulty encountered on attempting to resolve the contributions to the observed spectrum from each Q type. Figure 1.9 illustrates the different degree of resolution of Q^4 and Q^3 peaks in the 20 mol % alkali silicates.

(a) Li_2O–SiO_2

This system was first analysed using MAS NMR by Schramm *et al.* (1984) who interpreted their data in terms of a mixture of Q types corresponding to a statistical type distribution. Their treatment of the data has been criticized. Gladden *et al.* (1986b) and Grimmer *et al.* (1984) have shown spectra which are best interpreted as the sum of no more than two gaussians, i.e. a binary distribution.

(b) Na_2O–SiO_2

The observed ^{29}Si spectra from quenched glasses, over the compositional range 10–50 mol % Na_2O (Dupree *et al.*, 1984b; Grimmer *et al.*, 1984), could be fitted to a combination of one or two gaussians as predicted by the binary model. An excellent fit was obtained.

The ^{23}Na spectra of these glasses differed only in small variations in peak position from -2.3 ppm for 10 mol % Na_2O to $+2.8$ ppm for 50 mol % Na_2O. Large differences were produced on devitrification (*v.i.*).

(c) K_2O–SiO_2

Grimmer and Muller (1987) examined this system over the range 17.2–51 mol % R_2O and observed identical behaviour to the Na_2O–SiO_2 case.

(d) Rb_2O, Cs_2O

These were examined from 10 to 50 mol % R_2O (Dupree *et al.*, 1986a). At the lower R_2O content, the Q distribution follows the pattern exhibited by

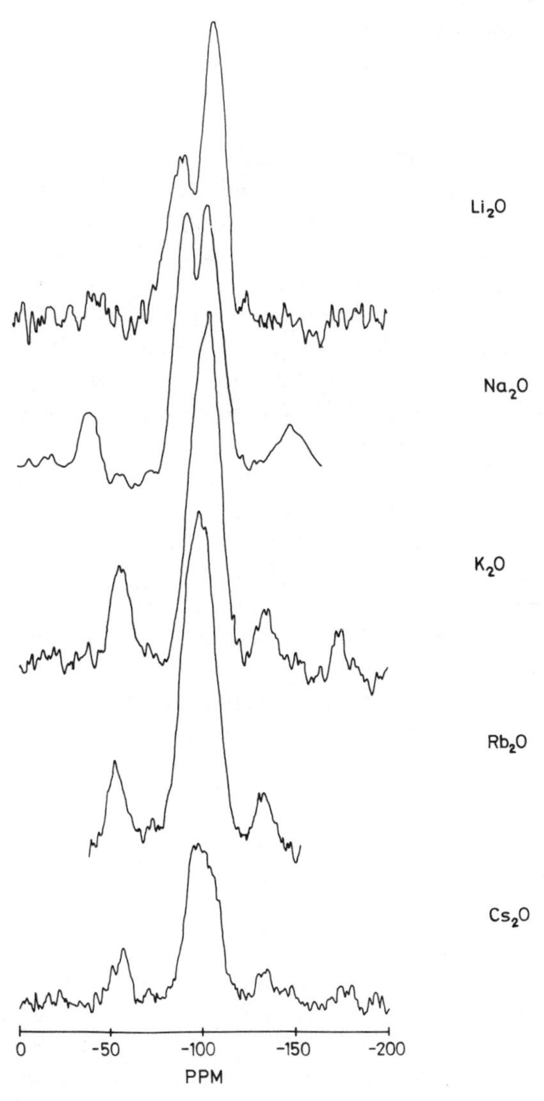

Fig. 1.9 ^{29}Si spectra for alkali-metal silicate glasses containing approximately 20 mol % alkali oxide (after Dupree *et al.*, 1986a).

Na_2O and K_2O. However, at the high concentrations >45 mol %, the spectra could not be fitted to the expected combination of one large Q^2 and one small Q^3 peak, nor to any other combination of two peaks only. They could be fitted to a combination of three narrow gaussians, $Q^3 + Q^2 + Q^1$, arising from a partial disproportionation $2\Delta q\ Q^2 \rightarrow \Delta q\ Q^1 + \Delta q\ Q^3$ such that below 50 mol % R_2O the ratio $Q^3:Q^2:Q^1$ is now given by $q3+\Delta q:(q2-2\Delta q):\Delta q$ and above 50 mol % by $\Delta q:(q2-2\Delta q):q1+\Delta q$, where $q2$ and $q3$ are the intensities predicted by the binary model. The degree of disproportionation $2\Delta q/q2$ amounts to ~32% for 46.7 mol % Rb_2O, ~79% for 48.5 mol % Rb_2O and ~70% for 51.1 mol % Cs_2O. These results indicate that the effect of the large cations is to favour a sheet silicate (Q^3) + dimer (Q^1) combination of silicon species rather than a mixture of chains and rings (Q^2). This structure may produce more free volume to accommodate the steric and coulombic requirements of Rb^+ and Cs^+.

The other unusual feature of the spectra for the high R_2O content glasses is the narrowness of the lines, implying a much smaller distribution of Si–O–Si bond angles in a system when these Q species are present.

Despite a small quadrupole moment, the linewidth of the resonances from ^{133}Cs in these glasses was large and the shape distorted in some instances. The chemical shift range for Cs is large and the peak width may simply reflect the local variation in site symmetry consequent on being in a glassy matrix; however, the distortion observed suggests that there are two different sites contributing to the spectra. Fitting of the peaks on this basis reveals two peaks which at low concentrations of Cs_2O are ~340 and 150 ppm wide respectively. On increasing the Cs_2O content, the lines move by ~+200 ppm and decrease in width to 220 and 70 ppm respectively. A decrease in width suggests an increasingly uniform environment which may be gained at the expense of a modification to the Q distribution of the silicons at high Cs_2O concentrations.

1.3.2 Other binary silicates

(a) $PbO–SiO_2$

This system is unusual in that glasses can be prepared to very high PbO contents ~90%. On the simple Zachariasen model of glass structure, this would not be possible if Pb^{2+} were behaving as a modifier ion. In the polymorphs of PbO, (PbO_4) square pyramidal units are formed and these units are generally considered to persist in the $PbO–SiO_2$ glasses, forming chain-like structures amongst which are distributed the remaining (SiO_4) tetrahedral units in various degrees of condensation. These glasses therefore provide the possibility of observing the full range of Q types in a glass. Bridging oxygens are detectable by XPS even at 70 mol % PbO where both the binary distribution and the unconstrained statistical distributions would predict that only Q^0, with no [bo] should exist.

Fujiu and Ogino (1984) interpreted static NMR spectra of ^{29}Si in $PbO–SiO_2$ as being composed of two lines, one relatively narrow at −107 ppm corresponding to Q^4 and one broad component at −79 ppm arising from a mixture of Q types with Q^2 predominating. The Q^4 peak decreases with increasing PbO and has disappeared at 60 mol %.

An MAS ^{29}Si study of glasses from 25 to 70 mol % PbO (Dupree *et al.*, 1987a) produced the following conclusions:

1. Below 30 mol % PbO, the number and disposition of [nbo] are consistent with a binary Q^4/Q^3 distribution indicating that, at this concentration Pb^{2+} may be considered as a traditional modifier ion;
2. From 30 to 40 mol % PbO, a peak can still be resolved in the spectrum at the Q^4 position, which may reflect a change to an unconstrained statistical distribution or may indicate the formation of Si–O–Pb bridging units which may be indistinguishable by NMR from Si–O–Si. XPS also indicates that there is more oxygen of [bo] type than would be expected at this stoichiometry;
3. From 40 to 65 mol % PbO, a statistical distribution of [nbo] is consistent with the observed spectra;
4. At 70 mol % PbO, the glass contains mainly isolated Q^0 units in a lead oxygen matrix.

(b) SiO_2–P_2O_5

Although silicon is usually four co-ordinated to oxygen, in stishovite a high-pressure silica polymorph, and in several crystalline silicates including SiP_2O_7 silicon is known to be six co-ordinated. A glass of composition $SiO_2P_2O_5$ was investigated (Weeding *et al.*, 1985) to see if octahedrally co-ordinated silicon was present. Only tetrahedrally co-ordinated silicon with a chemical shift of −110 ppm was found in the glass but upon devitrification at 1300°C, a narrow octahedral peak at −214 ppm and a broader line at −220 ppm were observed as well. Cubic SiP_2O_7 has six crystallographically distinct octahedral silicon sites, not all of which can be resolved in the NMR spectrum.

The ^{31}P spectra of the glass gave five lines at 15, −0.4, −12, −54 and −72 ppm, which were not assigned in Weeding *et al.* (1985) but presumably arise from end and chain groups in polyphosphate type units as well as from orthophosphate type units. An analysis of the anisotropy of these lines would be helpful in unravelling the spectra (Griffiths *et al.*, 1986).

Although octahedral silicon was not observed in this binary glass, we have observed this co-ordination in Na_2O–SiO_2–P_2O_5 glasses (*v.i.*).

(c) SiO_2–Al_2O_3

Roller-quenched SiO_2–Al_2O_3 glasses containing 10–50 wt % Al_2O_3 were investigated by Risbud *et al.* (1987). Samples containing more than ~15 wt % Al_2O_3 appeared to be phase separated into a SiO_2 rich phase containing ~10 wt % Al_2O_3 and a SiO_2-poor phase. The ^{27}Al spectra, although very broad, gave peaks at ~60 ppm, 30 ppm and 0 ppm of different relative intensity as the composition changed. The peaks at ~60 ppm and ~0 ppm are the well-known positions for tetrahedrally and octahedrally co-ordinated Al–O units. The peak at ~30 ppm was interpreted as 5-co-ordinated Al; a similar interpretation also based on the 5-co-ordinated Al chemical shift in andalusite was used to explain the peak at this position in anodic alumina films (Dupree *et al.*, 1985a).

The variation in relative intensity of these peaks as the composition is changed was interpreted as implying that Al(4) is charge compensated primarily by Al(5) in SiO_2-rich glasses and by Al(6) in the Si-poor phase. The spectrum for the composition with most Al_2O_3 apparently shows that the predominant co-ordination is Al(5), however.

1.4 TERNARY AND MINERAL GLASSES

1.4.1 Mixed alkali and alkaline-earth silicates

Since the effect of R_2O and $R'O$ additions is generally regarded as being the production of non-bridging oxygens, then a simple view might be that the chemical shift of Q^1, Q^2, Q^3 etc. should be independent of R and R′. This is only approximately true. There is a range of values of chemical shift for Q^3 and Q^2 depending on R and R′. Replacement of one modifier cation by another produces a linear change in chemical shift of a given Q species between the limits corresponding to each individual modifier (Dupree *et al.*, 1985b). Much grosser effects are observed for nuclei other than Si. If Na is substituted by other cations then a shift in peak position is observed with concentration although there is no detectable change in peak shape. For example, if Na in $20Na_2O \cdot 80SiO_2$ is gradually replaced by Cs, the peak position of Na moves from −3.5 ppm to +4 ppm at 75% substitution. In the case of 75% substitution by Ca the peak position approaches −16 ppm. This difference in response to substitution may reflect the different relative electronegativities of the substituents (Cs 0.86, Ca 1.04 compared with Na 1.01), or alternatively there may be a change in co-ordination preference.

1.4.2 Aluminosilicates

The role of Al^{3+} in glasses is generally accepted to be that of ‘intermediate’ – at least within certain composition limits. This means that it can substitute into the network structure and provide bridging oxygens. Since it does this by forming $(AlO_4)^-$ tetrahedral groups, then in a silicate system it acts as a ‘sink’ for modifier cations and non-bridging oxygens.

The presence of $(AlO_4)^-$ units will modify the environment of Si in several possible ways:

1. In replacing Si–O non-bridging oxygens by Si–O–Al
2. In transferring electronic charge from non-bridging oxygens to $(AlO_4)^-$ groups, resulting in preferential location of modifier cations near these groups.
3. In increasing the number of possible Si environments by substitution of neighbour Si by Al. We can now define the Si species as Qm/n where $m = 0$–4 bridging oxygens and n = number of Si–O–Al. Thus for example $Q3/2$ corresponds to $Si(OAl)_2(OSi)O^-$.

Lippmaa *et al.* (1981) showed that additional paramagnetic shifts occur when Si is substituted by Al in crystalline aluminosilicates. The shift is quite small and therefore more difficult to resolve in the case of the broad resonances obtained from glasses. Yet the difference in Si sites is observable although the effect of aluminium substitution on the shifts is often greater for the other nuclei, i.e. Na or Al.

1.4.3 Sodium aluminosilicates

Glasses of composition $(33.3 - x)Na_2O \cdot xAl_2O_3 \cdot 66.6SiO_2$ (Dupree *et al.*, 1985b) show more negative shifts for Si as x is increased. When $x = 0$ the sole silicon species is Q^3 and as Na_2O is gradually replaced by Al_2O_3 the Q^3 are replaced by $Q^4/1$ i.e. $Si(OSi)3O^- \rightarrow Si(OSi)_3(OAl)$. When $x = 16.67$, all the silicons should be of this type and the network is fully 3D linked. The change in chemical shift from Q^3 to $Q^4/1$ is 4 ppm – consistent with that observed in zeolites when a silicon in the second co-ordination sphere is replaced by aluminium. Although the ^{27}Al resonance shows little change with substitution the ^{23}Na is very sensitive. For 50% substitution in $33.3Na_2O \cdot 66.7SiO_2$ the peak shifts from $+0.3$ ppm to -18.5 ppm reflecting the change in association of Na^+ with SiO^- to AlO_4^-.

1.4.4 Calcium aluminosilicates

Engelhardt *et al.* (1985) reported ^{29}Si data for a series of glasses $CaO(0–70.1)Al_2O_3(0–38.3)SiO_2(7.7–100)$. The observed resonances could be described in terms of 1 to 3 different units of Qm/n depending on glass composition and differing from each other in one next nearest neighbour only, i.e. not a statistical distribution of Q types.

1.4.5 Alkali phosphosilicates

The introduction of P_2O_5 into binary alkali silicate glasses has a profound effect on their structure, particularly at high concentrations. Dupree *et al.*

(1987b,c) have studied the $R_2O:2SiO_2–P_2O_5$ system over a wide concentration range of P_2O_5; in small amounts P^{5+} acts as a 'scavenger' for alkali ions giving $P–O^-R^+$ units in preference to SiO^-R^+. This removal of R from the silicate network allows it to repolymerize – i.e. the number of [nbo] is reduced and Q^n units of higher n are formed. In the case of the disilicate glasses Q^4 silicons are formed from Q^3 when P_2O_5 is added. The consequent changes to the ^{29}Si spectrum can be seen in Fig. 1.10(a). The ^{31}P environment depends on the P_2O_5 concentration and the type of alkali metal oxide. At very low concentration and low atomic weight alkali metal the

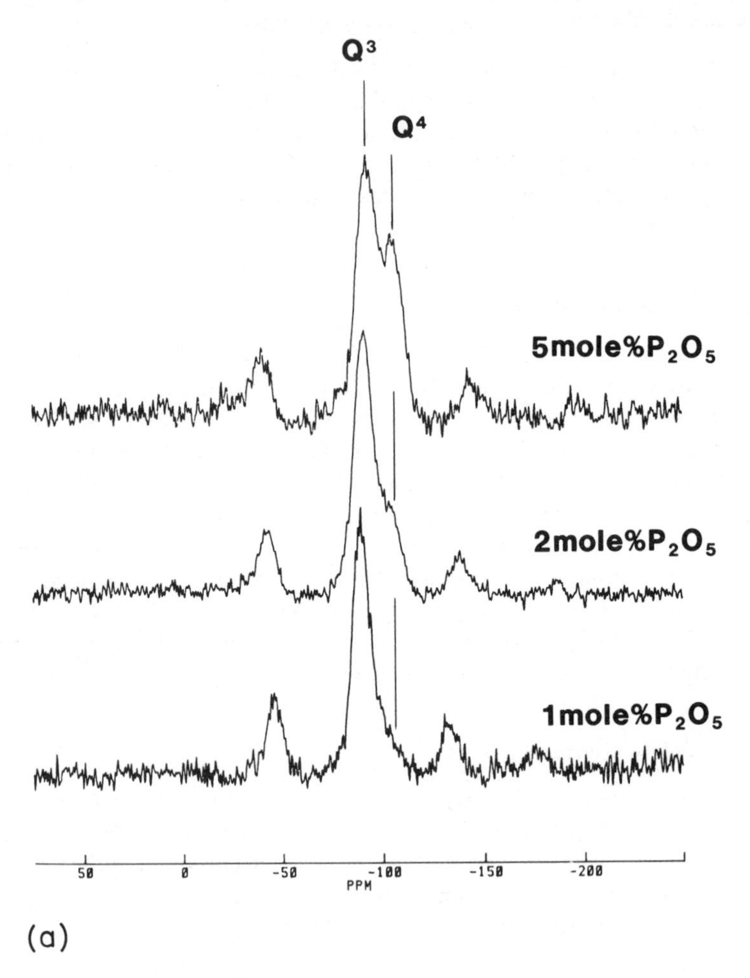

Fig. 1.10 (a) ^{29}Si resonance in a sodium disilicate glass containing small amounts of P_2O_5; (b) ^{31}P resonance in the same glass. I_0(ortho)-isotropic peak from orthophosphate unit, I_0(pyro) isotropic peak from pyrophosphate unit; all other peaks are spinning sidebands.

predominant species with a chemical shift of 10–15 ppm closely resembles that in alkali orthophosphates. As the P_2O_5 content and/or atomic weight increases then pyrophosphate-like environments (chemical shift ~2 ppm) occur (Fig. 1.10(b)). Since the effective scavenging power depends on the type of phosphate unit formed then this also determines the degree of repolymerization of the silicate network and the Q^4/Q^3 ratio observed from deconvolution of the MAS spectrum agrees well with that predicted from the known amounts of ortho- and pyrophosphate. As the P_2O_5 content is increased so is the degree of polymerization of both the silicate and phosphate networks, orthophosphate is replaced by pyrophosphate which in turn is replaced by metaphosphate (chemical shift ~ −15 to −20 ppm). At 30 mol % P_2O_5, the ^{29}Si spectrum of $Na_2O \cdot 2SiO_2$–P_2O_5 glass contains only

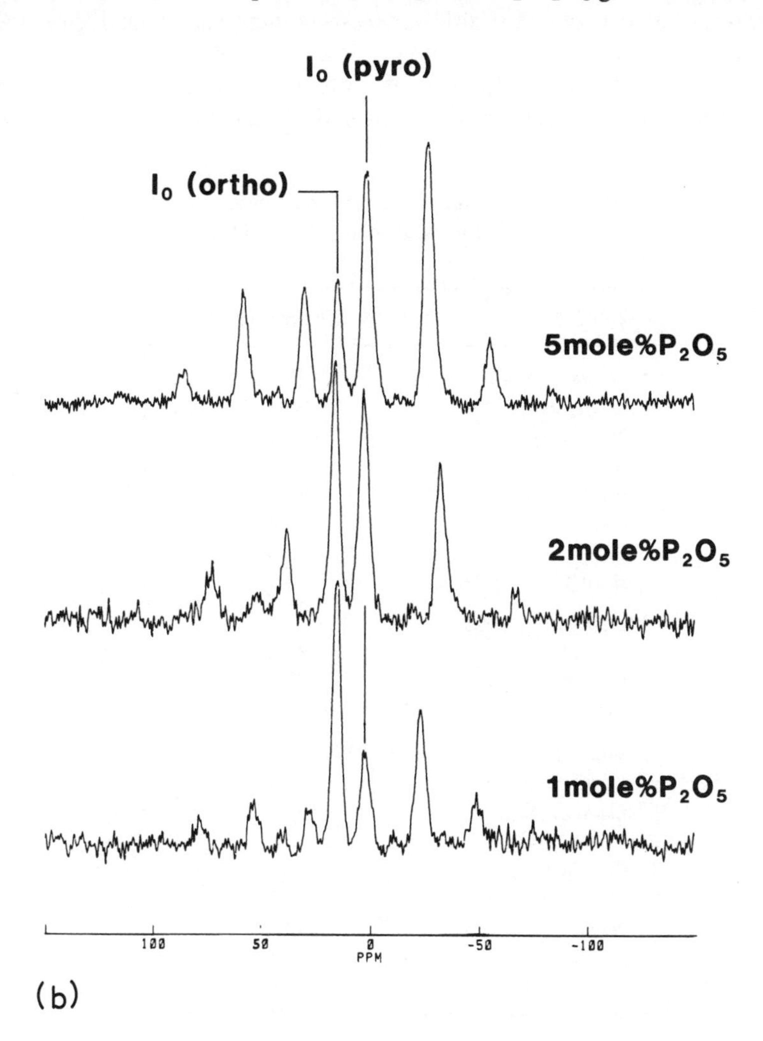

Q^4 units – however, at 40 mol % another peak appears at ∼ −213 ppm and grows in relative intensity as more P_2O_5 is added. This new peak originates from silicon in a six-co-ordinated environment as observed in a few crystalline compounds such as stishovite SiO_2 (−191 ppm), SiP_2O_7 (−214 ppm) and $Si_5O(PO_4)_6$ (−217 ppm).

1.4.6 Mineral glasses

The occurrence and microstructure of many minerals depends on the way in which glasses of magmatic origin cooled and devitrified. MAS NMR has been used as a tool by several groups of workers – notably Kirkpatrick *et al.* (1985, 1986), Murdoch *et al.* (1985) and Putnis *et al.* (1985) – to study the structure of glasses of composition corresponding to geological glasses and also their mineral derivatives. Table 1.1 is a compilation of the shifts for various binary and ternary glasses of synthetic origin (based on Murdoch *et al.*, 1985). In all cases the position of the peak maximum in the glass

Table 1.1 ^{29}Si chemical shifts in some mineral related glasses (based on Murdoch *et al.* (1985) with additions)

Type of glass	*Chemical shift* (ppm)
Disilicate glasses	
$Li_2Si_2O_5$	−90.5
$Na_2Si_2O_5$	−88.5
$K_2Si_2O_5$	−90.5
$Rb_2Si_2O_5$	−94.5
$Cs_2Si_2O_5$	−93.0
$SrSi_2O_5$	−92.5
$BaSi_2O_5$	−92.5
Metasilicate glasses	
$CaSiO_3$	−81.5
$CaMgSi_2O_6$	−82.0
$MgSiO_3$	−81.5
Aluminosilicates	
$CaAl_2Si_{12}O_{28}$	−107.1
$CaAl_2Si_6O_{16}$	−101.0
$CaAl_2Si_4O_{12}$	−95.6
$CaAl_2Si_2O_8$	−86.5
$CaAl_2SiO_6$	−82.3
$NaAlSi_3O_8$	−97.9
$NaAlSi_2O_6$	−92.8
$NaAlSiO_4$	−86.0

indicates the same dominant *Q* species as in crystalline materials. The authors observed that the chemical shift range covered by the glass peak increased as the size of the modifier cation decreased and charge increased. This they interpreted as reflecting the more disruptive nature of the highly polarizing cations resulting in the formation of a greater number of different *Q* species so that a localization of [nbo] is occurring in the glass. It should be noted that their half-width values and trends differ from those reported by other workers, where the trend is opposite to that claimed for disilicate glasses and in the case of the metasilicate glasses we would prefer to describe the Rb and Cs metasilicate data as being anomalous. The shape of the silicon resonance may reflect different Si–O–Si bond angles in addition to number of *Q* types. It will also be remembered that another feature of highly polarizing modifier cations is their tendency to induce phase separation which effectively duplicates the number of species present each of which, although of nominally the same type, has a slightly different environment.

In the case of the aluminosilicate glasses examined by Murdoch *et al.* (1985) the linewidth is determined mainly by the Si/Al ratio which they attribute to a manifestation of Lowenstein's aluminium avoidance principle. A secondary effect of polarizing power of the different modifier cations is also seen here.

1.4.7 Halide glasses

Several halide-containing glasses have been examined by wide-line NMR with particular reference to fast-ion conduction in these materials and the consequent motional narrowing of the ^{19}F lines (Bray *et al.*, 1983). High-resolution spectra of ^{19}F in quartz glass have been reported by Yonemari *et al.* (1986). Some MAS NMR data have been reported for the ^{23}Na resonance in fast-ion conducting haloborate glass-ceramics showing (a) three sodium sites, two of which may provide mobile Na^+, and (b) that chlorine plays an important part in the creation of these sites (Aujla *et al.*, 1987b).

The structure of glasses based on fluorozirconates is of considerable interest since the potential use of these materials as IR optical fibres relies on their being sufficiently stable to be capable of drawing into fibres of many kilometres length. In the fluorozirconate glasses, the unspun linewidth is of the order of 60–80 kHz. This cannot be narrowed simply by spinning and will require multiple pulse sequences to reduce the residual broadening. The width of the lines is, however, indicative of there being at least two different environments for fluorine in these glasses. Other nuclei which form constituents of fluoride glasses, such as Ba, Na, Al, La and Zr are capable of examination by this technique but all have quadrupole moments and therefore will exhibit broad lines, particularly in low symmetry environments.

1.4.8 Oxynitride glasses

The ^{29}Si resonance in a series of oxynitride glasses of general composition $Y_{1.04}Si_{1.27}Al_{1.27}O_{6-x}N_x$ ($0 \leq x \leq 1.2$) was investigated by Aujla *et al.* (1986). Distinct lines from the different $Si(O,N)_4$ tetrahedra were not resolved, but the data could be best fitted with a distribution of tetrahedral units having a shift change of 12.5 ppm for each replacement of an oxygen by a nitrogen and by assuming that each nitrogen enters the network producing 2($\pm$0.2) Si–N bonds and 1 Al–N bond (see Chapter 4 for further discussion of oxynitride glass structure).

1.5 DEVITRIFICATION

Glass-ceramics are derived from the controlled crystallization of glasses to give a material consisting of one or more crystal phases plus some residual glass depending on the starting composition and the heat treatment given.

During the crystallization process, molecular rearrangements occur to produce the appropriate crystalline structures. These structures are often themselves metastable polymorphs which under further heat treatment can transform to the thermodynamically stable crystal phases. These molecular rearrangements present a changing environment for the various nuclei in the glass and MAS NMR provides a technique for observing these various environments and following the crystallization process.

Several papers have been published in which the glassy and crystalline environments have been compared.

1.5.1 Magnesium aluminosilicates

Fyfe *et al.* (1986) studied the crystallization of a glass of the stoichiometric cordierite composition $Mg_2Al_4Si_5O_{18}$. The ^{29}Si resonance from the glass consists of a single broad peak but after 2 minutes at 1185°C, complete crystallization has occurred to give two multiple resonances based on chain and ring sites with peak splitting arising from the varying number of Al in the second co-ordination sphere. After prolonged heat treatment (~2000 hours) the initial crystallized phase has fully transformed to the thermodynamically stable orthorhombic polymorph in which Si/Al ordering has taken place so that the chain silicons are now completely Si(4Al) whereas the ring silicons are Si(3Al).

The ^{27}Al resonance was similarly examined. On crystallization, the resonance narrowed but remained at the position expected for tetrahedrally co-ordinated Al^{3+}. Ordering of the structure produces a small progressive change in chemical shift.

1.5.2 Magnesium aluminoborates

In this work (Dupree *et al.*, 1985c) the changes in the ^{27}Al spectrum during crystallization of a magnesium aluminoborate glass were used to determine which aluminium sites in the glass were involved in crystal formation. The parent glass $35MgO \cdot 20Al_2O_3 \cdot 45B_2O_3$ produced three distinct Al resonances which were assigned to $Al(OAl)_4$ at 58 ppm, $Al(OB)_4$ at 29 ppm (based on the findings of Muller *et al.* (1983a) for the $CaO–Al_2O_3–P_2O_5$ system) and $Al(OAl)_6$ at 1 ppm. The latter peak remained fairly constant in intensity over the composition range covered by the glass-forming region whereas the first two varied as the $B_2O_3 : Al_2O_3$ ratio changed, the second growing at the expense of the first as B_2O_3 is increased. On crystallization, the following changes to the spectrum are observed: (a) the $Al(OAl)_4$ peak diminishes in size, (b) the $Al(OB)_4$ peak moves to more positive chemical shift, (c) the $Al(OAl)_6$ peak also moves more positive. One interpretation of (a) and (b) is that both $Al(OAl)_4$ and $Al(OB)_4$ are being replaced by $Al(OAl)_{(4-x)}(OB)_x$ in the crystalline phase $Al_{18}B_4O_{33}$ which is produced.

1.5.3 Alkali silicates

(a) $Li_2O–SiO_2$

Schramm *et al.* (1984) related the bulk nucleation rates in $Li_2O–SiO_2$ glasses to the nature of the dominant Q species in the original glass. Their detailed analysis of Q distributions has been criticized (Gladden *et al.*, 1986b) and the situation is obviously complicated by phase separation.

The peaks observed in the spectra of the devitrified samples do not necessarily reflect the phase formation predicted by the thermodynamic phase diagram. For example, a $40Li_2O \cdot 60SiO_2$ glass might be expected to produce $Li_2O \cdot 2SiO_2$ and $Li_2O \cdot SiO_2$, i.e. Q^3 and Q^2 in the ratio 1 : 1. In practice, Q^2, Q^3 and Q^4 are obtained. Schramm *et al.* ascribe this to disproportionation during devitrification of the form $2Q^3 \rightarrow Q^2 + Q^4$. From the appearance of their spectrum it is obvious that the Q^4 represents residual glass and thus an alternative description of the devitrification of the glass is early nucleation of $Li_2O \cdot SiO_2$ (i.e. Q^2). Growth on these nuclei, involving diffusion of Li^+ to growth sites will leave a residual glass depleted in Li^+ and from which $Li_2O \cdot 2SiO_2$ can grow only in reduced quantities, depending on the level of remaining Li^+, to leave excess silicon as Q^4.

A more detailed examination of the crystallization behaviour of compositions around 33.3 mol % Li_2O has been carried out by Mortuza (1988) for materials with different levels of nucleating agent P_2O_5 q.v.

(b) Na_2O–SiO_2

The effect of devitrification on the Si and Na sites in Na_2O–SiO_2 glasses was studied by Dupree *et al.* (1984c). The spectra obtained are shown in Fig. 1.11. There are two points of interest here. Firstly, for the $Na_2O \cdot 2SiO_2$ phase formed, the crystal structure determined by XRD contains only single crystallographic sites for Si and for Na. However, NMR shows that there are two chemically distinct sites. Secondly, there are large differences in chemical shift between the sites in crystal and glass.

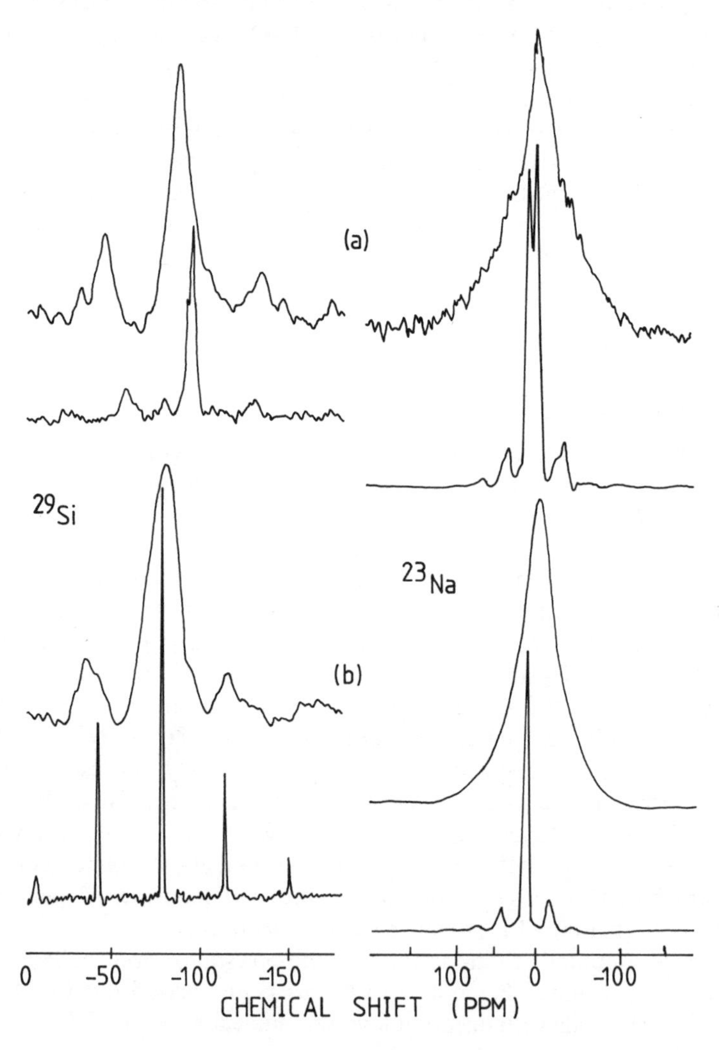

Fig. 1.11 ^{29}Si and ^{23}Na spectra for glassy (upper spectra) and devitrified (lower spectra): (a) $Na_2O \cdot 2SiO_2$ and (b) $Na_2O \cdot SiO_2$ (after Dupree *et al.*, 1984b).

(c) Rb_2O–SiO_2 and Cs_2O–SiO_2

The silicon spectra of glasses and the crystals formed from these systems were analysed as above in a similar fashion (Dupree *et al.*, 1986a). In many cases the single, 'mean' site in the glass is replaced by two or more resolved sites in the crystal where the requirements of 3D repeatability takes precedence over individual ion site preferences.

(d) PbO–SiO_2

The devitrification of glasses of the PbO–SiO_2 system and the resulting changes in NMR spectra have been studied by Dupree *et al.* (1987a) and Lippmaa *et al.* (1982). The crystallization behaviour is complex. Several separate phases can be obtained from a given composition by judicious choice of heat treatment (Figure 1.12(a)) and the molecular rearrangements during crystallization can produce intermediate species (Fig. 1.12(b)). The assignment of peaks to the individual Si environments is complicated (a) by the presence of Si–O–Pb bonds which cannot be simply approximated as Si-(nbo) as in the case of the alkali silicates and (b) by large deviations of the Si–O–Si angles from the usual dihedral value of ~140°.

The crystallization of alamosite $PbSiO_3$ from the 50 mol % glass is particularly complex in that there are several intermediate structures as shown (Fig. 1.12(b)). The first stage represents a rearrangement of the glass structure without crystallization. The Q^2 chains of the glass structure (broad peak at −85 ppm) have given rise to new species at ~ −77 ppm and ~ −98 ppm. Two possible explanations for this might be:

1. Partial disproportionation of Q^2 to give higher and lower Q species. The chemical shifts of these new species would suggest Q^0 for the −77 peak and Q^3 for the −98 peak. As crystallization proceeds, the Q^0 species rejoins the Q^2 chains to produce two crystal sites of −84 and −86.2 ppm and the excess PbO joins the Q^3 sites to give a Q^2 site of −94.7 ppm;
2. The Si–O–Si bond angle distribution in the glass is largely comprised of ~140° angles. In the crystal the three sites have differing Si–O–Si bond angles; see Fig. 1.12(c) where the peak positions assigned to these sites on the basis of their bond angles are indicated. As the Si–O–Si bond angle strongly affects chemical shifts it is possible to assign the NMR peaks to individual sites on the basis of their bond angle. The intermediate structure may represent the first stages of bond distortion whilst still in the glassy state. The species giving rise to the peak at −77 ppm could be a mobile Q^0 which on further heating recombines to give Q^2 and Q^3. It cannot be said whether these distorted sites are homogeneously dispersed or whether some phase separation occurs.

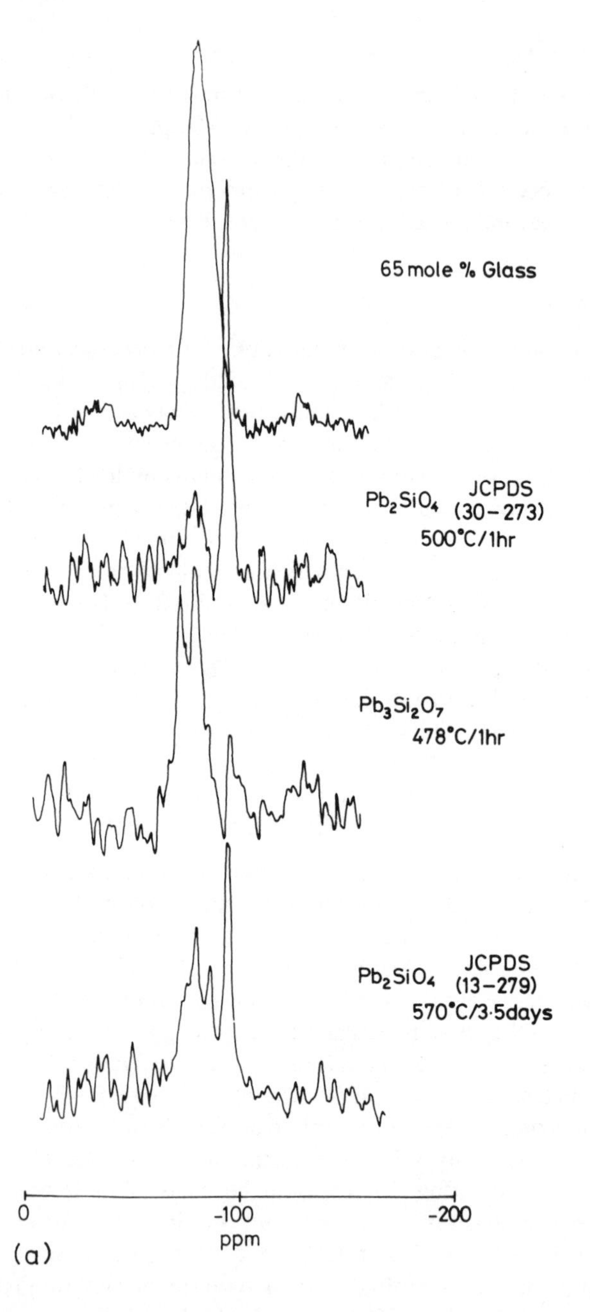

Fig. 1.12 (a) ^{29}Si spectra in 65% PbO 35% SiO_2 glass and after various heat treatments; (b) ^{29}Si spectra in a 48% PbO 52% SiO_2 glass after various heat treatments; (c) the three silicon sites in $PbSiO_3$ showing the Si–O–Si bond angles and the ^{29}Si position assigned to each site.

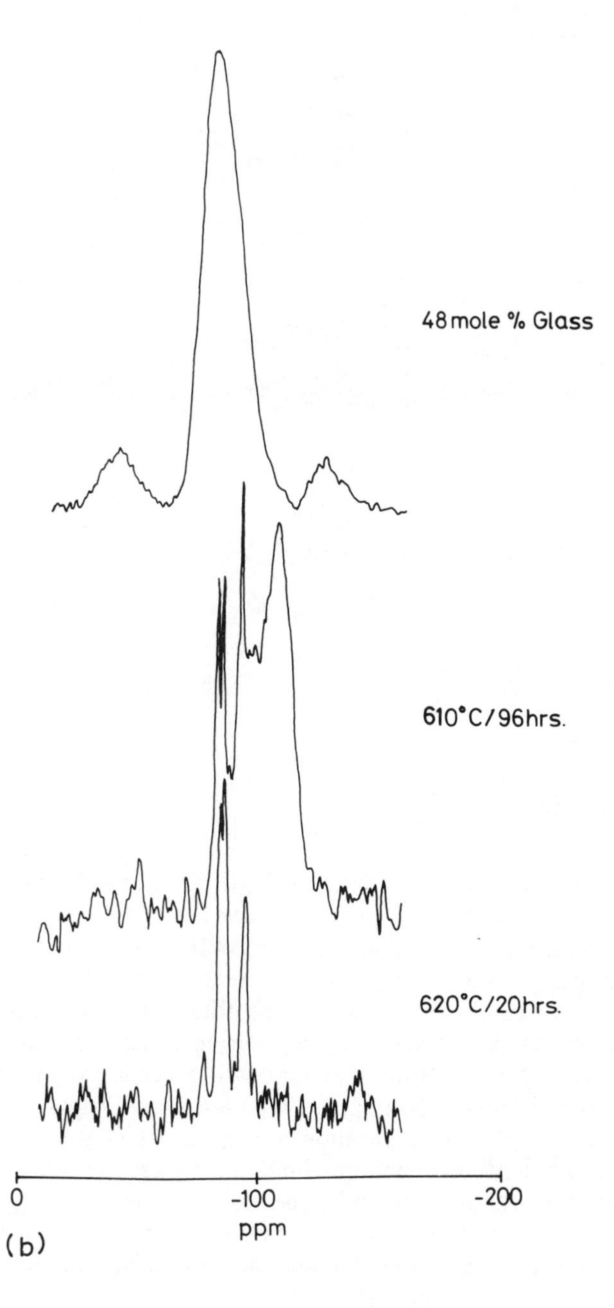
48 mole % Glass
610°C/96hrs.
620°C/20hrs.
0
-100
-200
ppm
(b)

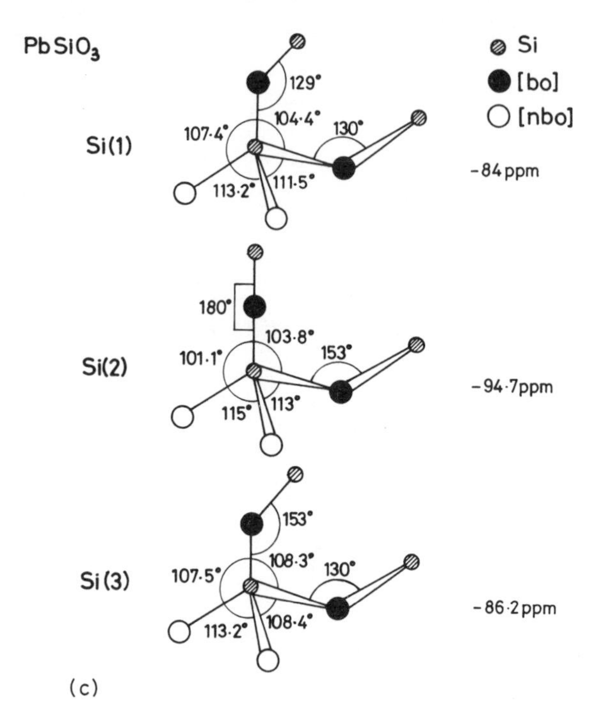

Fig. 1.12 *contd*

1.6 PHASE SEPARATION

Depending on composition, two or more glass phases may be thermodynamically more stable than a single phase at a given temperature. If two phases exist above the liquidus, then we have 'stable' phase separation and the two phases will be retained upon cooling to the glassy state. If the two phases exist below the liquidus only, then we have 'metastable' phase separation and a single phase glass can be obtained by rapid cooling. Heat treatment of this single phase can now bring about separation into two phases either by a nucleation and growth mechanism or by spinodal decomposition. For a review of phase separation see Chapter 6.

The presence of phase separation in a glass has consequences for the silicon and other nuclear environments. In the case of silicon, the total number of [nbo] is still dictated by the stoichiometry of the glass but the distribution of the [nbo] may be quite different from a single phase glass. If we neglect the interfacial sites, then the amount of each phase will depend on overall glass composition $xR_2O\cdot(1-x)SiO_2$ and on the composition of each phase. If one phase has the composition $aR_2O\cdot(1-a)SiO_2$ then the composition of the other is given by $(x-pa)R_2O\cdot(1-p-x+pa)SiO_2$ where p is the mole fraction of the first phase. Whilst any Q distribution is possible between the two phases, the typical product of phase separation of a glass containing initially Q^n and $Q^{(n-1)}$ is two phases in which one phase is enriched in Q^n and the other in $Q^{(n-1)}$, the limits being the pure Q types. The question now arises as to whether NMR can distinguish between a single, homogeneous $Q^n/Q^{(n-1)}$ distribution and two coexisting $Q^n/Q^{(n-1)}$ distributions in which the total amounts of Q^n and $Q^{(n-1)}$ may be the same. There are two ways in which the two systems differ which are detectable by NMR.

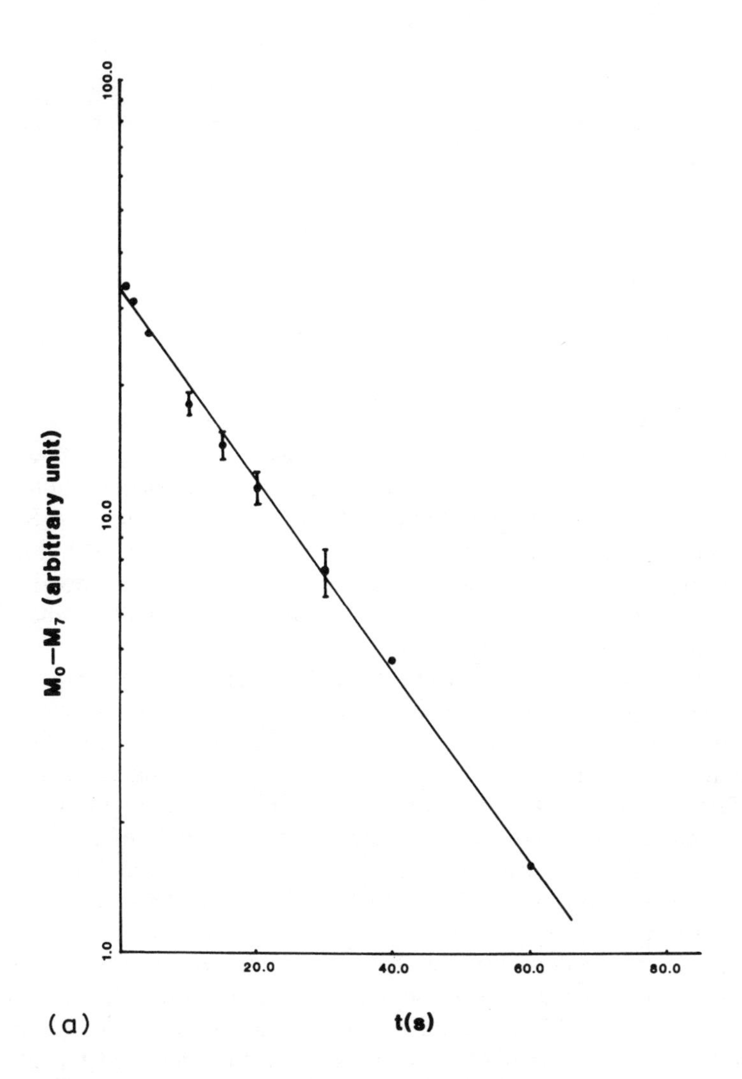

Fig. 1.13 ^{29}Si relaxation time T_1 plots for (a) $Na_2O{\cdot}2SiO_2$ glass, (b) $Li_2O{\cdot}2SiO_2$ glass possibly phase separated.

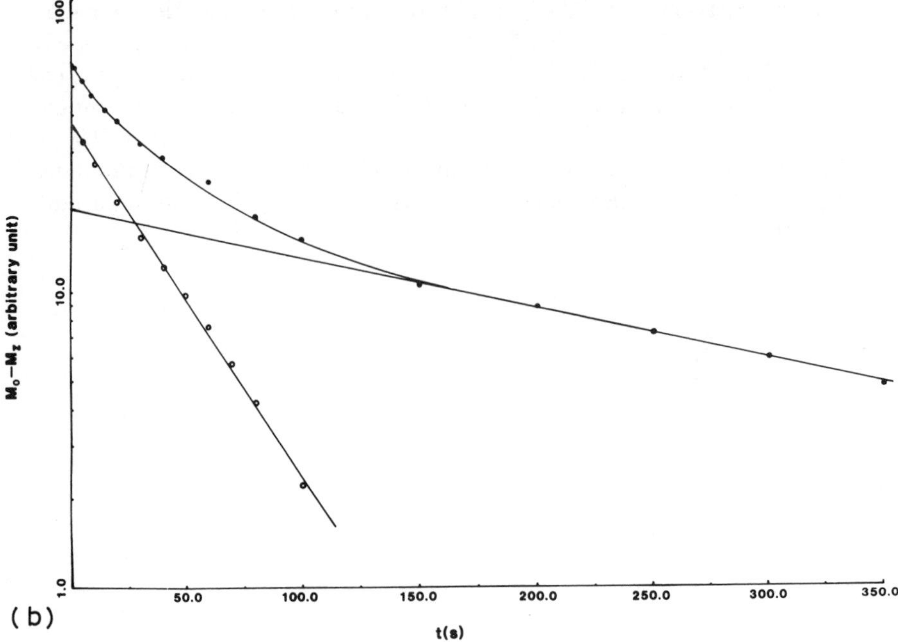

Fig. 1.13 *contd*

1. The chemical shift of a given Q^n species is an approximately linear function of the local composition, e.g. Q^3 in $15Na_2O \cdot 85SiO_2$ has a chemical shift of -94 ppm but in $40Na_2O \cdot 60SiO_2$ the shift is -87 ppm. Hence a mixture of two phases, each of which contains Q^3 but in very different compositions should produce two peaks. There is however the problem of resolving the two peaks when the typical half-widths of ^{29}Si resonances in glass are 10–15 ppm;
2. The relaxation times for ^{29}Si ($I = 1/2$) are very long and also very sensitive to environment. Therefore, if the relaxation times are measured for a phase separated glass in which Q^n and $Q^{(n-1)}$ occur in both phases, then four separate relaxation times are obtained. This exercise has been carried out (Mortuza, 1988) for the Li_2O–SiO_2 system. Figure 1.13 shows the intensity versus relaxation delay curves for Q^3 in lithium and sodium disilicate glasses. It can be seen that the graph is a straight line for Na_2O–SiO_2 material but curved for the Li_2O–SiO_2 system. This is because there is only a single Q^3 site in the Na_2O case where there is no phase separation, whereas in the Li_2O case there are probably two separate Q^n types, each contributing its separate relaxation time, the resultant of which is a curve. This may indicate phase separation. From these curves, not only can the individual relaxation times be obtained but also the quantities of Q^n type in each phase and hence the composition of each phase can be calculated.

1.7 CERAMICS

A wide variety of materials could be discussed under this heading and considerable work has been done on crystalline silicates (Smith, M. E., 1987; 1983, Kirkpatrick *et al.*, 1985) and on zeolites (Klinowski, 1984). However, we shall restrict this section to some aluminium compounds and to silicon carbide and nitride based ceramics.

1.7.1 Silicon oxynitrides

These form the basis of some well-known high-temperature ceramics and several compositions within the $\beta' Si_{3-x}Al_xO_xN_{4-x}$, $0 \leqslant x \leqslant 2$, systems have been investigated using both ^{29}Si and ^{27}Al (Dupree *et al.*, 1985d, 1988). The crystal structures were known to be predominantly based on differing arrangements of (Si, Al) $(O, N)_4$ tetrahedra. The terminal nitride Si_3N_4 has each nitrogen atom linked to three silicon atoms in planar arrays stacked parallel to the basal plane of a hexagonal symmetry unit cell. Two stacking sequences are observed for this material ABCD for α-Si_3N_4 and ABAB for β-Si_3N_4. In α-Si_3N_4 there are two silicon sites, the main difference being 0.0021 nm in one of the Si–N tetrahedron bond lengths. The ^{29}Si shift for β-Si_3N_4 occurs at -48.5 ppm whilst for α-Si_3N_4 the shifts are -48.0 ppm and -49.7 ppm. As x is increased from 0 to 2, no change in the ^{29}Si position occurs (see Fig. 1.14), indicating that throughout this range of compositions silicon is co-ordinated solely to nitrogen. A small increase in width occurs indicating a spread in nearest neighbour (n.n.) bond length or angle and/or substitution in next nearest neighbour (n.n.n.) sites. In a more oxygen-rich composition of the related β'' compound in which both magnesium and aluminium substitution are used to form $Mg_5AlSi_3O_{11}N$ two peaks at -63.1 ppm and -75.5 ppm are observed. The peak at -63.1 ppm must arise from SiN_3O tetrahedra as a peak at an identical position (see Fig. 1.14) occurs in Si_2N_2O whose structure is known to consist of SiN_3O tetrahedra linked by Si–O–Si bonds. The peak at -75.5 ppm must correspond to a more oxygen-rich co-ordination tetrahedron. The ^{27}Al spectra (Dupree *et al.*, 1988) are consistent with this picture of these materials but in addition show the previously unsuspected presence of small amounts of octahedrally co-ordinated $[AlO_6]$ aluminium in the higher oxygen content β' compounds and allow a model of the structure of the β' material to be suggested.

Several polytypoids of general composition $Si_{6-x}Al_{x+y}O_xN_{y+8-x}$ have been investigated as well as sialon X-phase with approximate composition $Si_3Al_6O_{12}N_2$ (Butler *et al.*, 1984; Klinowski *et al.*, 1984). The ^{27}Al spectra of the polytpoids show both octahedral $[AlO_6]$ and tetrahedral $[AlN_4]$ peaks with positions very similar to the octahedral peak in γ-Al_2O_3 and to AlN. Klinowski *et al.* (1984) reported two ^{29}Si peaks with a not very good signal-to-noise in their mostly 15R polytypoid; one at ~ -48 ppm

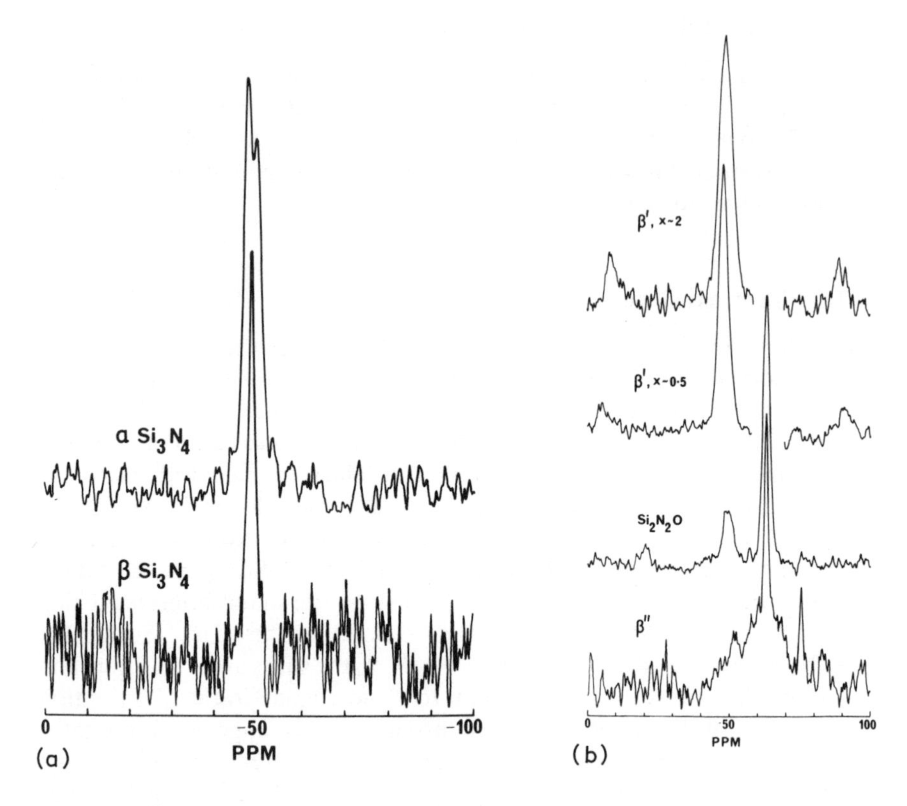

Fig. 1.14 ^{29}Si spectra of (a) α and β Si_3N_4, (b) the β' form of $Si_{3-x}Al_xO_xN_{4-x}$ for varying x, Si_2N_2O and β'' $Mg_xAl_{4-x}Si_2O_{4+x}$ (after Dupree *et al.*, 1985d).

which is clearly similar to Si_3N_4; the other smaller peak occurs at ~ -36 ppm. In our studies of a more single phase 15R sample only one peak at -48 ppm was visible. It is possible that the -36 ppm peak arises from other polytypoids 12H or 21R present in their sample; however, [SiN_4] tetrahedra have so far been observed only in the range -42 to -49 ppm. The one exception known to us is $Y_6Si_3N_{10}$ where non-bridging nitrogens must occur (Smith, 1987).

In the sialon X-phase ^{27}Al peaks were seen at 0.8 ppm and 67 ppm corresponding to [AlO_6] and [AlO_4] units respectively. However, because of large electric field gradients which are likely to occur at aluminium in mixed tetrahedra such as AlO_3N these could be present but contribute no signal; no estimate of the fraction of aluminium actually observed was given. No ^{29}Si signal could be observed in this sample – presumably because of the long relaxation time.

1.7.2 Silicon carbide

This has many different polytypes; the spectrum for the 6H polytype (hexagonal α-SiC) gives three ^{29}Si peaks at -13.9, -20.2 and -24.5 ppm corresponding to the three distinct silicon environments present. Similarly the ^{13}C gives three peaks for the 6H prototype whereas cubic β-SiC has only one ^{29}Si peak at -18.3 ppm (Finlay *et al.*, 1985). Silicon carbide, as used commercially, is usually a mixture of polytypes with some amorphous material present. The broad peak from the amorphous material can often dominate the spectrum because it contains traces of paramagnetic impurity which drastically reduces the spin lattice relaxation time of the ^{29}Si. Thus to ensure quantitative data long delays between pulses are required.

1.7.3 Aluminium oxides, hydroxides and phosphates

The chemistry of the Al–O–H ternary is complex. John *et al.* (1983) and Mastikhin *et al.* (1981) used MAS NMR to follow the changes in Al co-ordination and environment during the conversion of boehmite and bayerite to α-Al_2O_3 through the several intermediate phases. All the structures are based on the spinel structure with varying occupation of the tetrahedral and octahedral sites. This is reflected in the ^{27}Al spectra by peaks of differing intensity at ~ 60 ppm and ~ 5 ppm respectively. The exact position depends on field because of the second order quadrupolar shift. The fraction of octahedral aluminium is reported as 0.94 for boehmite, 0.84 for bayerite, 0.75 for δ-Al_2O_3 and γ-Al_2O_3, 0.65 for η-Al_2O_3, ~ 1.0 for θ-Al_2O_3 and 1.0 for α-Al_2O_3. The ratios were strongly influenced by surface area.

Muller *et al.* (1984) report the chemical shifts for tetrahedral and octahedral aluminium in aluminium phosphates. These occur at -13 to -21 ppm for octahedral and $+39$ ppm for tetrahedral aluminium. These values are shifted upfield by ~ 30 ppm compared to the same co-ordination in aluminium oxides, reflecting greater shielding in the presence of phosphorus.

1.7.4 Cements

The hydration of calcium aluminate ($CaO{\cdot}Al_2O_3$) cement was followed using ^{27}Al MAS NMR by Muller *et al.* (1984). They showed that hydration is accompanied by a change of aluminium co-ordination from tetrahedral to octahedral. By measuring the ratio of each co-ordination type, the hydration reaction could be followed as a function of time for different curing temperatures.

1.7.5 Spinels

The distribution of Al^{3+} between octahedral and tetrahedral sites in $MgAl_2O_4$ has been studied as a function of both temperature (Wood *et al.*, 1986) and stoichiometry (Dupree *et al.*, 1986b). In the former, the degree of inversion (the atomic fraction of Al in tetrahedral sites) increased from 0.02 at 700°C to 0.39 at 900°C. In the latter case, the incorporation of excess Al_2O_3 was shown to result in the occupation of tetrahedral sites by the excess Al at a rate consistent with a model in which the accompanying charge-balancing vacancies occupy octahedral sites only.

1.8 MATERIALS PRODUCED BY LOW-TEMPERATURE PROCESSES

1.8.1 Sol–gel

The technological importance of this technique of glass and ceramic preparation has, as yet, received little attention from the point of view of observation of changes in structure during the various stages in production – gellation, drying, sintering etc. Maciel *et al.* (1980) used cross-polarization and MAS of ^{29}Si to study a silica gel surface where Si–H cross-polarization is effective. They observed three peaks at -109.3, -99.8 and -90.6 ppm which they identified as $Si(OSi)_4$, $Si(OSi)_3OH$ and $Si(OSi)_2(OH)_2$ respectively.

1.8.2 Alumina gels

The environment of aluminium in alumina gels and in both single phase and diphasic mullite gels and their transformation products were studied by Komarneni *et al.* (1985). Both the gel prepared from boehmite and that from aluminium sec-butoxide gave spectra very similar to γ-alumina after heating to 500°C, i.e. both 4 and 6 co-ordinated aluminium were present although the ratio of tetrahedral to octahedral sites was apparently different, indicating perhaps that the vacancies in the structure are differently ordered in the two materials. Both mullite gels also gave spectra similar to γ-alumina after heating to $\sim$1050°C but on further heating to 1380°C showed a large increase in tetrahedral aluminium with a ratio approaching 1.25 : 1, the proper ratio for 3 : 1 mullite. Both tetrahedral and octahedral lines were much broader in the mullite material, indicating considerable disorder in this material and the material prepared from the single phase gel appeared to show two different tetrahedral Al environments. As no mention was made of whether all the aluminium atoms in the sample were contributing to the signal and because of the large electric field gradient present at distorted

Al sites in this type of material, as mentioned in the introduction, some caution is necessary in interpreting the tetrahedral to octahedral ratio in these materials.

1.8.3 SiO_x

Interest in the structure of *a*-SiO_x stems from the important role of oxide layers on device silicon. Several models of the structure of SiO_x have been proposed with the two main contenders being the random-bond (RB) model (Phillipp, 1972) and the random-mixture (RM) model (Temkin, 1975). The former assumes that Si–Si and Si–O bonds are statistically distributed throughout the SiO_x structure whilst the latter assumes regions of *a*-Si and *a*-SiO_2 with linear dimensions 10 Å and a significant amount of interphase material in which there are tetrahedral units Cn $Si(Si_{4-n}O_n)$ $0 \leq n \leq 4$ other than $Si(Si)_4$ C0 and $Si(O)_4$ C4. The predicted abundance of the different tetrahedral units Cn depends on the detailed arrangement. For macroscopic regions of *a*-Si, *a*-SiO_2, $C0 \sim C4 \sim 0.5$. Temkin assigned roughly equal volumes to *a*-Si, *a*-SiO_2 and interphase material, which gives $C0 \sim C4 \sim 1/3$ and $C1 + C2 + C3 \sim 1/3$. Ching (1982) refined this to $C0 = 0.191$, $C1 = 0.290$, $C2 = 0.167$, $C3 = 0.031$ and $C4 = 0.321$. Experimental evidence has been found for both models and seems to indicate that the method of preparation may dictate which structure is formed.

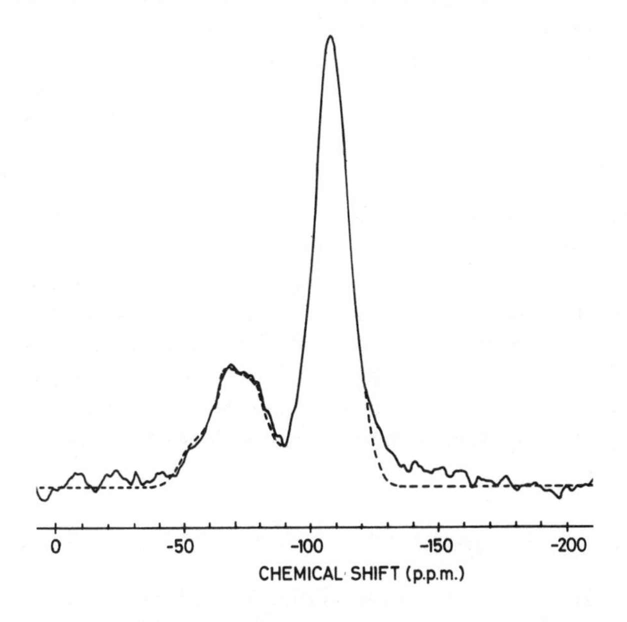

Fig. 1.15 ^{29}Si spectrum of SiO together with spectrum calculated from the Ching mixture model (shown dashed) (after Dupree *et al.*, 1984a).

Dupree *et al.* (1984a) investigated three commercial samples of SiO of greater than 99.99% purity but slightly oxygen-rich (x in $SiO_x = 1.1, 1.28, 1.25$). The ^{29}Si spectrum of all three were similar and an example (Fig. 1.15) shows one large broad approximately gaussian line at -109 ppm and a smaller, structured, line centred at -70 ppm. The line at -109 ppm is clearly from SiO_4 units and the other peak must be a mixture of C0–C3 units. The relative smallness of this peak could only be explained after quantitative experiment showed that slightly less than 50% of the silicon nuclei were being detected. ESR revealed a large spin density and nuclei within the 'wipe-out' radius of the unpaired electrons will not contribute to the NMR signal. The effect on each of the five contributing resonances depends on their location with respect to the unpaired electrons. Bond strength considerations indicate that Si–Si bonds break in preference to Si–O bonds and therefore the probability of a broken bond (unpaired electron) occurring near a specific Cn type increases as n decreases and the dashed line in Fig. 1.15 is the predicted spectrum obtained using the Ching model and applying a 50% total reduction for the C0–C3 units in the ratio 4 : 3 : 2 : 1.

1.9 CONCLUSION

MAS NMR is a technique with considerable potential for use in the understanding of glass and ceramic structures. Although its availability has been limited until the last few years it has provided considerable insight into the complex problems associated with these materials.

In view of the relative newness of the technique we have listed in Table 1.2 some isotopes of potential use in the investigation of glasses and ceramics. The table is divided into (i) $I = \frac{1}{2}$ nuclei (a) readily observable, (b) observable with difficulty (possibly requiring isotopic enrichment) and (ii) $I > \frac{1}{2}$ nuclei (a) readily observable in most environments, (b) readily observable only in relatively symmetric environments and (c) observable in very symmetric environments. It should be noted that for large Z isotopes the chemical shift anisotropy is likely to be large so that even for $I = \frac{1}{2}$ nuclei, lines are likely to be broad and relatively weak in glasses and split into many sidebands in anisotropic materials. The situation is exacerbated for large Z $I > \frac{1}{2}$ nuclei by the Sternheimer antishielding factor which increases rapidly with Z and acts to amplify the effective electric field gradient at the nucleus. This factor is greater than 100 for the last row of the periodic table.

ACKNOWLEDGEMENT

We would like to acknowledge the influence of Professor P.W. McMillan who recognized from the outset the utility of MAS NMR in studies of glass structure and was strongly supportive throughout the early stages of our work.

Table 1.2 Some isotopes of potential use in the investigation of glasses and ceramics

Isotope	*Sensitivity at natural abundance relative to* ^{29}Si
$I = 1/2$	
(a) Readily observable	
^{1}H	2.7×10^{3}
^{13}C	0.48
^{19}F	2.2×10^{3}
^{29}Si	1.00
^{31}P	180
^{77}Se	1.4
^{89}Y	0.32
^{113}Cd	3.6
^{119}Sn	12
^{125}Te	6.0
^{195}Pt	9.1
^{199}Hg	2.6
^{205}Tl	35
^{207}Pb	5.6
(b) Observable with difficulty or requiring isotopic enrichment	
^{15}N	1.0×10^{-2}*
^{57}Fe	1.0×10^{-3}*
^{103}Rh	8.5×10^{-2}
^{109}Ag	0.13
^{183}W	2.8×10^{-2}
$I > 1/2$	
(a) Readily observable	
^{7}Li	730
^{9}Be	37
^{11}B	360
^{17}O*	2.9×10^{-2}
^{23}Na	250
^{27}Al	560
^{51}V	1030
^{133}Cs	130
(b) Readily observable only in relatively symmetric environments	
^{25}Mg	0.73
^{33}S*	4.7×10^{-2}
^{37}Cl	1.7
^{39}K	1.3
^{43}Ca*	2.3×10^{-2}

Table 1.2 *continued*

Isotope	*Sensitivity at natural abundance relative to* ^{29}Si
^{45}Sc	820
^{55}Mn	470
^{59}Co	750
^{65}Cu	96
^{67}Zn	0.32
^{71}Ga	150
^{81}Br	133
^{87}Rb	133
^{93}Nb	1.3×10^3
(c) Observable in very symmetric environments	
^{49}Ti	0.44
^{53}Cr	0.23
^{61}Ni	0.11
^{73}Ge	0.30
^{75}As	69
^{87}Sr	0.51
^{91}Zr	2.9
^{95}Mo	1.4
^{105}Pd	0.68
^{115}In	910
^{121}Sb	250
^{127}I	260
^{135}Ba	0.89
^{137}La	160
^{209}Bi	380

* If isotopically enriched.

REFERENCES

Andrew, E. R. (1981a) *Int. Rev. Phys. Chem.*, **1**, 195.

Andrew, E. R. (1981b) *Philos. Trans. R. Soc. London, Ser. A*, **299**, 505.

Aujla, R. S., Dupree, R., Farnan, I. and Holland, D. (1987a) *Proc. 2nd Int. Conf. Effects of Modes of Formation on the Structure of Glass* (eds P. L. Kinser and R. A. Weeks) **99**, 53–54.

Aujla, R. S., Dupree, R., Holland, D. and Kemp, A. P. (1987b) *Mater. Sci. Forum* **19–20**, 147.

Aujla, R. S., Leng-Ward, G., Lewis, M. H. *et al.* (1986) *Philos. Mag. Lett.*, **54B**, L51.

Bray, P. J., Bucholtz, F., Geissberger, A. E. and Harris, I. A. (1982) *Nucl. Instrum. Methods* **199**, 1.

Bray, P. J. and Gravina, S. J. (1985) *Mater. Sci. Res.*, **19**, 1.

Bray, P. J., Hintenlang, D. E., Mulkern, R. V. *et al.* (1983) *J. Non-Cryst. Solids* **56**, 27.
Butler, N. D., Dupree, R. and Lewis, M. H. (1984) *J. Mater. Sci. Lett.*, **3**, 469.
Ching, W. Y. (1982) *Phys. Rev. B*, **26**, 6610, 6622.
Devine, R. A. B., Dupree, R., Farnan, I. and Capponi, J. J. (1987) *Phys. Rev. B*, **35**, 2305.
Dupree, R., Farnan, I., Forty, A. J. *et al.* (1985a) *J. Phys. (Paris).*, **C8**, 113.
Dupree, R., Ford, N. and Holland, D. (1987a) *Phys. Chem. Glasses*, **28**, 78.
Dupree, R., Holland, D., McMillan, P. W. and Pettifer, R. F. (1984) *J. Non-Cryst. Solids*, **68**, 399.
Dupree, R., Holland, D. and Mortuza, M. G. (1987b) *Nature (London)*, **328**, 416.
Dupree, R., Holland, D. and Mortuza, M. G. (1987c) *Phys. Chem. Glasses*, **29**, 18.
Dupree, R., Holland, D. and Williams, D. S. (1984a) *Philos. Mag.*, **50**, L13.
Dupree, R., Holland, D. and Williams, D. S. (1985b) *J. Phys. (Paris)*, **C8**, 119.
Dupree, R., Holland, D. and Williams, D. S. (1985c) *Phys. Chem. Glasses*, 50.
Dupree, R., Holland, D. and Williams, D. S. (1986a) *J. Non-Cryst. Solids*, **81**, 185.
Dupree, R., Lewis, M. H., Leng-Ward, G. and Williams, D. S. (1985d) *J. Mater. Sci. Lett.*, **4**, 393.
Dupree, R., Lewis, M. H. and Smith, M. E. (1986b) *Philos. Mag.*, **53A**, L17.
Dupree, R., Lewis, M. H. and Smith, M. E. (1988) *J. Appl. Cryst.* **21**, 109.
Dupree, R. and Pettifer, R. F. (1984a) *Nature (London)*, **308**, 523.
Engelhardt, G., Nofz, M., Forkel, K. *et al.* (1985) *Phys. Chem. Glasses*, **26**, 157.
Finlay, G. R., Hartman, J. S., Richardson, M. F. and Williams, B. L. (1985) *J. Chem. Soc. Chem. Commun.* 159.
Fujiu, T. and Ogino, M. (1984) *J. Non-Cryst. Solids*, **64**, 287.
Fyfe, C. A., Gobbi, G. C. and Putnis, A. (1986) *J. Am. Chem. Soc.*, **108**, 3218.
Gaskell, P.H. (1985) *J. Phys. (Paris)*, **C8**, 3.
Gladden, L. F., Carpenter, T. A. and Elliott, S. R. (1986a) *Philos. Mag.*, **B53**, L81.
Gladden, L. F., Carpenter, T. A., Klinowski, J. and Elliott, S. R. (1986b) *J. Magn. Reson.* **66**, 93.
Griffiths, L., Root, A., Harris, R. K. *et al.* (1986) *J. Chem. Soc. Dalton Trans.*, 2247.
Grimmer, A. R., Magi, M., Hahnert, M. *et al.* (1984) *Phys. Chem. Glasses*, **25**, 105.
Grimmer, A. R. and Muller, W. (1986) *Monatsh. Chem.*, **117**, 799.
Grimmer, A. R. and Radeglia, R. (1984) *Chem. Phys. Lett.*, **106**, 263.
John, C. S., Alma, N. C. M. and Hays, G. R. (1983) *Appl. Catal.*, **6**, 341.
Kirkpatrick, R., Kinsey, R. A., Smith, K. A. *et al.* (1985) *Am. Mineral.* **70**, 106.
Kirkpatrick, R. J., Oestrike, R., Weiss, C. A. Jr. *et al.* (1986) *Am. Mineral.*, **71**, 705.
Klinowski, J. (1984) *Prog. NMR Spectroscopy*, **16**, 237.
Klinowski, J., Thomas, J. M., Thompson, D. P. *et al.* (1984) *Polyhedron*, **3**, 1267.
Komarneni, S., Roy, R., Fyfe, C. A. and Kennedy, G. J. (1985) *J. Am. Ceram. Soc.*, **68**, C243.
Lippmaa, E., Magi, M., Samoson, A. *et al.* (1981) *J. Am. Chem. Soc.*, **103**, 4992.
Lippmaa, E., Samoson, A., Magi, M. *et al.* (1982) *J. Non-Cryst. Solids*, **50**, 218.
Maciel, G. E. and Sindorf, D. W. (1980) *J. Am. Chem. Soc.*, **102**, 7606.
Mastikhin, V. M., Krivoruchko, O. P., Zolotovskii, B. P. and Buyanov, R. A. (1981) *React. Kingt. Catal. Lett.*, **18** (1–2), 117.
Mortuza, G. M. (1988). PhD thesis, University of Warwick.
Muller, D., Berger, G., Grunze, I. and Ladwig, G. (1983a) *Phys. Chem. Glasses*, **24**, 37, 38
Muller, D., Grunze, I., Hallas, F. and Ladwig, G. (1983b) *Z. Anorg. Allg. Chem.*, **500**, 80.
Muller, D., Jahn, E., Ladwig, G. and Haubenreisser, O. (1984) *Chem. Phys. Lett.*, **109**, 332.

Muller, D., Rettel, A., Gessner, W. and Scheler, G. (1984) *J. Magn. Reson.*, **57**, 152.
Muller-Warmuth, W. and Eckert, H. (1982) *Phys. Rep.* **88**, 92.41.
Murdoch, J. B., Stebbins, J. F. and Carmichael, I. S. E. (1985) *Am. Mineral.*, **70**, 332.
Phillipp, H. R. (1972) *J. Non-Cryst. Solids*, **8–10**, 627.
Putnis, A. and Angel, R. J. (1985) *Phys. Chem. Min.*, **12**, 217.
Putnis, A., Fyfe, C. A. and Gobbi, G. G. (1985) *Phys. Chem. Min.*, **12**, 211.
Reimer, J. A., Murphy, P. D., Gerstein, B. C. and Knights, J. C. (1981) *J. Chem. Phys.*, **74**(2), 1501.
Risbud, S. H., Kirkpatrick, R. J., Taglialavore, A. P. and Montez, B. (1987) *J. Am. Ceram. Soc.*, **70**, C10.
Schramm, C. M., deJong, B. H. W. S. and Parziale, V. E. (1984) *J. Am. Chem. Soc.*, **106**, 4396.
Smith, K. A., Kirkpatrick, R. J., Oldfield, E. and Henderson, D. (1983) *Am. Mineral.*, **68**, 1206.
Smith, M. E. (1987) PhD Thesis, University of Warwick.
Temkin, R. J. (1975) *J. Non-Cryst. Solids*, **17**, 215.
Weeding, T. L., deJong, B. H. W. S., Veeman, W. S. and Aitken, B. G. (1985) *Nature (London)*, **318**, 352.
Wood, B. J., Kirkpatrick, R. J. and Montez, B. (1986) *Am. Mineral.*, **71**, 999.
Yonemari, S., Masui, A. and Nashiro, M. (1986) *Yogyo-Kyokai Shi*, **94**, 8.

2

X-ray absorption studies of glass structure

R. F. Pettifer

2.1 INTRODUCTION

X-ray absorption spectra have been known to be structurally sensitive since 1929; it has been appreciated that if the physics of the processes involved are understood then it should be possible to invert the spectra to give structural information in real space. The development of the necessary physics however took many years. The essential origin of the process was discovered very early on by Kronig (1931, 1932) but a lack of computational tools meant that a thorough test of the theory could not be undertaken. Developments in theory, largely as the result of progress in low energy electron diffraction, computation and experimentation (via the use of intense X-ray beams from synchrotron sources), caused a sudden reawakening of interest in hard X-ray spectra in the 1970s.

The first use, to the author's knowledge, of X-ray absorption spectroscopy in glass came when Nelson *et al.* (1962) compared the X-ray absorption edges of two crystalline forms of GeO_2, i.e. the hexagonal and tetragonal forms, with the glass, and showed conclusively that the glass was much closer in structure to the hexagonal form than the tetragonal. This was useful information but only qualitative.

In Fig. 2.1 we present a plot of the X-ray absorption spectrum of a well-known chalcogenide glass As_2Se_3. We can use this spectrum to illustrate several salient features of X-ray absorption spectra as a structural tool. At 11865 eV there is a sudden rise in the absorption which is caused by the X-ray photon achieving sufficient energy to ionize an arsenic atom by promoting an electron, from the As(1s) level, into the continuum states of the material. At the threshold of the edge, there is a large spike and this is caused by the electron not quite escaping but being trapped close to the arsenic atom. Within a few tens of electron volts from the edge there is a region of the spectrum now called XANES (X-ray absorption near edge structure) but formerly called Kossell structure. Beyond this region, from approximately

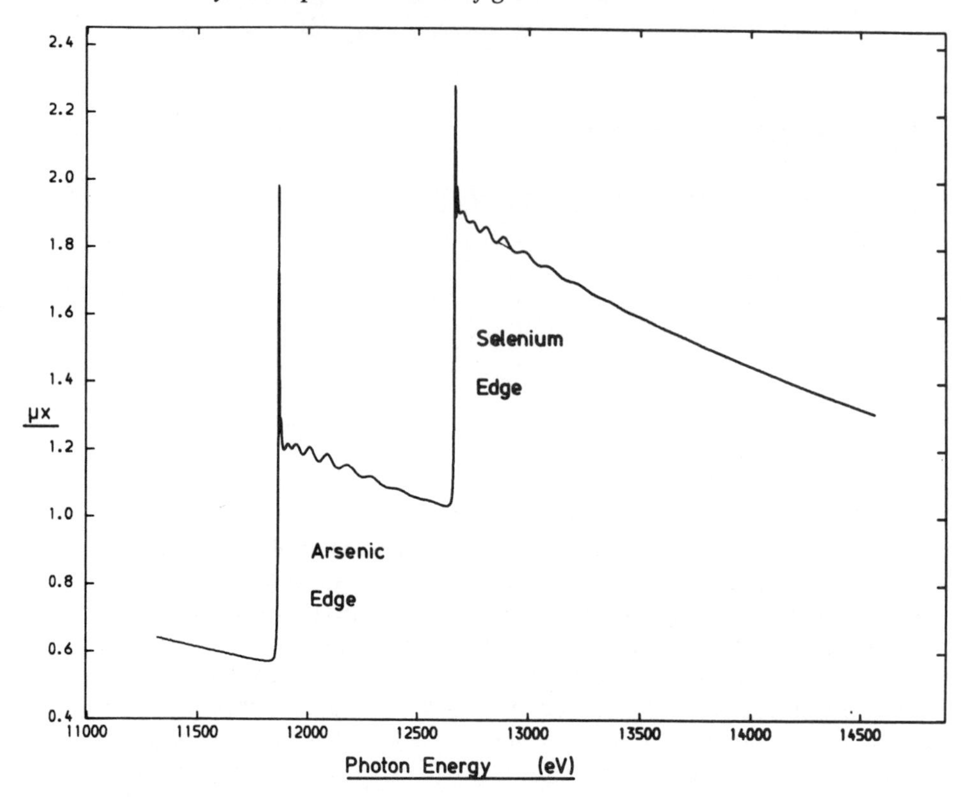

Fig. 2.1 The absorption thickness product (μx) as a function of X-ray energy. The two edges which are observed correspond to the As and Se K edges and result from excitation of 1s electrons to states of p-symmetry (dipole selection rule). The ripples above the edge in energy are EXAFS oscillations and may be observed over 1500 eV from the absorption edge.

50 eV onwards from the edge, we have a region called EXAFS (extended X-ray absorption fine structure). Both XANES and EXAFS contain structural information but the latter is easier to interpret. At 12654 eV in Fig. 2.1 there exists a second absorption edge caused by the onset of photo-ionization of selenium in As_2Se_3. One may observe that the size of the edge jump at the selenium edge is approximately 1.5 times that at the arsenic edge. This reflects the ratio of Se to As atoms in As_2Se_3, and can be used with refinement for compositional analysis.

The cause of the oscillations in the absorption coefficient is related to the scattering of the emitted photoelectron. The electron has wavelike properties and these waves are scattered by the surrounding atoms and this affects the matrix element for the transition, consequently changing the absorption coefficient. Thus, in X-ray absorption, we are dealing with an electron scattering phenomenon. As the electron is emitted from an atom and the scattering of the electron influences this emission, we can view the atom as a source and detector buried inside the material. As such the electron scattering is a local probe of the structure and this is a very important feature in glassy materials. The two main areas in which this aspect is important are in (a) multi-component systems in which the local structure surrounding a specific atom type can be determined and (b) in dilute systems, i.e. impurities. For example, in the celebrated chalcogenide glass STAG (Si–Te–As–Ge) it is possible to observe the bonding surrounding silicon, tellurium, arsenic and germanium separately. Conventional X-ray or neutron diffraction results in a pair correlation function which is related to the probability of observing two atoms separated by a given distance. The first peak in this function consists of contributions from Si–Si, Si–Te, Si–As, Si–Ge, Te–Te, Te–As, Te–Ge, As–As, As–Ge and Ge–Ge pairs. In contrast, a separate measurement of the X-ray absorption spectrum will provide Si–Si, Si–As, Si–Ge and Si–Te bonding data from the Si edge, together with the other partial pair correlation functions from the other edges. Clearly, four partial pair correlation functions place greater constraints on a structural model. For dilute systems, e.g. iron in silicate glasses, it is possible to study the environment of this metal using fluorescence EXAFS techniques at concentrations approaching parts per million.

To continue with our qualitative discussion of X-ray absorption spectra, it is wise to consider the types of electron scattering processes observed. Electron scattering is strong compared with neutron and to a lesser extent X-ray scattering. For the latter probes, it is only necessary to consider multiple scattering when atoms are arranged in a very regular manner, where the so-called dynamical theory becomes important. For electron scattering we must be more cautious owing to the strength of the scattering process. Figure 2.2 summarizes the relevant types of electron scattering which may be observed. Process 1 is scattering of the emitted photoelectron by the emitting atom itself. It is quite common especially in ionic materials and usually results in a sharp peak of absorption (called a white line) close to an absorption edge or in some cases below the edge. Whilst being essentially atomic in origin it may be perturbed by the symmetry of its neighbours and give useful results on the local structure. A particular example of the use of an edge feature is found in the examination of titanium in glasses, to be discussed later. Process 2 of Fig. 2.2 is a multiple scattering event which is only of significance in the low energy region of the spectrum, usually below 50 eV from an absorption edge. (We should note here that at the onset of absorption the emitted photoelectron has zero kinetic energy.) For photoelectrons with kinetic energies in excess of 50 eV large angle multiple scattering weakens and only processes 3 and 4 are relevant. For glasses process 4 where three atoms are collinear is very rare and can also be

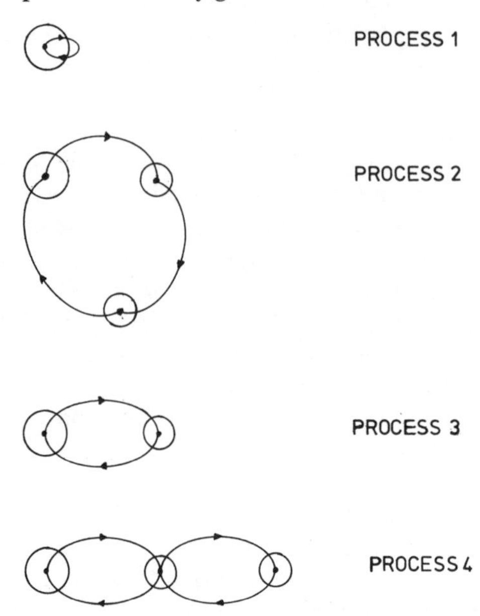

Fig. 2.2 Schematic diagram showing the various scattering processes present in X-ray absorption structure. The large circles represent the emitting atom whereas the smaller circles represent scattering centres. 1 is self-scattering of the electron by the emitting atom. This is relevant close to the edge. 2 is a multiple scattering path involving large angle scattering. For light elements this is only of relevance in the first 100 eV of photoelectron energy. 3 is dominant for glass structure and is the simple scattering process. 4 is a special strong multiple scattering event and expected to be absent for almost all glasses, but if such configurations exist can be present over all energies.

neglected. Consequently above 50 eV from an absorption edge simple single scattering dominates the EXAFS region.

In the next section of this chapter we examine the simple quantitative theory of EXAFS to develop further some of the strengths and weaknesses of the technique. In subsequent sections we examine some of the results obtained to date. Several reviews have appeared over the last ten years which relate to both the physics of X-ray spectroscopy and its applications. A short early review was written by the author (Pettifer, 1978) which covers the development of *ab initio* calculations originally formulated by Lee and Pendry (1975) with specific reference to glasses. A comprehensive review of both the theory and applications was published by Hayes and Boyce (1982) and a general review has been given by Lee *et al.* (1981). Of specific reference to glass are the reviews by Pettifer (1981), Gurman (1982), Brown *et al.* (1986) and Gurman (1987).

2.2 BASIC THEORY OF EXAFS

If we restrict our attention to the region of the X-ray absorption spectrum which coincides with kinetic energies of the photoelectron greater than 50 eV then the theory of the fine structure of the absorption process can be reduced to a simple understandable form. We start by defining a fine structure function

$$\chi(E) = \frac{\mu_3(E)\,x - \mu_1(E)\,x}{\mu_2(E)\,x} \tag{2.1}$$

$\mu_1(E)$, $\mu_2(E)$ and $\mu_3(E)$ are the X-ray absorption coefficients. $\mu_1(E)$ results from all absorption processes other than the K-shell excitation of interest. This is determined by observing the absorption below the X-ray absorption edge and assuming that this effect continues monatomically through the entire spectrum. Usually a functional form

$$\mu_1(E_x) = \mathrm{A}\lambda^3 + \mathrm{B}\lambda^4 \tag{2.2}$$

is assumed where λ is the wavelength of the X-ray. $\mu_3(E)$ is the observed absorption and $\mu_2(E)$ is an assumed absorption for an isolated atom, i.e. an

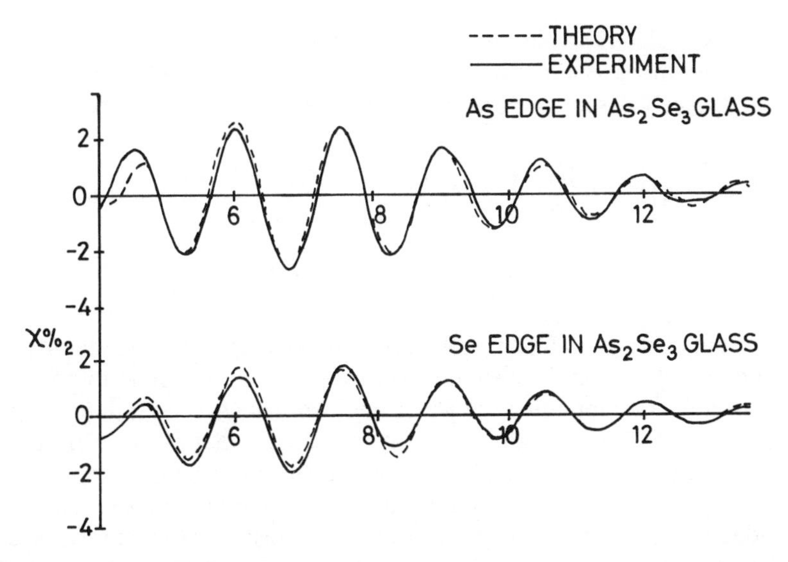

Fig. 2.3 The EXAFS fine structure extracted from the data of Fig. 2.1. Note that the data are plotted as a function of wave vector (k). This yields an even oscillatory pattern owing to the near linear dependence of phase changes with wave vector. The amplitude of the structure is roughly in the ratio 3/2 reflecting the three neighbours of As and two neighbours of Se. Backscattering of As and Se is roughly the same. The dotted line is the result of an *ab initio* calculation of the fine structure. Only the nearest neighbours are included.

edge which has zero fine structure. It is obviously impractical to measure this function and usually it is modelled by fitting a smooth polynomial through the fine structure oscillations. E, the argument of μ in (2.1) is the energy of the photoelectron which can be deduced by subtracting the X-ray energy at threshold E_0 from E_x, the X-ray energy which appears in (2.2). The net result of the application of (2.1) to the measured absorption is to extract the oscillations of fine structure from the spectrum and normalize them to the edge jump. We note that the thickness x of the specimen cancels in (2.1) and thus it is not necessary to know this parameter. This is of considerable help, however, in preparing specimens where, although the thickness is not needed absolutely, the material should be of uniform thickness. If the latter is not the case then the normalization (2.1) will yield a reduced $\chi(E)$ compared with the true value. We shall see in this section that this results in an incorrect assessment of co-ordination number. Figure 2.3 shows the extracted $\chi(E)$ functions from the absorption data of Fig. 2.1. This curve is the one most often presented in published literature, although the abscissa has been changed from the energy E of the photoelectron to k the wavevector. k can be defined by the equation

$$3.81k^2 = E \tag{2.3}$$

where E is in electron volts and k appears in Å^{-1}.

With the above definition of $\chi(k)$ an expression for this function has emerged largely based on the work of Lee and Pendry (1975) as follows:

$$\chi(k) = -[|S_0^2|] \sum_j \frac{[N_j]}{R_j^2} \left[\frac{|fj(\pi, k)|}{k} \right] \left[\exp \frac{(-2R_j)}{\lambda(k)} \right] [\exp(-2\sigma_j^2 k^2)] \times \{\sin[2kR_j + 2\delta_1(k) + \eta_j(k)]\} \tag{2.4}$$

Fortunately it is possible, with this approximate expression, to identify the meaning of each term. The summation over the index j refers to differing shells of atoms surrounding the absorbing element. A shell is defined by a collection of identical atoms sitting at the same radial distance from the absorber, e.g. 4 oxygen atoms surrounding silicon in SiO_2. The first five terms in (2.4) give rise to the amplitude modulation of the basic sine wave provided by term 6. The structural parameters present in this expression are N_j the co-ordination number, R_j the shell radius and σ_j^2 the variance of the emitter scatterer distance. It is possible to evaluate (2.4) from first principles and this has been done with success for several glasses by Pettifer (1978), and Gurman and Pettifer (1979). Alternatively one may use a structural standard to calibrate the non-structural terms in (2.4). We can list the terms in (2.4) giving their significance. $|S_0^2|$ is an amplitude reduction factor which represents that only a fraction of single photoelectrons are emitted from an excited atom. This factor is thought to be independent of wavevector k and take a numerical value ~ 0.7. The term

$$\left[\frac{|f(\pi, k)|}{k}\right]$$

dictates to a large extent the overall envelope of the fine structure. This factor has been calculated for the atoms oxygen, sulphur, selenium and tellurium by Pettifer and McMillan (1977) and the results are shown in Fig. 2.4. Focusing our attention on the oxygen scattering we can observe that the scattering is stringent close to the absorption edge and decays in an exponential manner. Unfortunately this means that the fine structure dies out rather quickly as a function of photoelectron energy. Consequently, absorption data of very high statistical accuracy needs to be obtained if a wide range of data is to be studied. This is a fundamental limit to the accuracy of the technique when dealing with light atom scatterers, e.g. oxide glasses.

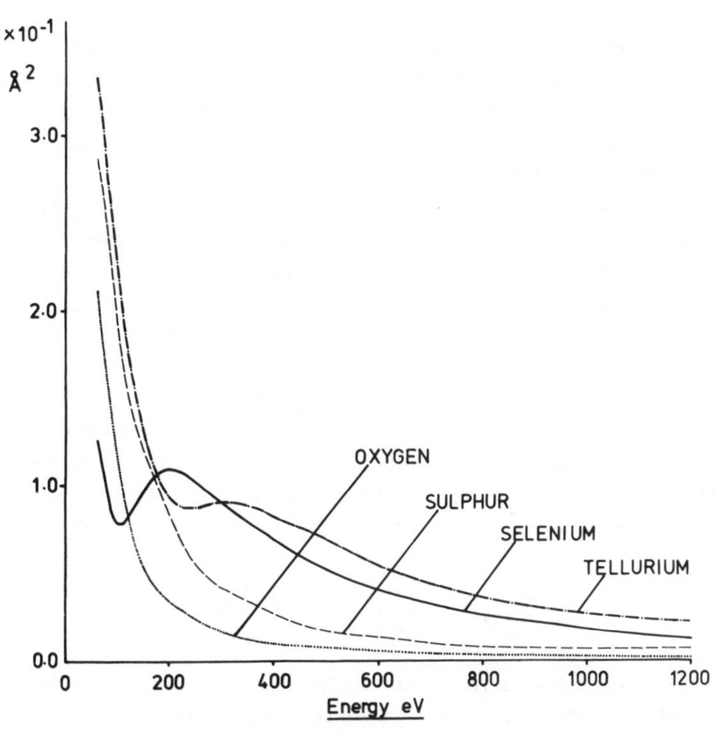

Fig. 2.4 Calculated amplitudes of backscattering for members of the chalcogenide family. For adjacent members of the periodic table the differences in amplitude are small. At high energies the scattering mirrors the size of the atom. Thus it is far easier to observe EXAFS at high energies for tellurides compared with oxide systems. Data from Pettifer and McMillan (1977).

The results of Fig. 2.4 are for atoms lying in the same column of the periodic table and exhibit distinctly different energy dependence. This fact can be used with good effect to determine the type of atom co-ordinating a particular atom in a solid. Similar calculations, however, have shown that adjacent atoms in the periodic table have very similar scattering behaviour and consequently it is very difficult to identify types under these circumstances. This is a similar problem to that encountered in X-ray diffraction. This means that the identification of neighbours, in GeSe glass for example, is impossible by both X-ray diffraction and EXAFS.

The term $[\exp(-2R_j/\lambda(k))]$ in (2.4) represents inelastic scattering of the photoelectron as it passes through the material. Inelastic losses are most serious in the low energy part of the spectrum.

Fortunately this energy dependence of the inelastic loss can be modelled quite successfully by replacing $\lambda(k)$ by

$$\lambda(k) = \frac{E_i}{(27.2k)} \tag{2.5}$$

where E_i is the imaginary component of the self-energy of the photoelectron and is set at a constant 4 eV. If the interpretation of the spectrum is to be made using *ab initio* techniques, both E_i and $(S_0)^2$ have to be determined by modelling the fine structure of a known crystalline standard prior to interpreting the fine structure of a glass.

Uncertainties in S_0^2 and experimental artefacts which reduce the fine structure (e.g. non-uniform specimen or contaminated X-ray beam) directly affect the amplitude and hence the co-ordination number found.

The term $[\exp(-2\sigma_j^2 k^2)]$ represents the effect of disorder in the shell radius. This disorder can be split into two components

$$\sigma_j^2 = r_{jT}^2 + \sigma_{js}^2 \tag{2.6}$$

if we assume that the distribution function of the positions of atoms is described by a gaussian probability density. The subscript T in 2.6 refers to the thermal vibrations which are also finite at 0 K owing to the zero point motion of the atoms. σ_{jT}^2 can be evaluated using the expression of Sevillano *et al.* (1979)

$$\sigma_{jT^2} = \frac{\hbar}{2m\omega_E}\coth\left(\frac{\hbar\omega_E}{2k_B T}\right) \tag{2.7}$$

where ω_E is the Einstein frequency and m is the effective mass of the emitter and scatterer system. Here we can make contact with other spectroscopic techniques in that ω_E frequently corresponds to an infra-red or Raman active mode. A successful comparison of σ_j^2 and Raman or infrared active modes has been made for a number of glasses by Pettifer (1978). Using (2.6) and temperature dependent measurements of the fine structure together with (2.7) it is possible to evaluate the additional static disorder σ_{js}^2 which may be present in glasses. For near-neighbours of glass-forming atoms this additional disorder is found to be negligible. This is quite understandable in that the free-energy differences between glass and crystal are usually very small and the compression of near-neighbour bonds is very expensive in terms of energy. The major difference between glasses and crystals comes from the second neighbour distances. Disorder of second-shell distances results from bond-bending rather than stretching and as such a great deal less costly in energy as these modes are comparatively soft for chalcogen (including oxygen) bonding.

In examining modifiers in oxide glasses and amorphous metals it is found that significant static disorder exists in the near-neighbour environment. Unfortunately this also means that the assumption of a gaussian pair distribution is suspect. Special care is necessary when applying (2.4) to these systems. Unfortunately this care is frequently not exercised in the published literature and yields grossly erroneous results.

The final term in (2.4) is responsible for the oscillations of the fine structure. It is the periodicity of the fine structure which yields the mean interatomic distance R_j. Together with this term there are two other components, resulting from a k-dependent shift of the phase owing to the photelectron climbing out of the emitter (δ_1) and a phase change on reflection from the surrounding atom ($\eta_j(k)$). These factors can be calculated, or the sum $2\delta_1 + \eta_j(k)$ can be calibrated from a known standard.

Although (2.4) has many terms we can rewrite it as:

$$\chi(k) = \Sigma_j A_j(k) \sin[2kR_j + \phi_j(k)] \tag{2.8}$$

where the amplitude modulation $A_j(k)$ and phase modulation $\phi_j(k)$ signature the number, type and disorder of shell j. This form immediately suggests the use of Fourier transform techniques to isolate scattering signatures at different radii. Unfortunately, owing to the $\phi_j(k)$ term this operation is not quite straightforward. Comparing (2.4) with (2.8) we find that

$$\phi_j(k) = 2\delta_1(k) + \eta_j(k) \tag{2.9}$$

and from *ab initio* calculations this can be approximated for some emitter scatterer pairs by a linear relationship

$$\phi_j(k) \simeq -2ak + b \tag{2.10}$$

Inserting this in (2.8) yields

$$\chi(k) = \Sigma_j A_j(k) \sin[2k(R-a) + b] \tag{2.11}$$

Thus a Fourier transform of $\chi(R)$ with respect to $2k$ yields δ-functions positioned at $R-a$ convolved with the Fourier transform of the envelope function. This would be the case if b were an integral multiple of 2π. In reality this is not the case and results in complex Fourier coefficients. To yield peaks at atom positions, the modulus of the Fourier transform is frequently plotted. For comparative purposes it is better to compare original data as the results of Fourier operation depend on the range and signal-to-noise ratio of the data.

This section is not intended as an exhaustive study of the physics of X-ray absorption, but merely to point out that there are many pitfalls which must be avoided prior to making structural conclusions. It has been said that EXAFS is a 'sporting technique' (C. D. Garner, private communication) although the local information provided is very valuable and quantitative when care is exercised in interpretation. For a greater understanding of interpretation the reader is urged to consult the article by Hayes and Boyce (1982).

2.3 GLASS STRUCTURE STUDIES BY EXAFS

2.3.1 Network formers

The most popular system for study via EXAFS in this class is GeO_2 and its derivatives. Several reasons exist for this choice, the main one being that the Ge K-edge lies in a wavelength range easily accessible to a spectrometer which has an air path. A more obvious choice would have been SiO_2 but the K-edge of Si lies at 1840 eV (6.74 Å) and as such requires stringent control of the specimen thickness and an all-vacuum instrument.

Nelson *et al.* (1962) were the first to conclude that the glass was most like the hexagonal form of GeO_2 which has four oxygens surrounding each germanium atom. The tetragonal form in which germanium is sixfold co-ordinated was not favoured. However, we mentioned in section 2.1 that this is not quantitative information and a more detailed examination was given later by Sayers *et al.* (1972). The latter authors gave a Ge–O distance of 1.60 Å compared with 1.65 Å for the hexagonal crystalline form. However, this must be a misprint as Smith and Isaacs (1964) found a Ge–O mean length of 1.74 in the quartz-like form whereas Baur (1956) reported a mean Ge–O bond-length of 1.88 (4–1.87, 2–1.91) in the tetragonal form. Later Cox and McMillan (1981) examined GeO_2 as part of a study of alkali germanate materials and found Ge–O distances of 1.71 Å. Comparison of these Ge–O bond-lengths, however, should not indicate the accuracy which can be achieved. The expected 0.03 Å difference is not structurally significant in that the interpretation was performed using *ab initio* calculations, i.e. specifically $\delta_{\mathrm{I}}(k)$ and $\eta_{\mathrm{II}}(k)$ in 2.4 were evaluated by solving Schrödinger's equation. Comparative techniques, when they can be applied, are usually far more accurate, and in very favourable situations are better than 0.003 Å in accuracy.

A more interesting point is that all three authors who worked on GeO_2 had evidence for Ge–Ge scattering from the second shell of their material.

Bearing in mind the restricted data range, excluding low photoelectron energy data, this indicates that the local environment in GeO_2 glass is well preserved with a narrow range about the mean Ge–O–Ge angle of 130°. Recently Okuro (1986) has quantified this spread in the Ge–Ge pairs distance. By triangulation we find that the mean bond angle is 130° with a static spread of angles spanning 11° centred on this value.

A similar study has been attempted by Greaves *et al.* (1981) on SiO_2; however, the data are of lower quality than that observed on the germanium spectra. Despite this the most interesting feature of their data was the observation of a scattering contribution from Si–Si pairs. Hence in these three-dimensional glasses sufficient correlation is preserved to observe the second shell scattering effect. From (2.1) we find that typical bond angle spreads (A–O–A) of 20° about a mean of 140° is insufficient to kill the EXAFS signal from the second shell of this tetrahedral glass. The observation of this signal probably results from the steric constraints imposed by the three-dimensionality of the material and is in contrast to the two-dimensional nature of some chalcogenide materials.

Despite the clear analogy between GeO_2 and SiO_2, even to the point that they both exhibit crystobalite, quartz and stishovite phases, the basic glasses differ when doped with an alkali. Cox and McMillan (1981) examined the Li_2O–GeO_2 system and found a progressive increase in the population of $GeO_{6/2}$ at the expense of $GeO_{4/2}$ species with increasing Li_2O content indicating that germanium may act in a modifier type role.

It is worth spending a few moments at this point on the perennial problem of a continuous random network versus a quasicrystalline description of glass networks. Both models are ill defined; however, if one adopts a conservative interpretation of these models in stating that a quasicrystalline network retains the typological graph of the parent crystal, then it is clear that merely an examination of first and second neighbours in a material is insufficient to distinguish between these extremes. Only by establishing the ring statistics in a material can a distinction be made. Some authors further confine this model to stating that small areas of a glass are crystal like in that they preserve their relative atomic positions. This can firmly be ruled out in some of the chalcogenide glasses, an example of which is shown in Fig. 2.3. Here the structure is simply explained by scattering from the near neighbours. The corresponding crystal shows strong evidence for second shell scattering which is reduced to at least 10% of the crystalline scattering contribution even at $k=5\,\text{Å}^{-1}$. Even in this simplistic model we can immediately assume that less than 10% of the material is a microcrystalline environment. Alternatively, if we adopt a gaussian pair correlation model for the second shell distribution we find (using $0.1=\exp(-2\sigma^2k^2)$ and converting the standard deviation into an angle by triangulation) that this absence of second shell structure is due to a spread of chalcogen bond angle $>\pm 7°$ from the crystalline case. The emphasis is thus on the ability of chalcogens (including oxygen) to accommodate a bond bend, without large penalties in terms of energy, to explain the stability of many glass systems.

An exception to the above discussion is the metallic glasses, both metal–metalloid and metal–metal systems. A large body of work has been performed on these systems which are difficult to characterize owing to the asymmetry of the distribution functions involved. The reader is referred to the works by Haensel *et al.* (1980) for further information. Unlike oxide and chalcogenide glass formers these materials exhibit disorder in the near-neighbour environments.

2.3.2 Intermediates in glasses

An interesting class of additives which can be investigated in dilute concentrations with EXAFS and XANES are the intermediates in which the role of these ions is uncertain but they may have a dramatic effect on the properties of the material.

An interesting case of the role of a transition metal in glass is that of TiO_2 in SiO_2. This is an important case in that some of the properties controlled by the addition of TiO_2 include the technologically important thermal expansion coefficient and nucleation. The environment of Ti in a silica matrix was first investigated by Sandstrom *et al.* (1980) using material which was made by flame hydrolysis. Later Greegor *et al.* (1983) extended this study to a wider range of compositions. Following these investigations Emili *et al.* (1985) investigated material which was formed using the sol-gel process. From the latter preparation technique it is possible to observe the environment as a function of heat treatment. Information concerning the environment of titanium comes from two sources. Firstly the XANES spectrum contains a characteristic feature, a pre-edge spike which is characteristic of Ti in tetrahedral symmetry. For this type of symmetry, admixture of oxygen 2p states into Ti 3d states is observed. This admixture yields a dipole allowed transition and consequently a peak in the absorption coefficient. Under O_h symmetry there is no mixing and consquently the absorption coefficient is low. This theory has of course been checked using model compound measurements. A series of Ti XANES spectra is shown in Fig. 2.5. The data are taken from the paper by Greegor *et al.* (1983).

It is clear that all spectra have a substantial pre-edge spike which clearly shows a compositional dependence. The maximum intensity of the pre-edge feature occurs for compositions between 1000 ppm and 7 mol % TiO_2. Below and above this range the pre-edge is weaker hence indicating O_h symmetry in these two regimes. At high concentrations this is understandable in terms of aggregation of Ti based clusters as precursors for precipitation of anatase. This is known to occur above ~ 10 mol % TiO_2. At low concentrations of TiO_2 the occurrence of octahedral sites is a surprise. Greegor *et al.* (1983) explain this fact on the basis of the Ti occupying 'holes'

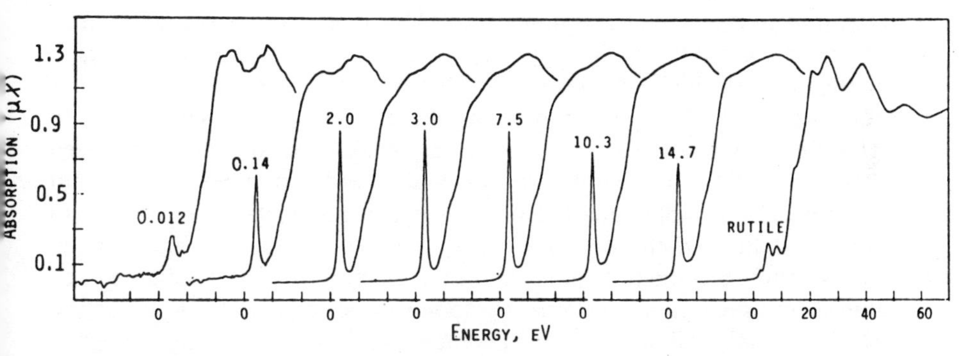

Fig. 2.5 The near-edge K absorption spectra of a series of TiO_2–SiO_2 glasses plus a standard material, rutile TiO_2. The composition of TiO_2 is labelled in wt %. The characteristic pre-edge peak is associated with T_d symmetry surrounding T_i. This sharp feature is absent in rutile which has local O_h symmetry. Hence the size of this feature maps the concentration of tetrahedral species.

in the silicon matrix. The present author would like to suggest, however, that the influence of water may well be crucial. It is known that water can be accommodated at concentrations of up to 1000 ppm in silica. The presence of this water will act as a strain relief mechanism in the glass as it creates non-bridging single bonded OH groups. It is highly likely that Ti getters these single bonded groups in regions of the glass structure which do not impose steric constraints. That is, at quasi-surface regions. This allows the preferred octahedral environment. When these 'surface states' are saturated Ti is forced into a tetrahedral geometry. All these studies agree that the dominant species in the intermediate concentration range is tetrahedral in symmetry. In contrast Emili *et al.* (1985) have shown that tetrahedral symmetry can be maintained in 15 mol % TiO_2 in SiO_2 provided the heat treatment of the sol-gel material is kept low (below 1000°C). The second set of information comes from the EXAFS region and this can be interpreted in the framework that was described in section 2.2. The results for both XANES and EXAFS agree.

Another glass additive which has an effect on nucleation is zinc which is believed to behave analogously to aluminium although the exact behaviour of aluminium is also the subject of great controversy. On the basis of the amphoteric nature of both zinc and aluminium where both octahedral and tetrahedral sites are found it seemed natural that both types of site could be found and these would explain the anomalies of macroscopic properties found as a function of composition particularly in the Na_2O–Al_2O_3–SiO_2 system or its zinc analogue. By studying a range of compositions of zinc-containing glasses $xNa_2O + yZnO + (1 - x - y)SiO_2$ Pettifer (1978) found that zinc retained an exclusively tetrahedral environment even when $y > x$.

A simple oxygen balance argument will show that the glass network can be composed of $ZnO_{4/2}$ and $SiO_{4/2}$ units at $y=x$. For $y<x$ the additional oxygen in the network can be incorporated via non-bridging oxygens on silica units (e.g. $SiO_{1+3/2}$ for a single non-bridging oxygen). For $y>x$ it is impossible to provide enough oxygen to maintain a simple corner linked system between tetrahedra. Thus the retention of fourfold co-ordination beyond this limit was a puzzle. This dilemma can be reconciled by noting that Lacy (1963) had suggested that tri-clusters may form. These are units whereby oxygen bridges three tetrahedral units rather than the normal two. Unusual though this seems such configurations are found in crystals. Indeed one of the parent crystals willemite Zn_2SiO_4 contains such triclusters and belongs to the phenocite (Be_2SiO_4) class of crystal. Such units would be expected to stiffen the network by increasing its connectivity and this is consistent with the increased viscosity observed when $y>x$. Further such rigid units may well act as nucleating centres and explain this aspect of the behaviour of zinc also.

To complete the analogy between the role of zinc and aluminium Waychunas and Brown (1984) have found a co-ordination number of four for a peraluminous glass ($Na_{0.75}Al_{1.2}Si_{2.9}O_8$), thus directly confirming Lacy's (1963) hypothesis.

Although we have directly discussed Ti, Al and Zn in oxide glasses, many others have been examined with EXAFS. The number of publications in this area is extensive and too numerous to give justice here. V, Fe, Zr, Yb, Pb, U, Th have all been examined in various compositions, brief reviews of the extensive work in this area are frequently to be found in the conference proceedings in the series *EXAFS and Near Edge Structure* which occur bi-annually.

2.3.3 Alkali and alkali-earth additives in glass

High-Z additives in glass are frequently found in asymmetric oxide cages and consequently the radical structure surrounding heavy elements is complex and cannot be modelled with a gaussian pair-correlation function. Thus, one of the basic assumptions of the simple theory outlined in section 2.2 is invalid.

This point also holds for alkali environments in glass. Consequently structural studies of these classes of material have proved inconclusive. Further the mobile nature of sodium, for example, and the thin specimens required for transmission measurements require great care in experimental technique. Greaves *et al.* (1981), Greaves (1981) and Greaves *et al.* (1984) examined Na environments in silicate glasses and concluded that the bond lengths were shorter and co-ordination numbers smaller than those determined by X-ray diffraction of the parent crystals. However, this work is in contrast to the study of McKeown *et al.* (1985) who observed a larger

Na–O mean distance than the previous study but also a lower co-ordination number, but these authors point out that the outer oxygen contribution is lost in the EXAFS experiment owing to the weak bonding and consequent damping of the signal via the $\exp(-2\sigma_j^2 k^2)$ term in (2.4).

Ca has been examined in a range of glasses by Geere (1983), Binstead *et al.* (1985) and Hardwick *et al.* (1985), although similar reservations hold for this work as for studies of the sodium environment it has been found that Ca can adopt a wide range of environments in glasses. With the continuing progress made in modelling asymmetric distributions many interesting features will be discovered.

2.3.4 Covalent glasses

Many covalent glasses have been measured using EXAFS and we therefore make a selection which is representative of the type of results which can be obtained. The system chosen is the As–chalcogenide glasses which have been prepared both by vapour deposition or by bulk glass techniques. Originally Sayers *et al.* (1972) examined evaporated As_2S_2, As_2Se_3 and As_2Te_3 but these data were taken with a conventional X-ray tube and suffered severely from signal to noise problems. Synchrotron radiation spectra of As_2Se_3 and As_2Se_2 bulk glasses which were air quenched were reported by Pettifer (1978). These data were analysed using the full theoretical analysis based on Lee and Pendry's (1975) spherical wave expansions of the scattered photoelectron. (This removes important approximations implicit in (2.4)). The results were consistent with full homopolar bonding of As in As_2S_3 whilst this had to be assumed for the case of As_2S_3 where the electron scattering from Se is indistinguishable from that of arsenic. In the same year Nemanich *et al.* (1978) examined the same material prepared by vapour deposition techniques and in contrast discovered evidence for homopolar bonding. More recently Lowe *et al.* (1986) have carefully reviewed the situation and also found that the irreversible photostructural effects in vapour deposited films was due to a homopolar–heteropolar transition. Further the bulk glass showed evidence for the reversible photostructural effects in the very weak changes in second shell scattering which was just visible in annealed bulk material. Also Yang *et al.* (1986) have found changes in the mean square relative displacements of atoms as a function of annealing conditions. This enabled these authors to determine an optimum temperature, anneal the glass and produce crystalline orpiment.

Over the past ten years many experiments have been performed on chalcogenide glasses and this has resulted in the examination of increasingly more subtle effects. Indeed some studies have suggested subtle co-ordination number changes as a function of sample preparation conditions. It is the author's opinion that many of these effects are at the limit of the spectroscopic systematic errors currently present in synchrotron radiation spectrometers. In future, greater attention will have to be paid to distortions of the spectra by (a) sample preparation, (b) the presence of contaminating harmonics in the X-ray beam and energy calibration of spectrometers.

Unlike evaporated material bulk glasses show little evidence of gross changes in their near-neighbour environment when transforming from the glass to the crystalline state. The only exceptions to this rule are Se (Hayes and Hunter, 1977), As_2Te_3 (Pettifer *et al.*, 1977), and *a*-As (Bordas *et al.*, 1977). All of these materials have a common feature in that in the crystalline state a degree of mesomeric bonding exists which relies on the translational symmetry to lock chains or layers together. In the glassy state the symmetry is lost resulting in a breaking of the mesomeric bonding and a resultant strengthening of the near neighbour bonding.

2.4 CONCLUSIONS

EXAFS has contributed a great deal to studies of local environments in glass, especially over the last ten years. The contribution to information concerning the topology of glass has, however, been slight. The technique should be considered as complementary to X-ray and neutron scattering and has a great similarity to other techniques such as Raman, infrared and Mössbauer studies which are also local probes. Fortunately, EXAFS is a quantitative technique interpretable into direct distances and disorders. This aspect sets the technique apart from the other local probes previously mentioned. In future it is clear that a greater understanding of the XANES spectra will increase the technique's power. The major role that this spectroscopy holds is in dilute and multicomponent glasses and in terms of its ability to examine the environment of almost every atom in the periodic table.

ACKNOWLEDGEMENTS

I wish to acknowledge the late Professor P. W. McMillan who took me on as a student and always showed faith in our work. I have also received much help from the theorists; in particular, Dr B. W. Holland, Professor J. B. Pendry and Dr S. Gurman have been very kind to me with their advice over the years. Also I would like to thank Professor A. J. Forty for tolerating my intransigence in pursuing some of this work.

REFERENCES

Baur, W. H. (1956) *Acta Crystallog.*, **9**, 515.

Binstead, N., Greaves, G. N. and Henderson, M. B. (1985) *Contrib. Mineral. Petrol.*, **89**, 103.

Bordas, J., Gurman, S. J. and Pettifer, R. F., reported in G. N. Greaves, S. R. Elliot and E. A. Davis (1979) *Adv. Phys.*, **28**, 49.
Brown, G. E. Jr., Waychunas, G. A., Ponader, C. W. *et al.* (1986) *J. Phys. (Paris)*, **C8**, 661.
Cox, A. D. and McMillan, P. W. (1981) *J. Non-Cryst. Solids*, **44**, 257.
Emili, M., Incoccia, L., Mobilco, S. *et al.* (1985) *J. Non-Cryst. Solids*, **74**(1), 129.
Geere, R. G., Gaskell, P. H., Graves, G. N. *et al.* (1982) *EXAFS and Near Edge Structure* (eds K. O. Hodgson, B. Hedman and J. E. Penner-Hahn), Springer-Verlag, Berlin, **27**, p. 256.
Greaves, G. N. (1981) *J. Phys. C*, **4**, 225.
Greaves, G. N., Binstead, N. and Henderson, C. M. B. (1984) *EXAFS and Near Edge Structure III*, Springer Proc. Phys., Berlin, **2**.
Greaves, G. N., Fontaine, A., Lagarde, P. *et al.* (1981) *Nature (London)*, **293**, 611.
Greegor, R. B., Lytle, F. W., Sandstrom, D. R. *et al.* (1983) *J. Non-Cryst. Solids*, **55**, 27.
Gurman, S. J. (1982) *J. Mater. Sci.*, **17**, 1541.
Gurman, S. J. (1987) *Extended X-ray Absorption Fine Structure* (ed. R. W. Joyner), Plenum Press, London, Chap. 6.
Gurman, S. J. and Pettifer, R. F. (1979) *Philos. Mag.*, **B40**, 345.
Haensel, R., Rabe, P., Tolkiehn, G. and Werner, A. (1980) *Proc. Nato Advanced Study Institute: Liquid and Amorphous Metals* (eds E. Lüsche and H. Cornful), Reidel, Sigthoff and Noordhoff, Netherlands, p. 459.
Hardwick, A., Wittaker, E. J. W. and Diakon, G. P. (1985) *Mineral Mag.*, **49**, 25.
Hayes, T. M. and Boyce, J. B. (1982) *Solid State Phys.*, **37**, 173.
Hayes, T. M. and Hunter, S. H. (1977) *The Structure of Non Crystalline Materials* (ed. P. H. Gaskell), Taylor and Francis, London, p. 78.
Kronig, R. de L. (1931) *Z. Phys.*, **70**, 317.
Kronig, R. de L. (1932) *Z. Phys.*, **75**, 468.
Lacy, E. D. (1963) *Phys. Chem. Glasses*, **4**, 234.
Lee, P. A., Citrin, P. H., Eisenberger, P. and Kincaid, B. M. (1981) *Rev. Mod. Phys.*, **53**, 769.
Lee, P. A. and Pendry, J. B. (1975) *Phys. Rev.*, **B11**, 2795.
Lowe, A. J., Elliot, S. R. and Greaves, G. N. (1986) *Philos. Mag.*, **B54**, 483.
McKeown, D. A., Waychunas, G. A. and Brown, G. E., Jr. (1985) *J. Non-Cryst. Solids*, **74**, 325.
Nelson, W. F., Siegel, I. and Wagner, R. W. (1962) *Phys. Rev.*, **127**, 6.
Nemanich, R. J., Connell, G. A. N., Hayes, T. M. and Street, R. A. (1978) *Phys. Rev. B*, **18**, 6900.
Okuro, M. (1986) *J. Non-Cryst. Solids*, **87**, 312.
Pettifer, R. F. (1978) *Trends in Physics*, 4th Gen. Conf. European Phys. Soc., Chap. 7, p. 522.
Pettifer, R. F. (1981) *EXAFS for Inorganic Systems*, Proc. Daresbury Study Weekend, DL/SCI/R17 Daresbury Laboratory.
Pettifer, R. F. and McMillan, P. W. (1977) *Philos. Mag.*, **35**, 871.
Pettifer, R. F., McMillan, P. W. and Gurman, S. J. (1977) *The Structure of Non-Crystalline Materials* (ed. P. H. Gaskell), Taylor and Francis, London, p. 63.
Sayers, D. E., Lytle, F. W. and Stern, E. A. (1972) *J. Non-Cryst. Solids*, **8–10**, 401.
Sandstrom, D. R., Lytle, F. W., Wei, P. S. P. *et al.* (1980) *J. Non-Cryst. Solids*, **41**, 201.
Sevillano, E., Meuth, H. and Rehr, J. J. (1979) *Phys. Rev. B*, **20**, 4098.
Smith, G. S. and Isaacs, J. B. (1964) *Acta Crystallogr.*, **17**, 842.
Waychunas, G. A. and Brown, G. E. (1984) *EXAFS and Near Edge Structure III*

(eds K. O. Hodgson, B. Heckman and J. E. Penner-Hahn), Springer, Berlin, p. 336.

Wong, J. (1981) *Metallic Glasses I* (eds H. J. Guntherot and H. Beck), Springer-Verlag, p. 45.

Yang, C. Y., Paesler, M. A. and Sayers, D. E. (1986) *J. Phys. (Paris),* C8, **47**, Suppl. 12, p. 391.

3

Volume nucleation in silicate glasses

Peter F. James

3.1 INTRODUCTION

The last twenty-five years or so have seen steady advances in the science and technology of glass-ceramics, materials prepared by the controlled crystallization of glass. Peter McMillan made many outstanding contributions to these advances and the publication of the first edition of his now classic textbook in 1964 remains a landmark in the development of the field.

The preparation of a glass-ceramic involves several stages. First, a glass is melted and formed into the appropriate shape. The glass article is then given a heat treatment schedule to nucleate and grow crystals in its volume until a material with the desired degree of crystallinity is produced. The kinetics of crystal nucleation and growth are thus critical in determining those compositions which can be formed into glasses reasonably stable towards devitrification, and which subsequently can be economically converted into fine-grained glass-ceramics by suitable heat treatment.

This chapter is concerned with crystal nucleation in glass, although no attempt is made to present a comprehensive discussion of the whole subject. Rather, certain topics, which are believed to be of particular interest, are highlighted. Throughout, the emphasis is on studies of 'simple' silicate systems involving quantitative measurements of volume nucleation kinetics. However, such studies are considered helpful in identifying the various factors influencing nucleation behaviour in glasses in general. Moreover, these factors also apply to the more complex compositions used in glass-ceramic manufacture.

After a general outline of the relevant theories, experimental studies of volume nucleation in various systems, in which the crystallizing phase has the same composition as the parent glass, are discussed. Both steady state and non-steady state nucleation are considered. The results form a remarkably consistent pattern and indicate that the nucleation in these systems is predominantly homogeneous. The more complex case when the crystallizing phase has a different composition from that of the parent glass is then discussed. Studies of heterogeneous nucleation on metallic particles and the roles of non-metallic nucleating agents are described. Finally, recent investigations of the effects of amorphous phase separation on crystal nucleation kinetics are reviewed.

3.2 SUMMARY OF CLASSICAL NUCLEATION THEORY

3.2.1 Steady state homogeneous nucleation

According to classical theory the rate of homogeneous steady state crystal nucleation (I) in a one-component supercooled liquid is related to absolute temperature T by the well-known expression (see, for example, Christian, 1975):

$$I = A \exp [-(W^* + \Delta G_D)/kT] \tag{3.1}$$

where W^* and ΔG_D are the thermodynamic and kinetic free energy barriers to nucleation respectively and k the Boltzmann constant. The pre-exponential factor A may be expressed as

$$A = 2n_v V^{1/3}(kT/h)(\sigma/kT)^{1/2} \tag{3.2}$$

where n_v is the number of atoms or, strictly, 'formula units' of the crystallizing component phase per unit volume of the liquid, V the volume per formula unit, σ the crystal–liquid interfacial free energy per unit area and h Planck's constant. In practice, the quantity A, which is typically 10^{41}–10^{42} m^{-3} s^{-1} may be treated as effectively constant over the temperature range of nucleation measurements and to a good approximation

$$A = n_v(kT/h) \tag{3.3}$$

The thermodynamic barrier W^* for a spherical nucleus is given by

$$W^* = 16\pi\sigma^3 V_m^2/3\Delta G^2 \tag{3.4}$$

where ΔG is the bulk free energy change per mole in crystallization and V_m the molar volume of the crystal phase (the free energy change per unit volume $\Delta G_v = \Delta G/V_m$).

To date no complete experimental test of the classical theory in absolute terms appears to have been performed in a condensed system. The main problem lies in evaluating the interfacial free energy σ which at present is extremely difficult to determine from independent experiments unrelated to nucleation kinetics. However, as will become evident, various predictions of the theory can be tested in terms of the experimental nucleation kinetics in silicate glasses.

To compare theory with experiment accurate data for the thermodynamic driving force, ΔG, are required. For a single component system at temperature T below the melting point T_m, ΔG is given by

$$\Delta G = -\Delta H_f(T_m - T)/T_m - \int_T^{T_m} \Delta C_p \, dT + T \int_T^{T_m} (\Delta C_p/T) \, dT \qquad (3.5)$$

where ΔH_f is the heat of fusion per mole and ΔC_p the difference in specific heats between the crystal and liquid phases at constant pressure at temperature T. If ΔC_p can be taken as independent of temperature from T_m to the temperature of interest (T), (3.5) reduces to

$$\Delta G = -\Delta H_f(T_m - T)/T_m - \Delta C_p[(T_m - T) - T \ln (T_m/T)] \qquad (3.6)$$

If ΔC_p is taken as zero, ΔG is given by the well-known approximate expression

$$\Delta G = -\Delta H_f(T_m - T)/T_m \qquad (3.7)$$

Finally, the following equation was obtained by Hoffmann (1958) assuming ΔC_p is an unknown constant:

$$\Delta G = -\Delta H_f(T_m - T)T/T_m^2 \qquad (3.8)$$

Both expressions (3.7) and (3.8) are usually only applicable for small undercoolings ($T_m - T$). Calorimetric measurements for alkali disilicate glasses (Takahashi and Yoshio, 1973) show that (3.7) and (3.8) overestimate and underestimate ΔG respectively, the errors being greater at high undercoolings.

It is usually assumed that the kinetic barrier ΔG_D can be expressed in terms of an effective diffusion coefficient D given by

$$D = (kT\lambda^2/h) \exp(-\Delta G_D/kT) \qquad (3.9)$$

where λ is a quantity of the order of atomic dimensions ('jump distance'). It is also usually assumed that D can be equated to the self-diffusion coefficient in the (one component) liquid. Furthermore, various authors have related D to the viscosity of the liquid (η) by the Stokes–Einstein relation

$$D = kT/3\pi\lambda\eta \qquad (3.10)$$

From (3.1), (3.9) and (3.10)

$$I = (Ah/3\pi\lambda^3\eta) \exp(-W^*/kT) \qquad (3.11)$$

Substituting for A from the approximate expression (3.3) we have

$$I = (n_v kT/3\pi\lambda^3\eta) \exp(-W^*/kT) \qquad (3.12)$$

Hence, if experimental data for I, η and ΔG are available a plot of $\ln(I\eta/T)$ against $1/\Delta G^2 T$ should produce a straight line with the slope and intercept yielding σ and A respectively. The analysis of results by this procedure will be discussed later.

3.2.2 Non-steady state homogeneous nucleation

In practice, an equilibrium size distribution of crystal embryos in the supercooled liquid is not achieved immediately at a given temperature. As a result a steady state nucleation rate I is only approached gradually, the effect being characterized by an induction time which increases with fall in temperature. In the case of glasses transient nucleation becomes of particular importance near the glass transformation temperature.

An analytical treatment has been given by Kashchiev (1969) who showed that the transient nucleation rate I' at time t can be expressed as an infinite series.

$$I'/I = 1+2\sum_{n=1}^{\infty}(-1)^n \exp(-n^2 t/\tau) \qquad (3.13)$$

where τ is an induction time and n an integer. The number of nuclei $N(t)$ at time t is

$$N(t)/I\tau = t/\tau - \pi^2/6 - 2\sum_{n=1}^{\infty}[(-1)^n/n^2]\exp(-n^2 t/\tau) \qquad (3.14)$$

For $t > 5\tau$ this reduces to the simple linear relation

$$N(t) = I(t - \pi^2\tau/6) \qquad (3.15)$$

It can be shown (James, 1974) that

$$\tau = (16h\lambda^2\sigma/\pi^2 V^2 \Delta G_v^2)\exp(\Delta G_D'/kT) \qquad (3.16)$$

where $\Delta G_D'$ is an activation free energy which may be identical to the kinetic barrier ΔG_D in (1.1), although this remains to be established. If $\Delta G_D'$ is equated to the activation free energy for diffusion in the liquid, as above, and if (3.10) applies we have

$$\tau = (48\sigma\lambda^5/\pi V^2\Delta G_v^2)\,\eta \qquad (3.17)$$

i.e. τ is proportional to the viscosity. However, as we shall see this is not necessarily true in practice.

3.2.3 Heterogeneous nucleation

In view of the subsequent discussion it is appropriate to consider briefly the steady state heterogeneous nucleation rate on a flat substrate in the supercooled liquid. This is given (Christian, 1975) by

$$I_{het} = n_s(kT/h)\exp[-(W_{het}^* + \Delta G_D)/kT] \qquad (3.18)$$

where n_s is the number of atoms or 'formula units' of the liquid in contact with the substrate per unit area and

$$W_{het}^*/W^* = f(\theta) = (2 - 3\cos\theta + \cos^3\theta)/4 \qquad (3.19)$$

where θ is the contact angle between the crystalline nucleus and the substrate and $f(\theta) < 1$ for $0 \leqslant \theta \leqslant \pi$. The rate of heterogeneous nucleation per unit volume of the liquid will depend on the total surface area of catalysing substrate per unit volume and thus on the state of dispersion of the substrate in the liquid, including the particle size distribution.

Another important factor is the curvature of the substrate. In general, W^*_{het} for convex and concave (to the liquid) substrates will be greater and smaller respectively than the corresponding value for a flat substrate. For example, for a substrate in the form of a spherical particle of radius R in the liquid, Fletcher (1958, 1959) has shown that

$$W^*_{het} = W^* g(\theta, x) \tag{3.20}$$

where $g(\theta, x)$ is a function given by Fletcher; $x = R/r^*$, r^* being the critical nucleus radius for homogeneous nucleation. In effect, this theory predicts that the substrate curvature (hence R) only has a large effect on W^*_{het}, and hence on the heterogeneous nucleation rate, when x is less than 10, and particularly for the range $x = 1$ to 10. In practice, r^*, from experimental results for silicate glasses (James, 1982) is $\simeq 1$ nm for the temperature range of interest. Thus for a value of R below about 10 nm the heterogeneous nucleation rate for the convex substrate is expected to show a sharp decrease relative to the value for the flat substrate. As a result very fine convex particles below a certain size are not expected to act as efficient nucleation sites.

3.3 EXPERIMENTAL STUDIES IN 'SIMPLE' ONE-COMPONENT SYSTEMS

3.3.1 General features and non-steady state effects

The most commonly observed form of nucleation in glass-forming systems is on the surface. It is probably heterogeneous in origin and is generally sensitive to the chemical or mechanical condition of the glass surface. Internal or volume nucleation, is much more rarely observed and is often only achieved by adding nucleating agents. However, certain glass systems exhibit volume nucleation without deliberate additions and it is probable that such nucleation is homogeneous. Here the discussion will be restricted to 'simple' stoichiometric compositions, which are defined in the present context as those in which the crystallizing phase has the same composition as the parent glass, i.e. they are effectively single component systems. Initially only 'self nucleation' in the volume is discussed, i.e. nucleation without the addition of nucleating agents. It will be evident that the available data, most of which have only been obtained in the last decade show a clear and interesting pattern.

A number of simple compositions exhibit volume nucleation, including $Li_2O \cdot 2SiO_2$ (LS_2), $BaO \cdot 2SiO_2$ (BS_2), $3BaO \cdot 5SiO_2$ (B_3S_5), $Na_2O \cdot 2CaO \cdot 3SiO_2$ (NC_2S_3), $2Na_2O \cdot CaO \cdot 3SiO_2$ (N_2CS_3), $Na_2O \cdot SiO_2$ (NS) and $CaO \cdot SiO_2$ (CS). All these are one component, or effectively one component, systems. Of these, lithium disilicate has been the most intensively studied by various authors (see James, 1982, 1985 for a full list of references), mainly because the nucleation rates can be conveniently measured, but also because the LS_2 crystal phase is an important constituent of many glass ceramics. Another reason for the importance of this system is that detailed ΔG data are available. Figure 3.1 shows the steady state nucleation rate for LS_2 plotted against temperature. This plot illustrates the general features of volume nucleation in all the simple systems. The 'bell-shaped' curve is of the general form predicted by the classical nucleation theory. The maximum nucleation rate ($4.25 \times 10^9\,m^{-3}\,s^{-1}$) occurs at about 454°C, which is within the transformation range of the glass. Nucleation occurs in the range 425–530°C, which is at high undercoolings below the liquidus at 1034°C. Outside this range volume nucleation is negligible, although surface nucleation can be observed at somewhat higher temperatures. It should also be mentioned that the crystal growth rate curve for LS_2 shows a maximum at a much higher temperature than the nucleation curve. The growth rates also become increasingly smaller as the temperature approaches the transformation range, and are negligible below 425°C.

Before proceeding it is appropriate to consider briefly the experimental method used to obtain the nucleation rates. At higher temperatures, usually well above the nucleation maximum, a single stage heat treatment can be used. The number of crystals per unit volume, N_v, can be readily determined by reflection optical microscopy of polished and lightly etched sections through the glass samples. In some cases it is more appropriate to use scanning electron microscopy of surface sections or transmission optical or electron microscopy of thin sections, depending on the magnitude of the nucleation rates involved. At lower nucleation temperatures a double stage heat treatment is commonly employed. After the nucleation treatment at T_N the glass is given a short heat treatment at a higher growth or development temperature (T_G) to grow the crystals to observable dimensions for counting by optical microscopy. The temperature T_G is chosen for a more rapid growth rate but a negligible nucleation rate. The validity of this procedure depends on the assumption that after nucleation the glass contains an assembly of nuclei, some of which will have grown into small crystals, the large majority not redissolving on heating to the second stage. Of particular importance is the choice of growth treatment and growth temperature, which should not be too high. However, the method has been investigated in detail and justified (see James, 1982) and produces accurate and reproducible results for nucleation rates provided it is used correctly.

A typical optical micrograph of LS_2 glass given a double stage treatment

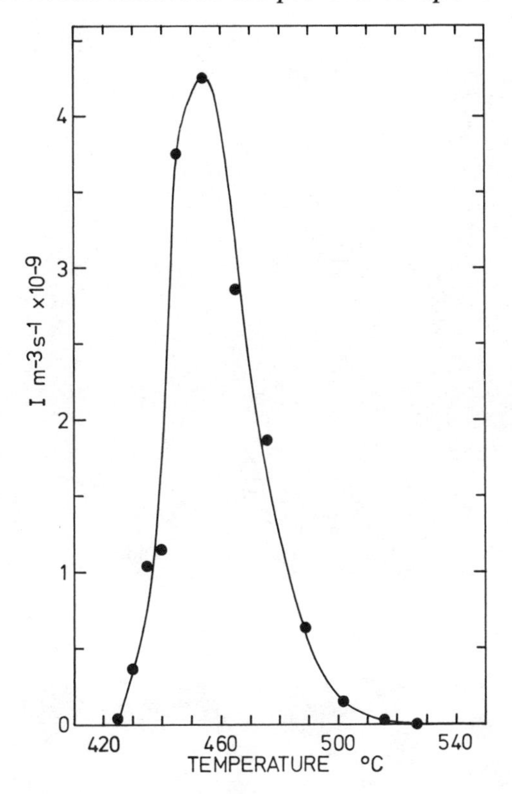

Fig. 3.1 Steady state nucleation rate (I) as a function of temperature for a glass close to lithium disilicate in composition (33.1 mol % Li_2O). After James (1974).

shows LS_2 crystals at a fairly advanced stage of development (Fig. 3.2). From transmission electron microscopy (James and Keown, 1974) of the earlier stage of growth, small plate-like crystals form from the initial nuclei. These develop branches, eventually forming a tight cluster of platelet crystals with the approximate overall shape of a prolate ellipsoid, as shown in Fig. 3.2(a). It is relatively straightforward to determine the number of such 'spherulites' per unit volume, as indicated above, and this can be equated to the number of original nuclei. Such spherulitic growth is also observed in the other systems, although the 'developed' particles are usually close to spherical in shape and the detailed morphologies of the early growth vary considerably between systems. Figures 3.2(b) and (c) show optical micrographs of spherulites in BS_2 and NC_2S_3 glasses.

Typical plots of nucleation density N_v against time for LS_2 glass are shown in Fig. 3.3, at 440°C and 476°C. At lower temperatures pronounced non-steady state nucleation behaviour was observed, as shown by the 440°C curve. The nucleation rate increased with time up to about 12 h and then

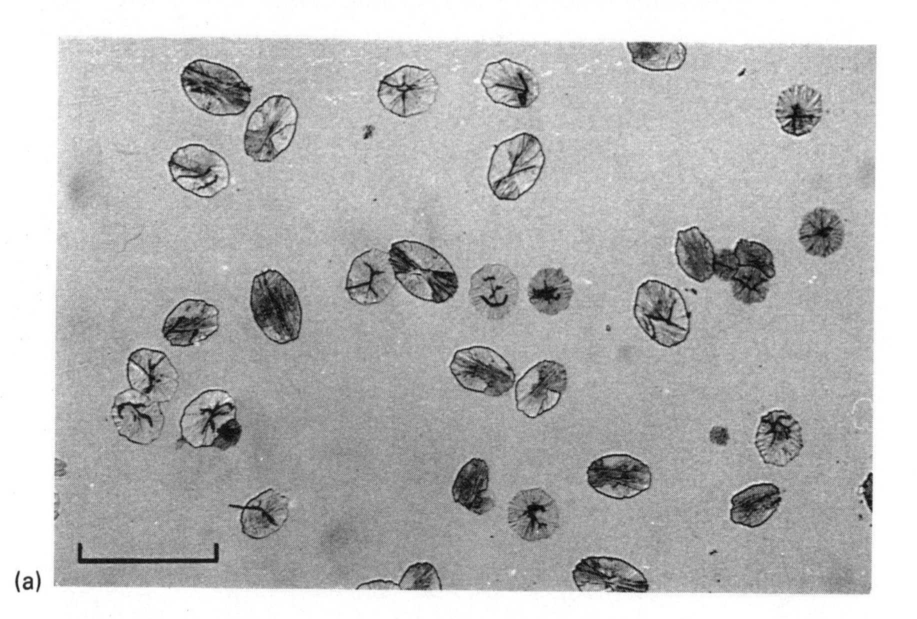
(a)

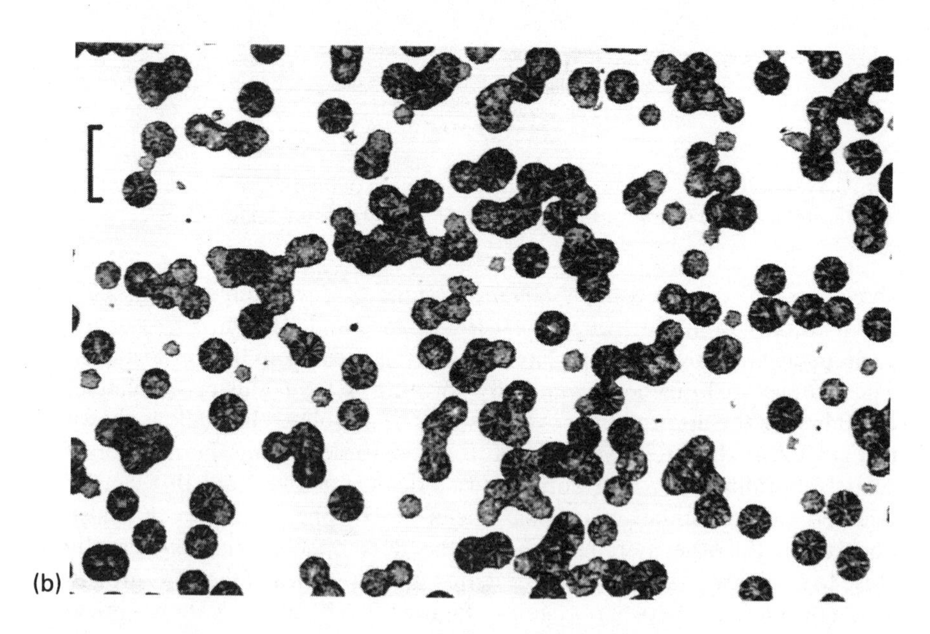
(b)

Fig. 3.2 Reflection optical micrographs of polished and lightly etched sections through glass samples: (a) $Li_2O \cdot 2SiO_2$ glass heat-treated at 476°C for 1 h followed by a growth treatment at 600°C for 15 min; bar denotes 100 μm; (b) $BaO \cdot .2SiO_2$ glass heated at 868°C for 10 min; bar denotes 200 μm; (c) $Na_2O \cdot 2CaO \cdot 3SiO_2$ glass heated at 654°C for 75 min; bar denotes 40 μm.

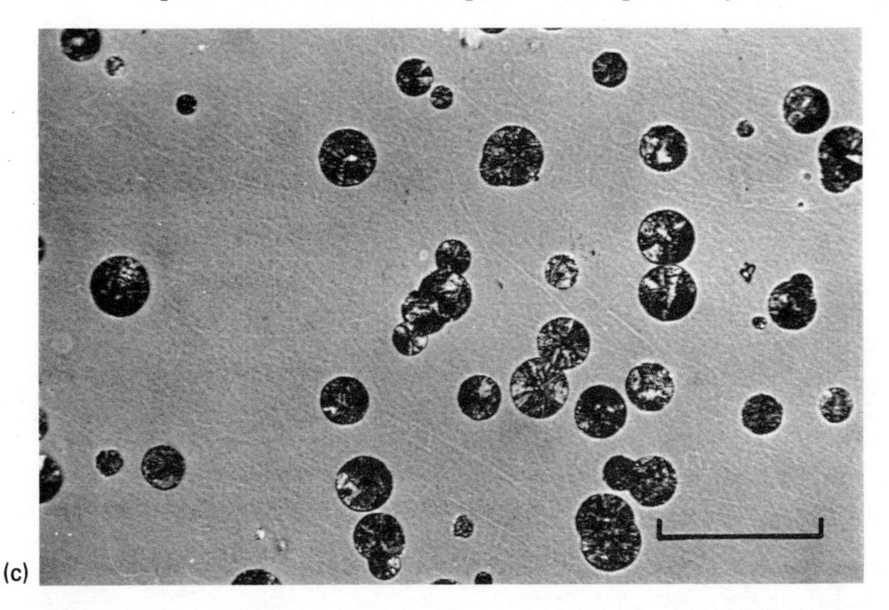
(c)

became constant, the form of the curve in Fig. 3.3. being very similar in general form to that predicted by the Kashchiev theory (expression (3.14)). The induction time, τ, determined from the intercept on the time axis of the N_v against time plot (using (3.15)) (see Fig. 3.3), was 8 h at 440°C, compared with only 13 min at 476°C. Analysis of many such curves (James, 1974) showed that the form of the N_v against time curves were accurately described by the Kashchiev theory. The temperature dependence of the induction time was also described by an Arrhenius type relationship, in accord with (3.16). However, approximate calculations of absolute values of τ using (3.17) (based on the Stokes–Einstein relation) and viscosity data were about an order of magnitude smaller than the experimental values. This may be considered reasonable in view of the limitations of the Stokes–Einstein relation, which is subject to an uncertainty of at least an order of magnitude in absolute terms.

Induction times for volume nucleation have also been observed in the NC_2S_3 (Gonzalez-Oliver and James, 1980b) and N_2CS_3 (Kalinina *et al.*, 1980) systems. As for LS_2, plots of $\log(\tau\Delta_v^2)$ against $1/T$ gave straight lines in accordance with (3.16). The slopes of such plots give $\Delta H'_D$, since $\Delta G'_D$ may be expressed as $\Delta H'_D - T\Delta S'_D$ where $\Delta H'_D$ and $\Delta S'_D$ are the activation enthalpy and entropy respectively. However, the $\Delta H'_D$ values obtained (or ΔH_τ) were not identical to the corresponding activation energies for viscous flow (ΔH_η) indicating that τ may not be proportional to viscosity as suggested by (3.17). There is clearly scope for further work in this area. Non-steady state nucleation effects have also been demonstrated in the crystallization of a multicomponent glass (Gutzow *et al.*, 1977).

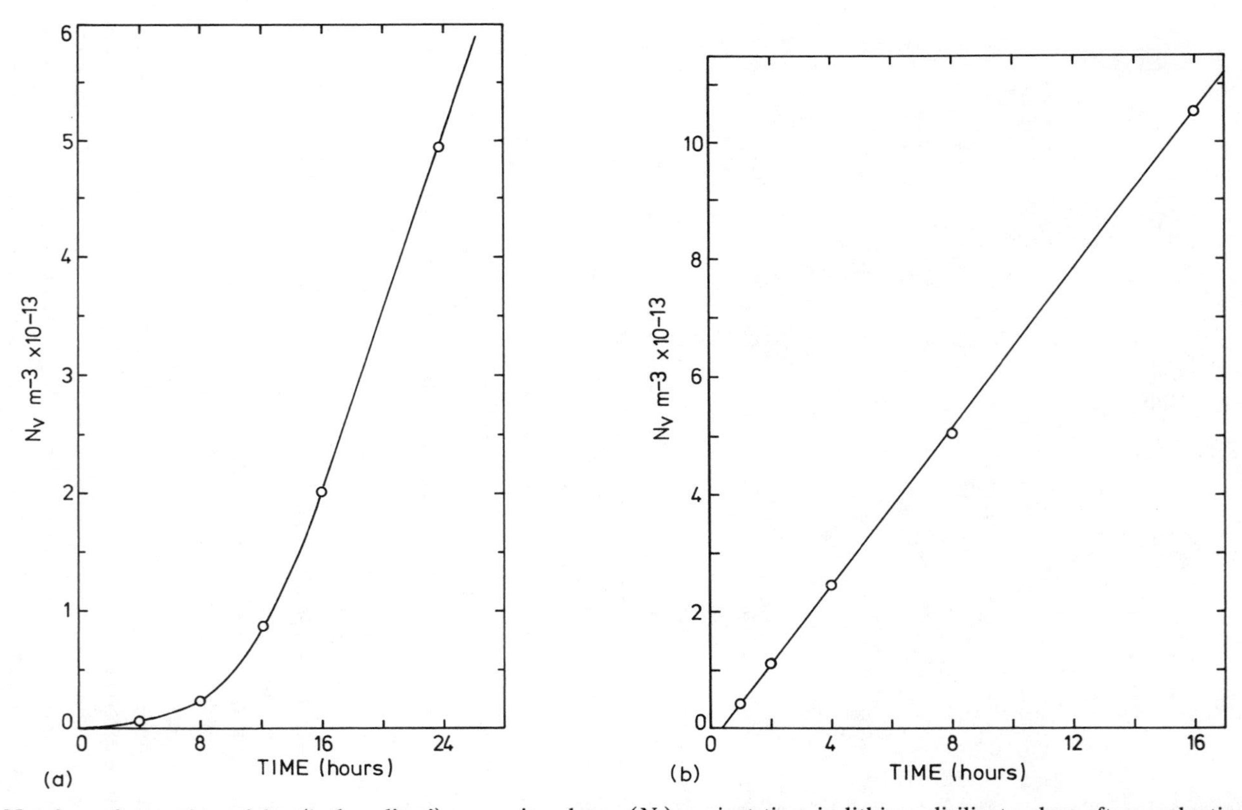

Fig. 3.3 Number of crystal particles ('spherulites') per unit volume (N_v) against time in lithium disilicate glass after nucleation (a) at 440°C and (b) at 476°C. Growth (development) temperature 560°C. After James (1974).

Kelton *et al.* (1983) recently performed a detailed theoretical treatment of transient nucleation in a one-component condensed system. They obtained an exact closed form expression for the time dependent cluster populations during nucleation and growth and developed a numerical simulation by computer which modelled directly the evolution of the cluster distribution and the nucleation rate in the transient regime. The results of the simulation were in good agreement with the approximate analytical treatment of Kashchiev, mentioned previously, which was shown to give a good description of the nucleation behaviour. Kelton and Greer (1986) have extended the numerical method to the non-isothermal case of a fast cool from the melt and showed that, in some cases, transient effects have a critical effect on glass formation.

3.3.2 Comparison of steady state nucleation rates in $Li_2O \cdot 2SiO_2$ with theory

Rowlands and James (1979) performed a detailed analysis of the nucleation data of James (1974) for LS_2. A plot was made of $\ln(I\eta/T)$ against $1/\Delta G^2 T$ using the viscosity (η) data of Matusita and Tashiro (1973) and published ΔG data. A good straight line fit was obtained over most of the temperature range (445 to 527°C) as predicted by (3.12), the slope yielding an interfacial energy σ of 190 mJ m^{-2}. The classical theory thus appeared to give a good description of the temperature dependence of I over most of the range, but the description was poor at lower temperatures (425 to 440°C). Moreover, the experimental value of the pre-exponential factor A from the intercept of the plot was about 20 orders of magnitude higher than the theoretical value of A (about 5×10^{41} m^{-3} s^{-1}) for LS_2. Almost identical conclusions were reached by Neilson and Weinberg (1979) who independently carried out a similar analysis using the same experimental data of James (1974).

Recently, the viscosity of LS_2 glass has been accurately redetermined (Gonzalez-Oliver, 1979; Zanotto and James, 1985) yielding values much higher than the previous data (a factor of 10 or more depending on temperature). Figure 3.4 shows the plot of $\ln(I\eta/T)$ against $1/\Delta G^2 T$ using the original nucleation data of James (1974), the ΔG data from Takahashi and Yoshio (1973) and the *new* viscosity data. Clearly a good straight line is obtained over a fairly wide temperature range. In fact, comparison with the original plot of Rowlands and James (1979) shows that a much better fit is obtained at lower temperatures. It would be tempting to state that this indicates a good agreement with classical theory. However, the departure from linearity still present at lower temperatures may be real and not caused by experimental errors, as will become evident later. The slope of the plot in Fig. 3.4 gives $\sigma = 207$ mJ m^{-2}, similar to the previous value although somewhat higher. The experimental value of A from the intercept is still much higher than the theoretical value (by a factor of over 10^{25} times).

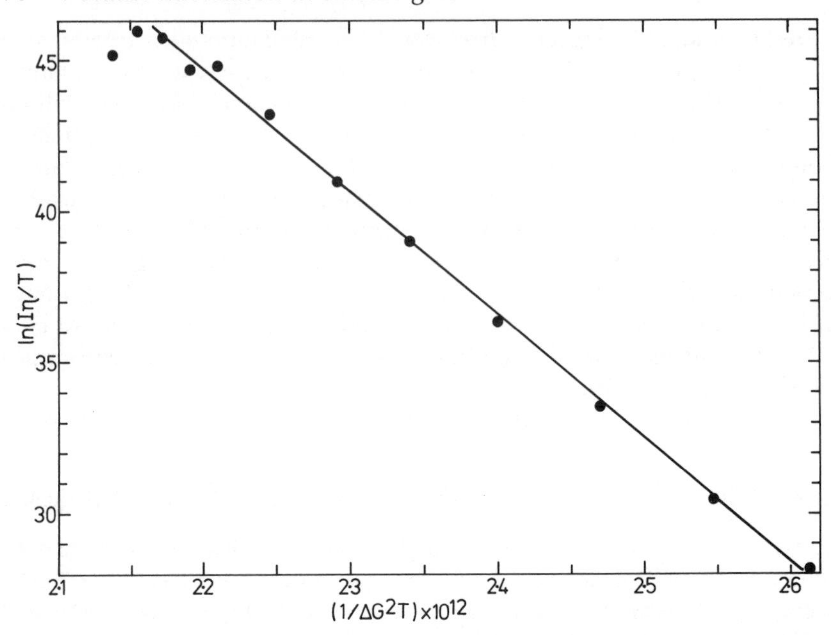

Fig. 3.4 Plot of ln $(I\eta/T)$ against $1/\Delta G^2 T$ for $Li_2O \cdot 2SiO_2$ glass. I in $m^{-3}\ s^{-1}$, η in $N\,s\,m^{-2}$, ΔG in $J\ mol^{-1}$, and T in K. After James (1985).

In addition to redetermining the viscosity of LS_2 glass, Zanotto and James (1985) recently obtained new data for the crystal nucleation rates in this system. Efforts were made to minimize impurity levels in the glasses by careful choice of very pure starting materials because of the possible effects these might have on the nucleation rates. They also re-examined all the available nucleation results for LS_2 glass. The nucleation rates in the new very pure LS_2 glass were slightly lower than those of James (1974) by a fairly constant factor, independent of temperature, but, in general, the new results were in excellent agreement with the previous data taken as a whole. The small differences in nucleation rates might be attributed to slight differences in base compositions used. Other possible reasons are discussed by Zanotto and James (1985).

Plots of ln $I\eta/T$ against $1/\Delta G^2 T$ were constructed for the new and previous nucleation data but almost identical results to Fig. 3.4 were obtained in all cases, the slopes and intercepts giving very similar σ and A values to those stated above. The effect of possible errors in the measured ΔG data were also considered. However, even allowing for reasonable variations in ΔG, the experimental A was still much higher than the theoretical value, and σ was hardly altered.

Possible reasons for the apparent discrepancy between the theoretical and experimental pre-exponential factors were discussed by Rowlands and

James (1979) and Zanotto and James (1985). First, it was most unlikely that the disagreement could be caused by heterogeneous nucleation. The high undercoolings below the liquidus necessary to observe internal nucleation and the consistency of the data from various investigators strongly suggest that homogeneous nucleation is predominant. This is strongly supported by the work of James *et al.* (1978), who found that LS_2 glass melted in platinum showed no significant difference in nucleation behaviour from LS_2 glass melted under completely platinum-free conditions in a silica crucible, indicating that heterogeneous nucleation on platinum particles, possibly introduced during melting, was negligible for the conditions used. They also compared LS_2 glass melted from ordinary purity batch materials with LS_2 glass melted with very high purity materials and found that minor impurities in the levels present had at most only a small effect on nucleation, as indicated later by the work of Zanotto and James (1985).

Heterogeneous nucleation is also very unlikely, since according to theory, the pre-exponential factor A for heterogeneous nucleation is expected to be many orders of magnitude smaller than that for homogeneous nucleation (not greater, as observed).

Other possibilities, such as transient nucleation effects and experimental errors in the nucleation rates have been considered but it was concluded that these could not account for the observed discrepancy in the values of A.

Since an error in the free energy (ΔG) data sufficiently large to explain the discrepancy appears unlikely, there are two remaining possibilities. The first is that the use of the Stokes–Einstein equation to relate the nucleation kinetic term (diffusion term) and viscosity may be incorrect. In partial justification of this procedure it may be noted that diffusion coefficients calculated from measured viscosities, using the Stokes–Einstein equation, agreed with measured diffusion coefficients of oxygen in silicate glasses to within an order of magnitude (Oishi *et al.*, 1975). Also the apparent activation enthalpies ΔH_D for oxygen diffusion were close to those for viscous flow. Furthermore, experimental growth rates at high undercoolings in one-component systems can be predicted to within an order of magnitude or so using the Stokes–Einstein equation (Uhlmann, 1971). However, in view of the induction time results discussed earlier, there remains some uncertainty in the use of the Stokes–Einstein equation for the analysis of nucleation rates.

The other remaining possibility, considered by the present author to be the most plausible, is to invoke a temperature dependent interfacial energy, σ, as first suggested for LS_2 by Rowlands and James (1979). If A is assumed to be given by theory (expression (3.2)) σ_T can be calculated by substituting I, ΔG and η into (3.11) (W^* is given by (3.4)). The value of σ_T is then obtained by successive approximation since it also enters the more accurate expression for A (expression (3.2)). The values of σ_T and W^*/kT calculated in this way by James (1985) to fit exactly the nucleation data are shown in

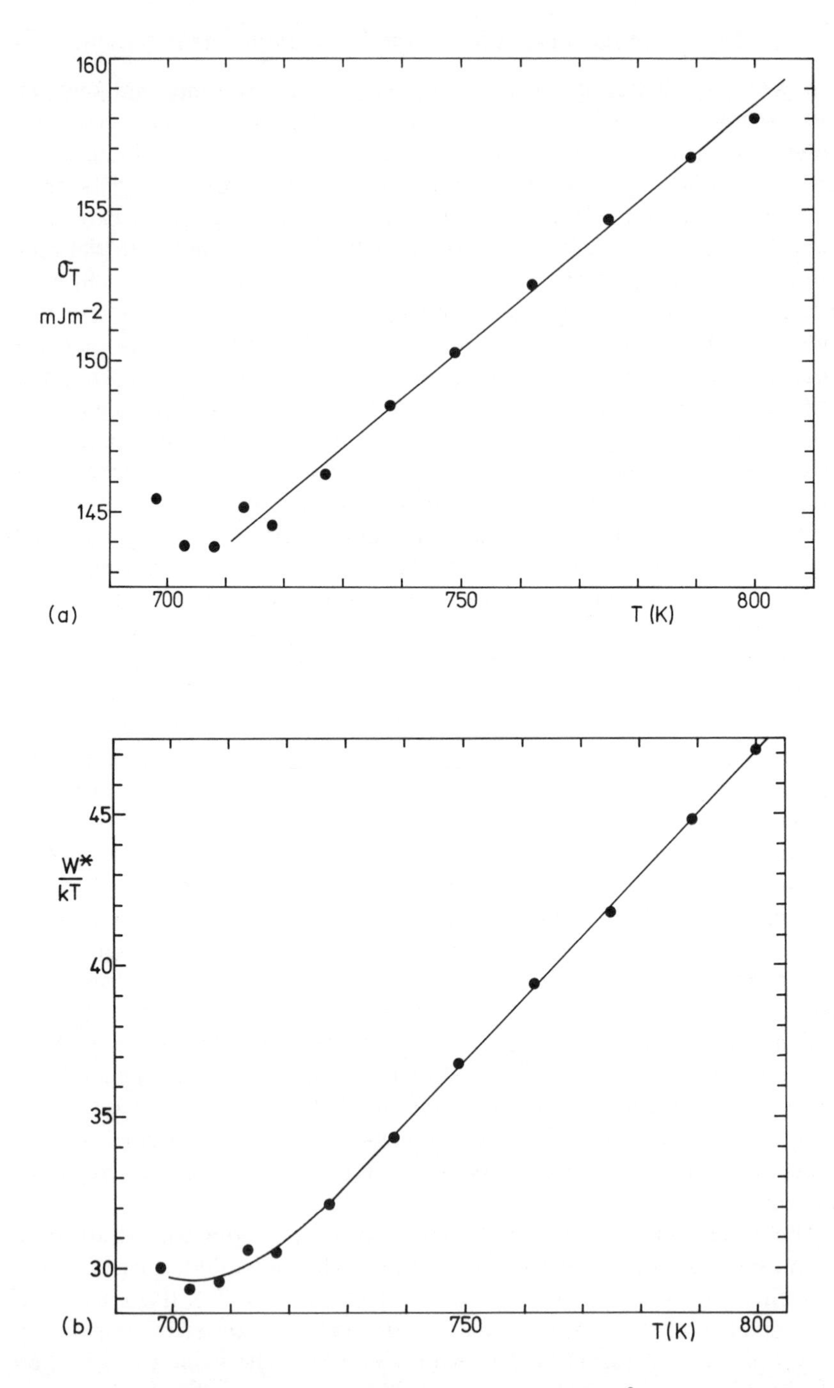

Fig. 3.5 (a) Crystal–liquid interfacial energy σ_T (mJ m^{-2}) as a function of temperature T (K) for lithium disilicate (LS_2). (b) W^*/kT against T for LS_2 (W^* is the thermodynamic free energy barrier to nucleation). After James (1985).

Fig. 3.5(a) and (b). In the nucleation range σ_T increases with temperature from about 144 to 158 mJ m^{-2} and over most of the range a linear relation applies. As expected the value of W^*/kT (assuming the theoretical A) increased with temperature (Fig. 3.5(b)) varying from 29 to 47 in the range of nucleation measurements.

It is clear that the σ_T values are significantly lower than the σ value (207 mJ m^{-2}) obtained from the earlier plot of $\ln (I\eta/T)$ against $1/\Delta G^2 T$. Also, at first sight, the temperature dependent σ_T might appear to be inconsistent with the good straight line plot over most of the range in Fig. 3.4. That no inconsistency exists was demonstrated by James (1985), who calculated the slope of the plot in Fig. 3.4 ($d \ln (I\eta/T)/d(1/\Delta G^2 T)$) at a given temperature T assuming the linear relation $\sigma_T = a + bT$ where a, b are constants, and was able to determine the 'apparent σ' (σ_a) from this slope. The details of the calculation will not be given here. At 454°C, $\sigma_a = 207.4$ mJ m^{-2} and $\sigma_T = 146.6$ mJ m^{-2} and at 527°C, $\sigma_a = 209.2$ mJ m^{-2} and $\sigma_T = 158.4$ mJ m^{-2}. Thus σ_T had increased by 8% whereas σ_a had increased by less than 1% in this range. Also the slope changed only slightly (< 3%), explaining the good straight line obtained (within experimental error). The important point is that an apparent good straight line plot, as in Fig. 3.4, does not necessarily imply a constant, temperature independent σ. It could hide an appreciable linear variation of σ with T and indicate a higher interfacial energy than the true value. Moreover, the A factor determined from the intercept of the line would be much higher than the true value.

A similar analysis to that above was first used by Turnbull (1952) to explain a discrepancy in the theoretical and experimental A values for the nucleation of supercooled mercury (the experimental value was 10^7 times the theoretical value). Miyazawa and Pound (1974) also found that for homogeneous nucleation in supercooled gallium the experimentally derived value of A was 10^6 times the value calculated from classical theory and showed that this could be accounted for by a σ that increased linearly with rise in temperature. Spaepen and Turnbull (1976) have discussed the possible effect of a temperature dependent σ on metallic glass formation, and Spaepen (1975) proposed a model for the liquid–crystal interface in metallic systems in which σ is largely entropic in origin and increases approximately linearly with rise in temperature. The result for LS_2 glass also indicates a large negative entropy contribution to the interfacial free energy σ_T. The reason for this behaviour is unclear at present, although the increase in σ_T with temperature may arise from a sharpening of the liquid–crystal interface and be associated with the increase in critical nucleus size with rise in temperature (Fokin *et al.*, 1981).

3.3.3 Results for $Na_2O \cdot 2CaO \cdot 3SiO_2$ and $BaO \cdot 2SiO_2$

Apart from LS_2 the most complete data available are for the compositions $Na_2O \cdot 2CaO \cdot 3SiO_2$ (Gonzalez-Oliver and James, 1980b) and $BaO \cdot 2SiO_2$

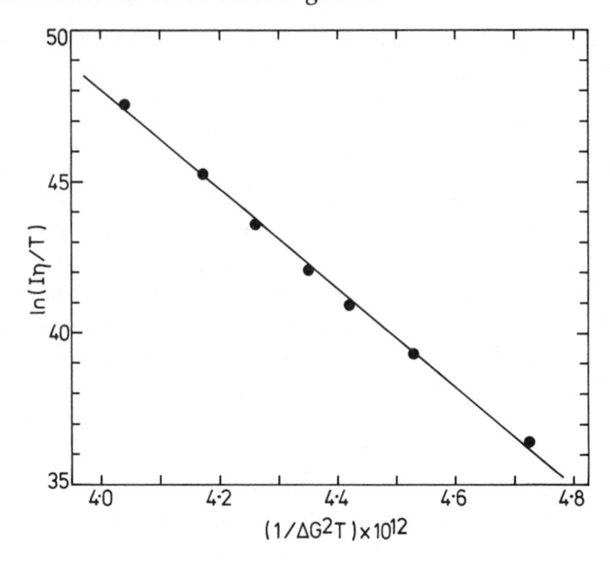

Fig. 3.6 Plot of ln $(I\eta/T)$ against $1/\Delta G^2 T$ for $BaO \cdot 2SiO_2$ glass. I in $m^{-3}\,s^{-1}$, η in $N\,s\,m^{-2}$, ΔG in $J\,mol^{-1}$ and T in K. After James (1985).

(James and Rowlands, 1979; Zanotto and James, 1985). Both are congruently melting systems with maximum nucleation rates of $5.60 \times 10^{11}\,m^{-3}\,s^{-1}$ (NC_2S_3) and $1.87 \times 10^{12}\,m^{-3}\,s^{-1}$ (BS_2) at 595 and 700°C respectively. For both compositions plots of ln $(I\eta/T)$ against $1/\Delta G^2 T$ gave straight lines in accordance with theory (a plot for BS_2 is given in Fig. 3.6). The ΔG data were calculated using measured ΔH_f values and were therefore subject to more uncertainty than the ΔG for LS_2. However, the experimental values of A determined from the intercepts of the 'nucleation plots' were again very high compared with theory. This was the case even when reasonable allowance was made for ΔC_p using (3.6). Once more a temperature dependent σ was shown to provide a reasonable explanation of the discrepancy with theory (James, 1985). The calculated σ_T and W^*/kT for $BaO \cdot 2SiO_2$ are plotted against temperature in Fig. 3.7.

3.3.4 Comparison of 'simple' systems and overall picture

As mentioned earlier, volume nucleation without addition of nucleating agents is relatively rare and only a few systems have been investigated quantitatively. In contrast, surface nucleation is readily observed in most compositions. Recently, James (1985) has collected the available information on those systems exhibiting volume nucleation. A summary of this information is given in Tables 3.1 and 3.2. The results for $2Na_2O \cdot CaO \cdot 3SiO_2$ are from Kalinina *et al.* (1980); for $Na_2O \cdot SiO_2$ from

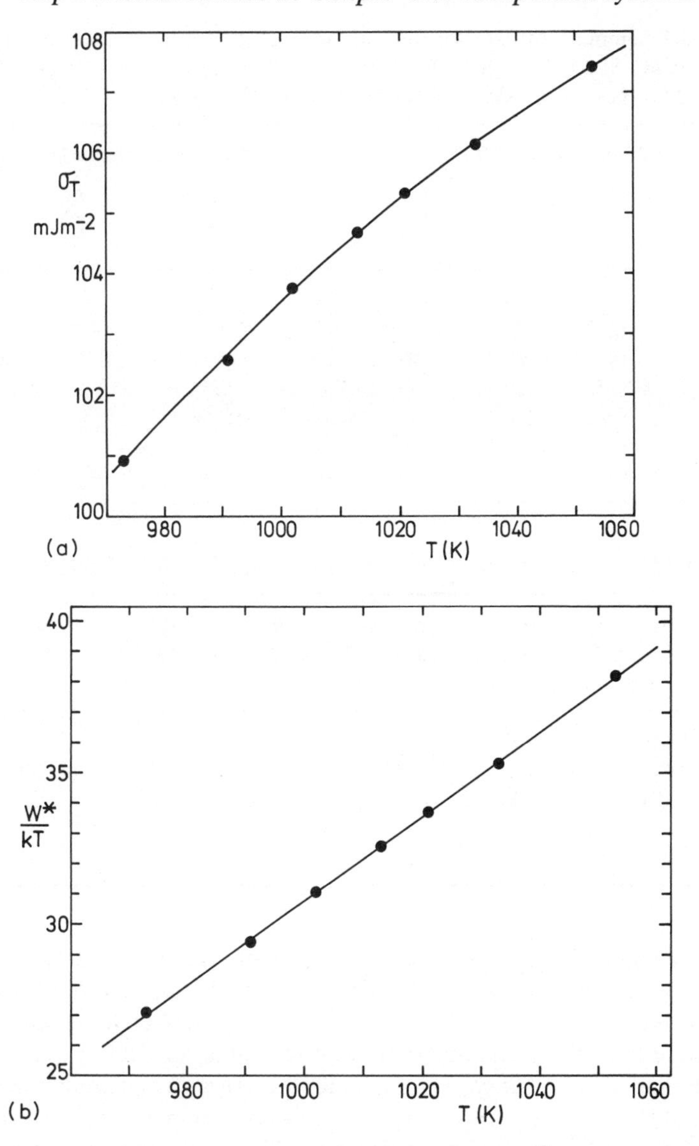

Fig. 3.7 (a) Crystal–liquid interfacial energy σ_T (mJ m^{-2}) as a function of temperature T (K) for barium disilicate (BS_2). (b) W^*/kT against T for BS_2. After James (1985).

Filipovich *et al.* (1975); for $3BaO \cdot 5SiO_2$ from Freiman *et al.* (1972) and for $CaO \cdot SiO_2$ from Wang and James (1984). In Table 3.1 all the melting points (T_m) are congruent except for LS_2, N_2CS_3 and B_3S_5. Although crystalline LS_2 melts incongruently the LS_2 composition is completely liquid 1°C above the incongruent melting point. Thus the theoretical congruent melting point

Table 3.1 Summary of data for nucleation in simple silicates: $Li_2O{\cdot}2SiO_2$ (LS_2); $Na_2O{\cdot}2CaO{\cdot}3SiO_2$ (NC_2S_3); $BaO{\cdot}2SiO_2$ (BS_2); $2Na_2O{\cdot}CaO{\cdot}3SiO_2$ (N_2CS_3); $Na_2O{\cdot}SiO_2$ (NS); $3BaO{\cdot}5SiO_2$ (B_3S_5); and $CaO{\cdot}SiO_2$ (CS)

System	*Temperature (°C)*				*Reduced temperatures T (K)*		
	T_m	T_{max}	T_d	T_g	$\frac{T_{max}}{T_m}$	$\frac{T_d}{T_m}$	$\frac{T_g}{T_m}$
LS_2	1034	454	535	454	0.556	0.618	0.556
NC_2S_3	1289	595	730	565	0.556	0.642	0.536
BS_2	1420	700	846	689	0.575	0.661	0.568
N_2CS_3	≃1175	505	–	470	0.54	–	0.51
NS	1089	460	–	410	0.54	–	0.50
B_3S_5	≃1433	700–725	–	≃690	0.57–0.59	–	0.56
CS	1544	790	–	757	0.58	–	0.57

Table 3.2 Nucleation parameters in various systems (for key see Table 3.1)

System	σ_T ($mJ\,m^{-2}$) (*at* $T = T_{max}$)	σ_a ($mJ\,m^{-2}$)	α *from* σ_T *at* $T = T_{max}$	α' (*from* σ_a)	W^*/kT (*at* $T = T_{max}$)	W^*/kT (*at* $T = T_d$)
LS_2	146	207	0.31	0.45	32	49
NC_2S_3	108	182	0.25	0.42	30	53
BS_2	101	136	0.40	0.54	27	52
N_2CS_3	–	–	–	–	~25	–
NS	173	–	0.36	–	32	–

must be almost identical with the incongruent one and the system can be treated as effectively one component with a melting point at 1034°C. For N_2CS_3 and B_3S_5 the effective metastable congruent melting points were estimated from the phase diagrams, and the values in Table 3.1 are subject to only minor uncertainty. T_{max} is the temperature of maximum nucleation rate. For the present purposes T_g is taken as the temperature for a viscosity of 10^{12} Pa s (10^{13} poise). The temperature T_d is that estimated for a volume nucleation rate of 1 cm^{-3} s^{-1} (or 10^6 m^{-3} s^{-1}). The latter is chosen arbitrarily as the 'just detectable' limit. It should be noted that certain systems such as Al_2O_3–SiO_2, which are not included in Table 3.1, also show volume nucleation but no quantitative nucleation data or even values of T_{max} are available at present.

A striking feature of Table 3.1 is the remarkably consistent values of T_{max}/T_m for the seven glass compositions. They are all in the range 0.54 to 0.59 in spite of the large variation in T_m. T_d/T_m (0.62 to 0.66) and T_g/T_m

(0.50 to 0.57) also show small variations. T_{max}/T_m for each composition was either slightly greater than, or equal to T_g/T_m. The values of T_{max}/T_m and T_d/T_m are plotted against T_m in Fig. 3.8 for comparison. It is interesting to note that in the present silicate systems the onset of homogeneous nucleation occurs at reduced undercoolings $(T_m - T_d)/T_m$ of 0.33 to 0.38. These are much higher than the corresponding values of 0.15 to 0.25 that have been observed in 'droplet' nucleation studies of metals, alkali halides, organic liquids and other systems (Turnbull, 1950; Jackson, 1966).

In Table 3.2, σ_T and W^*/kT were calculated as described previously using (3.11) and assuming A was given by theory (expression (3.2)). The values of W^*/kT show a striking similarity being ~ 30 at $T = T_{max}$ and ~ 50 at $T = T_d$.

The quantity α in Table 3.2, the 'Turnbull parameter', is given by

$$\alpha = \sigma(N_A V_m^2)^{1/3}/\Delta H_f \tag{3.21}$$

where N_A is Avogadro's number. The quantity $\sigma(N_A V_m^2)^{1/3}$ is defined as the 'molar interfacial energy' (Turnbull, 1950) and α is dimensionless.

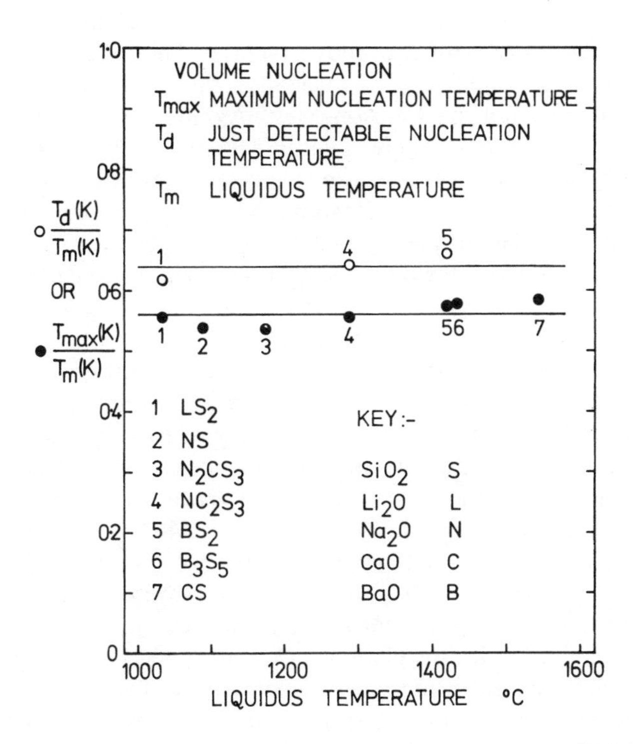

Fig. 3.8 Plots of T_d/T_m and T_{max}/T_m against liquidus temperature (T_m) for various stoichiometric silicate systems. T_d and T_{max} are the 'just detectable' temperature for volume nucleation and the temperature of maximum nucleation rate respectively.

Nucleation experiments using droplets (Turnbull, 1956; Jackson, 1966) indicate α values between 0.25 and 0.5 for a range of materials, with $\alpha \sim 0.4$ to 0.5 for metals and $\alpha \sim 0.3$ typical for most non-metals. In Table 3.2 the α values were obtained using the σ_T results at $T = T_{max}$ in (3.21). The values denoted α' were obtained from the σ_a results (using the slopes of the 'nucleation' plots). The former values using σ_T appear to correspond closest to the 'droplet' nucleation studies for non-metals.

The remarkably consistent pattern of the results in Tables 3.1 and 3.2 and the high undercoolings required for nucleation provide further support, in addition to the evidence given previously, that the nucleation observed in most, if not all, of these silicate systems is predominantly homogeneous. Clearly the highest volume nucleation rates in these simple systems occur in the vicinity of T_g or just above. In addition, the striking constancy of T_{max}/T_m and T_d/T_m indicates that all stoichiometric silicate systems may behave in a similar manner. Thus the observed ratios may be used to 'predict' approximate values of T_{max} or T_d in systems where only T_m is known, and thus provide a guide to the temperature ranges where volume nucleation should occur.

It is relevant to consider data on other simple systems which are not apparently consistent with the above picture. Thus Klein *et al.* (1977) deduced apparent volume crystal nucleation rates in a $Na_2O \cdot 2SiO_2$ glass indirectly by means of the Johnson–Mehl–Avrami (JMA) equation using the times required to reach a given volume fraction of crystallinity and the known crystal growth rates. The results appear to indicate good agreement with classical homogeneous nucleation theory both in the temperature dependence of I and in the pre-exponential factor. However, it now appears that this agreement was fortuitous since only surface nucleation has been observed in this system by various authors (see James, 1985).

$Na_2O \cdot 2SiO_2$ is one common example of a system where no volume nucleation has been reported. $K_2O \cdot 2SiO_2$ and anorthite ($CaAl_2Si_2O_8$) glasses are other examples but there are many other simple compositions in the same category which appear to exhibit only surface crystallization. This general conclusion, that homogeneous crystal nucleation is only observed in certain silicate compositions, is supported by recent studies of geologically important systems (Berkebile and Dowty, 1982). No homogeneous nucleation was observed at temperatures from 400°C to the liquidi in a number of compositions in the diopside–anorthite–forsterite system. Instead, various types of heterogeneous nucleation occurred on, for example, container walls, the glass surface and on cracks in the glass.

Let us briefly consider the conditions likely to favour the occurrence of homogeneous nucleation. The glass transformation temperature T_g was taken above as the temperature corresponding to a fixed viscosity (10^{12} Pa s). Therefore, at T_g the values of D and ΔG_D (from (3.9) and (3.10)) may be regarded as essentially constant and independent of composition, $T = T_g$ corresponding to a given level of atomic transport in the glass.

At T_g the driving force (from (3.7)) is given approximately by

$$\Delta G_v = (-\Delta H_f/V_m)\,[1 - (T_g/T_m)] \qquad (3.22)$$

If we neglect the interfacial energy σ for the moment, a greater tendency towards volume nucleation (i.e. a higher nucleation rate) will be favoured by a higher magnitude of ΔG_v, and hence on this simple argument, as low a value of T_g/T_m as possible.

If we now return to the data, Table 3.1 indicates that just detectable nucleation is expected at reduced temperatures of about 0.62–0.66. The T_g/T_m values range from 0.5 to 0.57. A particularly high T_g/T_m, perhaps greater than 0.6, could indicate low or undetectable volume nucleation since T_g would be higher than the temperature range expected for just detectable nucleation. The T_g/T_m values for $Na_2O \cdot 2SiO_2$ and anorthite are relatively high (about 0.64 and 0.61 respectively) supporting this argument. However, the value for $K_2O \cdot 2SiO_2$ is ~ 0.58, only slightly higher than the values in Table 3.1. Thus a high T_g/T_m is insufficient as a sole criterion for the absence of volume nucleation and other factors (such as σ) must be considered.

More generally, no volume nucleation will occur if W^*, which decreases with increasing undercooling, is too high. If it is too high even at T_g (or just below) no nucleation will be observed since the steep rise in viscosity in the transformation range will provide an effective 'cut-off', an effect accentuated by the induction times (non-steady state behaviour) in this range. A simple calculation (using (3.11)) indicates that $W^*/kT \leqslant 40$ is required for just detectable volume nucleation at $T = T_g$. Achievement of this condition will depend on the values of both ΔG_v and σ (expression (3.4)). Volume nucleation will be favoured by a high ΔG_v and low σ. A high ΔG_v will, in turn, be favoured by a high $\Delta H_f/V_m$ and a high T_m (see (3.7)). Not surprisingly candidate compositions for homogeneous nucleation will probably tend to have a high T_m or liquidus temperature and will often be near the limiting compositions for glass formation (the 'glass-forming boundaries'). The same conditions should also favour surface nucleation. However, a combination of factors involving ΔG_v, σ and T_g are involved. Thus the presence of volume nucleation in $Li_2O \cdot 2SiO_2$ but its apparent absence in $Na_2O \cdot 2SiO_2$ and $K_2O \cdot 2SiO_2$ may be ascribed to the greater ΔG_v in the case of $Li_2O \cdot 2SiO_2$ (Matusita and Tashiro, 1973).

The available results show that relatively high undercoolings are required for volume nucleation and that the nucleation rates range from high to undetectable depending on composition. In contrast, surface nucleation appears to occur in virtually all systems and usually can be observed at smaller undercoolings (Strnad and Douglas, 1973). However, there have been comparatively few quantitative studies of surface nucleation, perhaps because of the difficulties in obtaining reproducible data as a result of the sensitivity of the phenomenon to surface condition. There is clearly a need for such studies in view of its likely dominant effect on glass-forming ability in many compositions.

3.4 EFFECT OF GLASS COMPOSITION ON NUCLEATION KINETICS

James (1982) has given a detailed discussion of the more complex case when the crystallizing phase has a different composition from that of the parent glass, i.e., when the system is no longer 'simple' or one component. The effects of parent glass composition on nucleation kinetics were considered within the framework of the classical theory in terms of changes in ΔG, σ and ΔG_D, these quantities varying with both composition and temperature.

If the crystal nucleus is assumed to have the same composition as the equilibrium precipitating phase, the bulk free energy change ΔG ('driving force') at a given temperature can be determined in principle, provided detailed free energy versus composition data are available. However, such detailed information rarely, if ever, exists. In the case of a simple binary system, for example a simple eutectic with two crystallizing phases (α and β) and little or no solid solution effects, the driving force for nucleation of phase α will be a maximum when the initial liquid composition is at or near that of the phase α (Fig. 3.9). In these circumstances, and neglecting the influence of σ and ΔG_D, the nucleation rate for phase α will be a maximum. An approximate expression for ΔG in a simple binary system assuming ideal mixing is

$$\Delta G = -\Delta H_f(T_L - T)/T_L \tag{3.23}$$

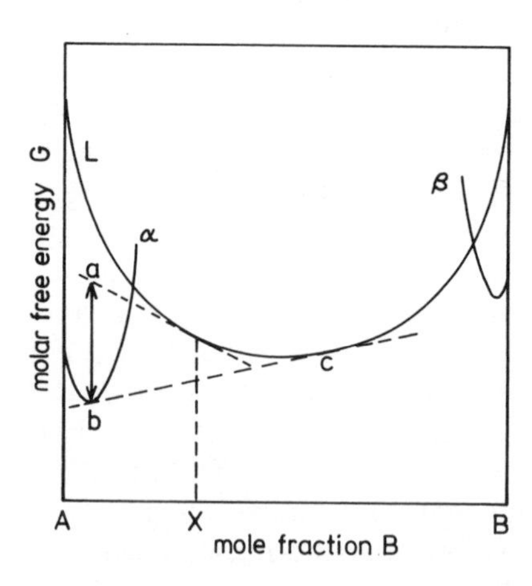

Fig. 3.9 Schematic plot of Gibbs free energy per mole against composition for a simple binary system (A–B) at a temperature below the melting points of the crystalline phases α and β. The thermodynamic driving force (ΔG) for crystallization of the equilibrium phase α from the liquid (L) composition at X is given graphically by the distance *ab*. A tangent is drawn to the liquid free energy curve L at point X and *ab* is the height of this tangent above the free energy curve for phase α. Points *b* and *c* are the equilibrium compositions for α and L respectively.

where T_L is the liquidus temperature (compare with (3.7)). Thus, changes in liquidus temperature can be a useful guide to changes in ΔG.

If appreciable solid solution effects occur the situation may be quite different from that described above. The free energy–composition relationship for the solid phase α is broadened and no longer exhibits as sharp a minimum as shown in Fig. 3.9. In these circumstances, the driving force ΔG may no longer be a maximum when the composition of the nucleating phase is the same as that of the final equilibrium phase. It is thus difficult to predict the composition of the phase that first precipitates since this will also be governed by kinetic considerations involving the interfacial energy σ and the kinetic barrier ΔG_D.

In most cases the highest nucleation rates are expected for a liquid composition which is the same as that of the equilibrium precipitating phase, since in this situation ΔG will tend to be a maximum, σ a minimum and hence W^* a minimum (expression (3.4)). However, this may not hold in the presence of a solid solution, as just described, or if ΔG_D varies with composition. Changes in ΔG_D can be assessed from viscosity measurements (assuming (3.9) and (3.10)) but at present there is no method of determining σ independent of nucleation studies.

Relatively few quantitative experimental studies of the effects of glass composition on nucleation rates have been carried out. Burnett and Douglas (1971) examined the Na_2O–BaO–SiO_2 system including the barium disilicate–sodium disilicate join. The region of glass-ceramic formation, i.e. the region of high volume nucleation rates was adjacent to the practical limit of glass formation and to the barium disilicate (BS_2) stoichiometric composition. The critical temperature for the onset of rapid nucleation of BS_2 crystals was found to decrease as the glass composition was shifted away from the BS_2 composition and the liquidus temperature was reduced. This observation can be explained by the associated reduction in ΔG (expression (3.23)).

A detailed study of nucleation kinetics was also made in the Li_2O–BaO–SiO_2 system, particularly for the $Li_2O \cdot 2SiO_2$–$BaO \cdot 2SiO_2$ section, which may be regarded as essentially a simple binary eutectic with primary phases lithium disilicate and barium disilicate (Rowlands, 1976; James and Rowlands, 1979; James, 1982). The I against T curves were determined for compositions on the section. These showed the familiar shape (as in Fig. 3.1) with a maximum (I_{max}). The highest nucleation rates were observed for the BS_2 and LS_2 compositions. I_{max} for the barium disilicate crystal phase decreased as the glass composition was moved away from BS_2 and as the liquidus temperature T_L was reduced. Similarly, I_{max} for the lithium disilicate crystal phase decreased as the glass composition was moved away from LS_2, also with a reduction in T_L. No volume nucleation of either phase was observed near the eutectic composition. Analysis of the nucleation rates using available thermodynamic data indicated that the variation of ΔG with composition was a crucial factor. However, variations in ΔG_D and also (probably) in σ were required to explain the behaviour completely.

Another system which has been studied intensively is Na_2O–CaO–SiO_2. The glass compositions NC_2S_3, N_2CS_3, NS and CS all exhibit volume nucleation, although the nucleation rates in the CS glass (Wollastonite composition) are very low. Some of the results for these compositions, which are all on the CS–NS (50 mol % SiO_2) join have been described earlier. The regions of glass formation and volume nucleation were studied by Strnad and Douglas (1973). Regions on the ternary diagram around the NC_2S_3 and N_2CS_2 compositions showed volume nucleation, and the glass-forming boundary was at about 50 mol % SiO_2, the exact position depending on cooling rate. Outside these regions on the low silica side only surface crystallization was observed. The surface and volume nucleation kinetics were investigated in a number of compositions and the results showed that the onset of surface nucleation occurred at lower undercoolings than in the case of volume nucleation.

A detailed study of the effects of small changes in composition around NC_2S_3 on nucleation rates and viscosities was made by Gonzalez-Oliver (1979) and Gonzalez-Oliver and James (1980b). Six compositions were

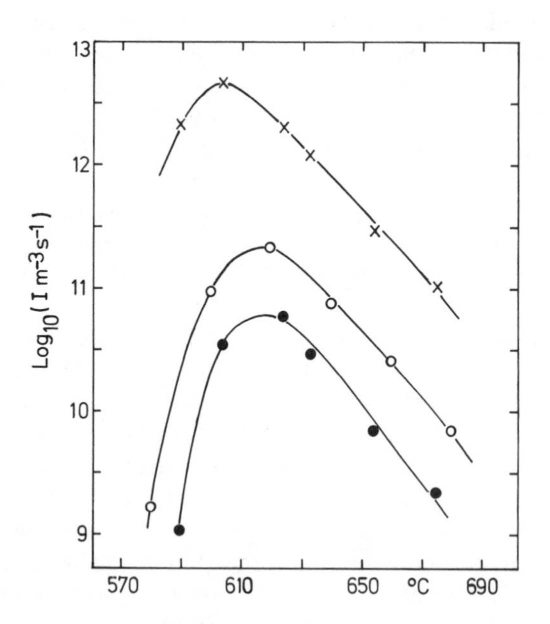

Fig. 3.10 Plots of $\log_{10} I$ against temperature for a NC_2S_3 glass composition (16.7 Na_2O, 33.3 CaO, 50 SiO_2 mol %) (points ○), for a 17.7 Na_2O, 32.9 CaO, 49.4 SiO_2 mol % glass (points ×) and for a 15.7 Na_2O, 33.7 CaO, 50.6 SiO_2 glass (points ●). I (in $m^{-3}\,s^{-1}$) determined from N_v/t where N_v is the number of crystal spherulites present at time t (40 min). After Gonzalez-Oliver (1979).

examined each containing 1 mol % more or 1 mol % less of Na_2O, CaO or SiO_2 than the base NC_2S_3 glass. The highest nucleation rates (and largest decrease in viscosity compared with the base glass) was observed for the composition with 1 mol % *more* Na_2O, the rates being about a factor of ten greater than in the base glass. The lowest rates were observed for the composition with 1 mol % *less* Na_2O. These results are shown in Fig. 3.10. In general, the compositions with more SiO_2, or less SiO_2, than the base glass showed lower and higher rates respectively than the base glass at all temperatures. A particularly interesting result was that several of the compositions had higher nucleation rates than the base glass, which is surprising if it is assumed that ΔG is a maximum for the latter. However, these glasses had lower viscosities than the base glass, which might indicate that a decrease in ΔG_D (as suggested by the decrease in viscosity) had overridden any possible reduction in ΔG for these compositions. The results also showed that viscosity changes alone were insufficient to explain the effects on nucleation and that variations in W^* with composition (for a given temperature) were also certainly involved. Another complicating factor, which will not be considered here, is the occurrence of appreciable solid solution in the NC_2S_3 crystalline phase (Moir and Glasser, 1974). The possible effects of a solid solution on ΔG have been mentioned above.

It is clear that viscosity measurements are a useful aid in interpreting the effects of composition on nucleation kinetics, particularly when ΔG (or even T_L) data are available. However, it should be emphasized that in most multicomponent systems the almost complete lack of thermodynamic information, not to mention complicating factors such as solid solution behaviour and the occurrence of metastable phases, are major obstacles to further progress. This point has also been considered in recent reviews by Tashiro (1985) and Uhlmann (1985).

3.5 HETEROGENEOUS NUCLEATION: EXPERIMENTAL RESULTS

Some of the earliest glass-ceramics were prepared by precipitating metallic particles in glasses, these particles then acting as sites for nucleation and growth of the major crystalline phase(s). (See, for example, Stookey, 1959.) The precipitation of metals such as copper, silver or gold is achieved by heat treatment of the glass (to which a reducing agent is added) or in the photosensitive process by irradiating the glass (usually containing CeO_2) with ultraviolet light followed by heat treatment. This forms the basis of the well known process of photochemical machining developed by Corning. The uses of metallic nucleating agents have been fully reviewed by McMillan (1979) and Berezhnoi (1970). A detailed discussion of heterogeneous nucleation in glasses has also been given by Gutzow (1980).

Some aspects of the theory of heterogeneous nucleation have been described above. For a metallic nucleating agent to be effective the factor $f(\theta)$ (3.19) needs to be as low as possible, and hence the contact angle θ between crystal nucleus and metal substrate should be low. It is also generally considered that the lattice disregistry between crystal and substrate should be small. However, Gutzow and Toschev (1971) compared the catalysing efficiencies of Au, Ag, Rh, Pt and Ir in promoting heterogeneous nucleation in a $NaPO_3$ glass and concluded that the efficiency was not apparently related to the degree of lattice disregistry. They suggested that a more important factor might be mechanical strains produced at the substrate–glass interface arising from thermal expansion differences between glass and metal. Gutzow (1980) from a theoretical treatment concluded that the nucleating efficiency could be correlated with the molar heat of fusion of the metal.

Earlier it was stated that from classical theory spherical particles below a certain size (about 10 nm) are not expected to act as efficient nucleation sites. This is supported by experimental evidence. Thus Maurer (1959) found that heterogeneous nucleation of lithium metasilicate on gold particles in a silicate glass did not occur until the particles were larger than 8 nm. He offered a different explanation from the 'Fletcher effect' mentioned above, and suggested that nucleation was inhibited by strains between the crystalline embryos and the underlying gold particles when the latter were less than 8 nm. Gutzow *et al.* (1968) also found that crystallization only occurred on gold particles above a certain size in $NaPO_3$ glass.

Another important factor is substrate condition. Theory predicts that sharp depressions or cavities in a substrate are particularly effective sites for heterogeneous nucleation. This was demonstrated by Burnett and Douglas (1971) who studied the nucleation of barium disilicate crystals at the interface between a soda–baria–silica glass and a Rh/Pt crucible. Nucleation from the crucible wall occurred at lower supercoolings than observed for volume nucleation. Also the supercoolings for a scratched crucible wall were much lower (by 200°C) than for a normal (smooth) wall.

One of the few quantitative studies of both homogeneous and heterogeneous nucleation kinetics on the same glass composition, $Na_2O \cdot 2CaO \cdot 3SiO_2$, has been performed by Gonzalez-Oliver and James (1980a, 1982). Glasses of this composition containing 0.2 and 0.46 wt % platinum were compared with the platinum-free glass. In the platinum-containing glass platinum particles were precipitated during the melting stage. In the 0.46 wt % Pt glass the number of particles was approximately 7×10^5 mm^{-3}, the diameters ranging from 0.3 to 3.3 μm with an average of 1.2 μm. On heat treatment heterogeneous nucleation of NC_2S_3 crystals occurred on the platinum particles, in addition to the internal (probably homogeneous) nucleation which was observed in the absence of platinum.

This was clearly demonstrated by transmission electron microscopy of thin sections at 100 to 1000 kV. Frequently several individual crystals nucleated on the same platinum particle as shown in Fig. 3.11. An important result was that the contact angle θ between the crystals and platinum particles could be directly measured. A value of about 100° was given by Gonzalez-Oliver and James (1980a) who also discuss the difficulties associated with the measurement. As far as we are aware this is one of the few cases where such a measurement has been made.

Gonzalez-Oliver and James (1982) measured the nucleation kinetics using optical microscopy and showed that Pt could considerably enhance the volume nucleation of NC_2S_3 glasses particularly at temperatures above the maximum in the rate of homogeneous nucleation (at about 595°C). However, in a quite different glass ($10Na_2O \cdot 10CaO \cdot 80SiO_2$ mol %), containing 0.3 wt % Pt, no heterogeneous nucleation was observed on the Pt after heat treatment. In another experiment 0.5 wt % Ag_2O was added to a NC_2S_3 glass and a high density of metallic particles (12.5 nm diameter)

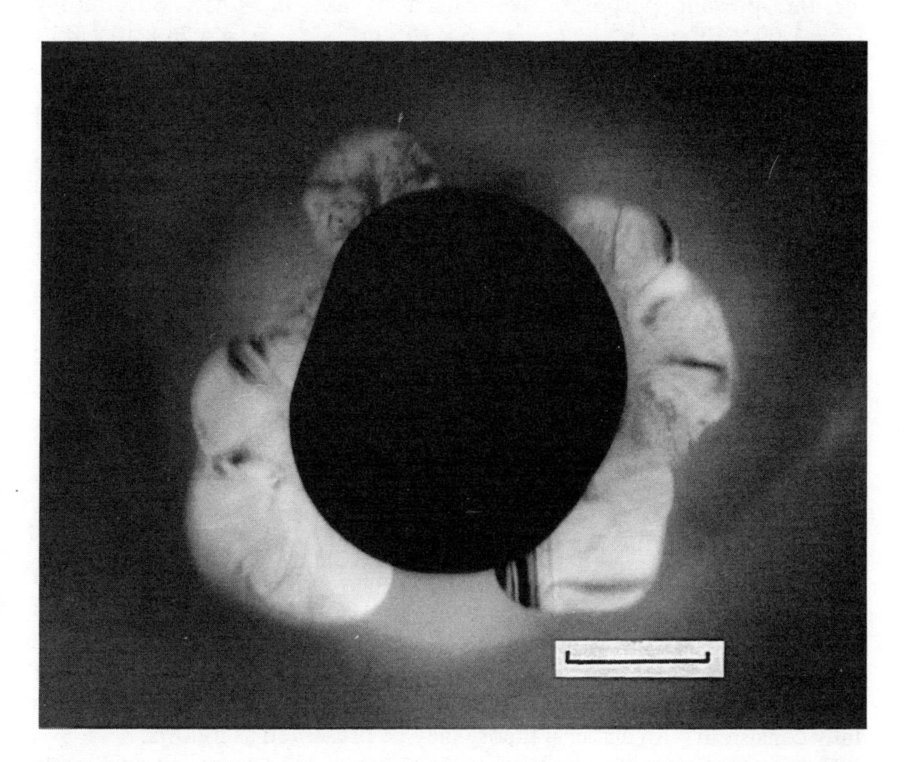

Fig. 3.11 Transmission electron micrograph (100 kV) of thin film of NC_2S_3 glass containing 0.46 wt % platinum. Glass heat treated at 596°C for 6 h 21 min prior to preparing thin film by ion-beam machining. The bar denotes 0.52 μm. After Gonzalez-Oliver and James (1980a).

were precipitated, but again no heterogeneous nucleation could be produced.

Another interesting feature of the previous investigation was the direct experimental determination of I_{het} the heterogeneous crystal nucleation rate per unit area of the platinum substrate. This was obtained from electron microscopy by estimating the total number of NC_2S_3 crystals nucleated on approximately spherical particles of known size after a given nucleation time at a known temperature. Using this value, the experimental homogeneous nucleation rate I_{hom} at the same temperature and the contact angle θ, the values of W^*_{hom} and W^*_{het} can be calculated by the following novel method. From (3.1), (3.3) and (3.18)

$$I_{het}/I_{hom} = (n_s/n_v)\,[\exp\,(W^*_{hom} - W^*_{het})/kT] \tag{3.24}$$

Here the reasonable assumption is made that the diffusional (kinetic) process is independent of the type of nucleation (i.e. $\Delta G_D^{hom} = \Delta G_D^{het}$). The approximate expression for A is also used (3.3) which is adequate for the present purposes. We can also take $n_s \simeq n_v^{2/3}$. Substituting $I_{het} = 1.09 \times 10^8\,m^{-2}\,s^{-1}$, $I_{hom} = 5.6 \times 10^{11}\,m^{-3}\,s^{-1}$, $n_v = 4.76 \times 10^{27}\,m^{-3}$, from the original paper we obtain at $T = 869\,K$

$$W^*_{hom} - W^*_{het} = 12.7\,kT \tag{3.25}$$

Note, this result is independent of any assumptions about ΔG or σ and the viscosity is not introduced. The only assumptions are that $\Delta G_D^{hom} = \Delta G_D^{het}$ and that classical theory applies. It is also worth pointing out that the calculation is not sensitive to the exact value used for I_{het}. For example, an order of magnitude increase in I_{het} gives $W^*_{hom} - W^*_{het} = 15\,kT$, an increase of only 18%. The uncertainty in I_{het} is not expected to be larger than this.

Taking the experimentally estimated contact angle (θ) of 100° we have from (3.19)

$$f(\theta) = W^*_{het}/W^*_{hom} = 0.63 \tag{3.26}$$

Gonzalez-Oliver and James (1982) calculated W^*_{hom} using (3.4) and the σ value derived from the slope of the ln $(I\eta/T)$ against $1/\Delta G^2 T$ plot (see section 3.3.4). In this way they obtained $W^*_{hom} = 141\,kT$ at $T = 869\,K$, giving $f(\theta) = 0.91$ from (3.25). This is not in agreement with the value from the contact angle measurement. However, as explained earlier (section 3.3.2) the method used to derive σ does not give the true value if σ is temperature dependent. Therefore it is now considered that these values of W^*_{hom} and $f(\theta)$ are too high. Moreover, a more plausible argument which is fully consistent with classical theory may be presented as follows.

First, returning to the results given in (3.25) and (3.26) we obtain directly at $T = 869$ K that

$$W^*_{hom} = 34.3\,kT \text{ and } W^*_{het} = 21.6\,kT \qquad \text{(Method 1)} \tag{3.27}$$

Secondly, let us use a different approach (Method 2) to calculate the values of W^*_{hom} and W^*_{het}. If we assume the kinetic barrier ΔG_D can be calculated from the viscosity η, that A is given by theory (expression (3.3)) and $\lambda^3 \simeq n_v^{-1}$ to a good approximation, we have, from (3.12)

$$I_{hom} = (n_v^2\, kT/3\pi\eta) \exp(-W^*_{hom}/kT) \tag{3.28}$$

Substituting numerical values for I_{hom} (5.6×10^{11} m^{-3} s^{-1}), η etc. we obtain

$$W^*_{hom} = 28.5\, kT \text{ at } T = 869\,\text{K}$$

This method is the same as that previously described in section 3.3.2 except that the approximate value for A (expression (3.3)) is used rather than the more rigorous expression of (3.2), to be consistent with the calculation for Method 1. The more rigorous expression gives $W^*_{hom} = 29.6\,kT$ (see also Table 3.2), only a little higher than the above result.

In the same way, from (3.9), (3.10) and (3.18)

$$I_{het} = (n_v^{5/3}\, kT/3\pi\eta) \exp(-W^*_{het}/kT) \tag{3.29}$$

Substituting numerical values for I_{het} (1.09×10^8 m^{-2} s^{-1}), η etc. we obtain

$$W^*_{het} = 15.8\, kT \text{ at } T = 869\text{ K}$$

Summarizing, at $T = 869$ K

$$W^*_{hom} = 28.5\, kT \text{ and } W^*_{het} = 15.8\, kT \qquad \text{(Method 2)} \tag{3.30}$$

These values are in good agreement with the results of Method 1 (3.27) given above. Uncertainties in the measured values of I_{het} and the contact angle θ could easily account for the small differences in values from the two methods. For example, a contact angle of 94° would give perfect agreement for the two methods, and such a value is within the error limits of the experimental measurement.

These results are of great interest since they show that within the limits of experimental error, the predictions of the classical theories of homogeneous and heterogeneous nucleation are correct. Moreover the results indicate that the expressions for the pre-exponential factors in the theories are at least approximately correct and that the use of viscosity to calculate the kinetic barrier is also approximately valid (to within, say, about an order of magnitude). Although this can be considered only a partial vindication of classical theory since the thermodynamic driving force ΔG and the interfacial energy σ do not enter the calculations, the approach used appears promising and indicates that further experimentation on similar lines would be worthwhile.

3.6 NON-METALLIC NUCLEATING AGENTS

A 'nucleating agent' may be broadly defined as an addition which either causes volume crystallization (after suitable heat treatment) of a composition which would not otherwise show it, or, if the composition already exhibits low volume nucleation, increases it. The use of metallic particles as nucleating agents has been briefly described previously. Apart from these the most important nucleating agents, vital in the preparation of practical glass-ceramics are oxides, such as TiO_2, P_2O_5 and ZrO_2 and various fluorides and sulphides. The function of nucleating agents, particularly in the aluminosilicate based systems has been discussed in detail by McMillan (1979) and Beall and Duke (1983) and only a brief discussion, with examples, is appropriate here.

In spite of their importance the roles of non-metallic nucleating agents are often obscure. Also their action is specific to certain systems. Thus, although TiO_2, P_2O_5 and ZrO_2 are well known nucleants in various aluminosilicate based systems, they have been found to decrease nucleation rates when added to the $Na_2O \cdot 2CaO \cdot 3SiO_2$ glass composition (Gonzalez-Oliver, 1979). Similarly, small additions of TiO_2 to the $BaO \cdot 2SiO_2$ composition reduced nucleation significantly (Hillig, 1962; Isard *et al.*, 1978).

A nucleating agent may have various roles. It may increase homogeneous crystal nucleation rates by increasing ΔG, or by reducing σ, or by increasing the diffusion rate (reducing ΔG_D), as already discussed in Section 3.4. Alternatively, it may precipitate, either on its own or in combination with other components in the glass, and initiate heterogeneous nucleation of a major phase. For example, in various systems TiO_2 forms compounds with other oxides, which precipitate as a high density of fine crystals, the process usually involving prior amorphous phase separation. These crystals then act as heterogeneous sites for crystallization of the glass (McMillan, 1979; Beall and Duke, 1983).

Oxide nucleating agents, such as TiO_2, are used in amounts ranging typically from 2 to 20 wt %. As pointed out by McMillan (1979) when the larger concentrations are employed, the use of the term 'nucleating agent' is questionable since the addition becomes a major component in the glass.

Detailed studies of the role of P_2O_5 in various systems have been reviewed by McMillan (1979) and James (1982). One system in which P_2O_5 is particularly effective as a nucleating agent is $Li_2O–SiO_2$. In a recent study (James, 1982) nucleation rates of lithium disilicate in a $33.3Li_2O \cdot 65.7SiO_2 \cdot 1P_2O_5$ (mol %) glass were compared directly with a $Li_2O \cdot 2SiO_2$ base glass. The nucleation rates in the P_2O_5 containing glass were much greater than in the base glass at higher temperatures (1000 times greater at 500°C), but the effect became less with decreasing temperature and below 450°C the nucleation rates converged to similar values. Amorphous phase separation was also observed in the P_2O_5 glass. However, this had at most a relatively minor effect on the crystal nucleation since another glass of composition $36Li_2O \cdot 62.9SiO_2 \cdot 1.1P_2O_5$ (mol %), which did not exhibit phase separation, gave an I against temperature curve close to the first P_2O_5 glass. Incorporation of 1 mol % P_2O_5 in the glass affected the viscosity by only a small amount compared with the large effect on I. This indicated that the main effect of P_2O_5 was to reduce W^* rather than ΔG_D (expression (3.11)). Moreover the decrease in W^* could be attributed to a lowering of the interfacial energy σ. An alternative explanation (James and McMillan, 1971) is that precipitation of very fine crystallites of Li_3PO_4 had occurred which in turn initiated growth of lithium disilicate crystals. Evidence in favour of such a heterogeneous mechanism is given by Headley and Loehman (1984), who showed that Li_3PO_4 crystals were precipitated from a Li_2O–Al_2O_3–SiO_2 based glass after various heat treatments in the region 800–1000°C. Moreover, lithium disilicate, lithium metasilicate and cristobalite were observed to crystallize by epitaxial growth on the Li_3PO_4 crystals. Although there was no direct evidence from TEM for the presence of Li_3PO_4 crystallites in the work of James (1982) the crystallites may have been too fine to detect by TEM after the heat treatments used, which were carried out at temperatures in the range of high nucleation rates (450–550°C). Further work may settle this question.

Another particularly effective 'nucleating agent' is 'water', which is present in most glasses in small amounts as either hydroxyl groups or molecular water. Commercial glasses, for example, contain typically 0.01 to 0.03 wt % H_2O. It is well known that small concentrations of water can increase nucleation and growth kinetics and reduce the viscosity of silicate and other glasses (for reviews see McMillan, 1979; James, 1982). In practice, the melting conditions and nature of the starting materials may influence the water concentration and hence the crystallization kinetics. Such factors must therefore be carefully considered when performing kinetics studies in glasses. The effects of water content are also highly relevant to crystallization studies of glasses prepared by the sol-gel process, as recently discussed by Zarzycki (1982). These glasses often contain higher hydroxyl levels than glasses prepared by conventional melting.

A systematic study of the influence of water content on the crystal nucleation and growth kinetics in silicate glasses was made by Gonzalez-Oliver *et al.* (1979). The two base compositions $Li_2O \cdot 2SiO_2$ and $Na_2O \cdot 2CaO \cdot 3SiO_2$, with water contents ranging from 0.019 to 0.136 wt % and 0.007 to 0.04 wt % respectively, were investigated. Different water concentrations were introduced by bubbling the melts with steam, wet air and dried air. The crystal nucleation and growth rates increased markedly with water content whereas the viscosities of the glasses decreased. Figure 3.12 shows the nucleation rates in the LS_2 glasses containing 0.136 wt % and 0.02 wt % H_2O (L5 and L1 respectively). Gonzalez-Oliver *et al.* (1979) concluded that the main effect of water was to lower the kinetic barrier ΔG_D. From (3.11) we can write, for a given temperature,

$$\ln I_2 = \ln I_1 + \ln (\eta_1/\eta_2) + (W_1^* - W_2^*)/kT \qquad (3.31)$$

where the subscripts 1 and 2 refer to glasses L1 and L5 respectively. This

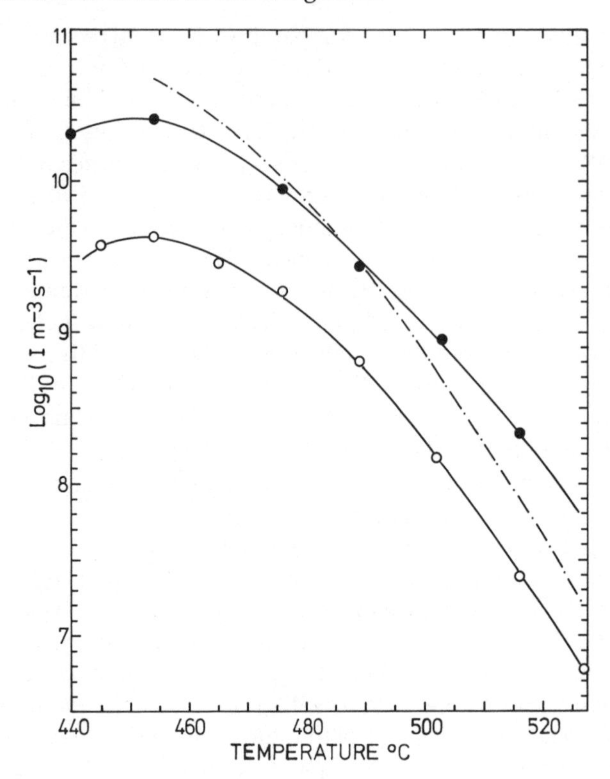

Fig. 3.12 Plots of $\log_{10} I$ against temperature for lithium disilicate glasses with 0.136 wt % H_2O (glass L5) (points ●) and 0.02 wt % H_2O (glass L1) (points ○). I in $m^{-3}\ s^{-1}$. Data from James (1974) and Gonzalez-Oliver *et al.* (1979). For explanation of dotted curve see text.

assumes that A is constant and that ΔG_D can be expressed in terms of viscosity. The dotted curve in Fig. 3.12 is $\ln I_1 + \ln(\eta_1/\eta_2)$ and represents the nucleation rates predicted for L5 if water affects only the viscosity (and hence ΔG_D) and not W^* (i.e. $W_1^* = W_2^*$). The values of η_1 and η_2 at each temperature were calculated from the data given in Table 3.3. The agreement between the predicted curve and the actual curve in Fig. 3.12 is very close around 490°C and below but a deviation occurs, particularly at higher temperatures, indicating a significant change in W^* at these temperatures. Thus the original conclusion of Gonzalez-Oliver *et al.* (1979) appears to be correct for a restricted range of temperatures only. Outside this range the change in W^* is most likely attributed to a change in the interfacial energy σ since there was a negligible difference in the liquidus temperatures of L1 and L5, indicating a negligible change in ΔG. The results suggest that above 490°C σ is decreased by the presence of water in the glass.

Table 3.3 Fulcher parameters for lithia–silica glasses L1 and L5 (see text) from Gonzalez-Oliver (1979)

	A	B (°C)	T_0 (°C)
Glass L1	−2.47	3281	222.7
Glass L5	−2.76	3745	175.2

where $\log_{10}\eta$ (Pa s) $= A + B/(T - T_0)$

It should be noted that the above comparison between L1 and L5 was made at temperatures above 450°C to avoid complications from non-steady state nucleation effects. Also, the conclusions obtained here are broadly the same as those of Zarzycki (1982) who performed a similar analysis of the results for L1 and L5.

Gonzalez-Oliver *et al.* also concluded that for the soda–lime–silica glasses the water affected both ΔG_D and W^*. In addition they found that the kinetic barriers to crystal growth were lowered by water content for both the lithia and soda–lime glasses, increases in growth rates corresponding closely to reductions in the viscosities.

These results indicate that the influence of water content on nucleation kinetics in silicate glasses is two-fold. First, incorporation of water results in a rupturing of silicon–oxygen bonds and formation of Si–OH bonds with a consequent reduction in viscosity, and in the kinetic barrier ΔG_D. Gonzalez-Oliver *et al.* (1979) suggested that the hydroxyl content may lead to an increase in the oxygen ion diffusion which may be the rate determining process in nucleation and growth. Secondly, the hydroxyl content can alter W^*, probably by changing the interfacial energy σ.

It is interesting to note that small additions of fluorides, such as NaF, appear to have similar effects to water addition, i.e. to reduce viscosity and increase nucleation and growth kinetics (see, for example, James 1982). For further discussion of nucleating agents reference is made to the reviews given previously.

3.7 EFFECTS OF AMORPHOUS PHASE SEPARATION ON CRYSTAL NUCLEATION KINETICS

Many glasses exhibit amorphous phase separation (APS) prior to crystal nucleation and growth during the heat treatment schedule required to convert them to glass ceramics. It is well known that such separation may aid subsequent crystallization by producing a phase with a greater tendency to crystallize than the initial glass. From the foregoing discussion it is clear that APS is not essential to produce volume crystal nucleation in glasses since compositions such as LS_2 and NC_2S_3 crystallize internally but do not phase separate. Nevertheless, in some compositions, nucleating agents such as TiO_2 or P_2O_5 increase the tendency towards APS. Consequently, some debate remains on the role of APS in glass ceramic formation, in spite of numerous studies (see McMillan, 1979; James, 1982; Craievich *et al.*, 1983).

The possible effects of APS on crystal nucleation may be broadly classified as 'compositional' or 'interfacial' and have been discussed in detail by James (1982) and Ramsden and James (1984a). In the former case, it is assumed that crystal nucleation occurs within one of the amorphous phases and is determined by its composition. For example, composition changes resulting from APS may affect W^* and ΔG_D for homogeneous nucleation (see previous discussion in section 3.4). In the 'interfacial' case, crystal nucleation occurs preferentially (for various reasons) at the interface between the phases, i.e. heterogeneously.

Recently, the effects of APS on crystal nucleation in the $BaO–SiO_2$ system have been intensively investigated (Ramsden and James, 1984a,b; Zanotto, *et al.*, 1986) and it is only possible here to summarize the main conclusions of these studies. The $BaO–SiO_2$ system is ideal for a 'model' study since it has an extensive region of sub-liquidus immiscibility and exhibits volume nucleation of the barium disilicate crystal phase. Also the nucleation rates are relatively high but not too high to be conveniently measured by optical microscopy. The phase diagram is shown in Fig. 3.13. The nucleation rates were measured for a wide range of temperatures in a series of $BaO–SiO_2$ glasses in the range 25.3 to 33.1 mol % BaO, i.e. straddling the immiscibility boundary. Thus some of the glasses showed APS and others, including the 33.1 mol % BaO glass, did not. The crystal nucleation kinetics were correlated with quantitative studies of the amorphous phase separation using small angle X-ray scattering (SAXS) and transmission electron microscopy (TEM). SAXS and replica or thin foil TEM provided information on the average diameter, number and surface area of amorphous droplets in the glasses. It should be mentioned that in most cases, after the primary crystal nucleation heat treatments, the glass samples were given a short development treatment (typically at 810–830°C for 10 to 30 min) to grow the crystals to observable dimensions for counting by optical microscopy. The technique was discussed earlier (section 3.3.1).

In these studies, the APS was found to have a marked influence on crystal nucleation and it was shown conclusively that almost all the observations could be explained in terms of the composition of the baria-rich phase. The highest nucleation rates were observed in the 33.1 mol % BaO glass close to the BS_2 composition, which was just outside the immiscibility region (as may be expected from theoretical considerations as shown in section 3.4). Lower rates were found in the glasses with lower BaO contents including those exhibiting phase separation. However, APS had a marked but indirect effect on crystal nucleation because of the accompanying shift in composition of the baria-rich phase in which crystal nucleation occurred. Thus, for a given

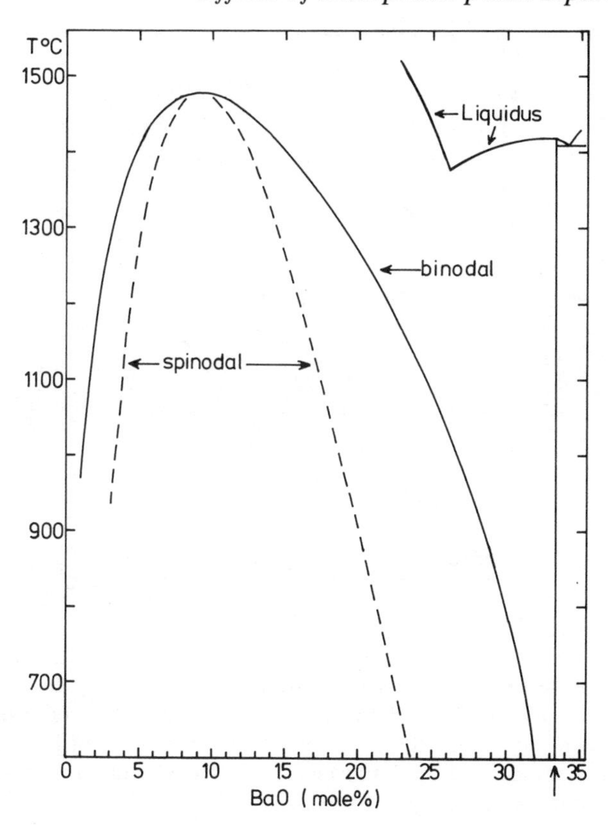

Fig. 3.13 Part of the $BaO–SiO_2$ phase diagram. Sub-liquidus liquid–liquid immiscibility boundary (binodal) and spinodal curve from Haller *et al.* (1974). $BaO \cdot 2SiO_2$ composition indicated by ↑.

temperature (such as 700°C) APS and crystal nucleation occurred simultaneously over an extended period and the crystal nucleation rate increased with time. This was manifested in a curved plot of N_v (number of crystals per unit volume) against time, the plot becoming linear at longer times as the nucleation rates approached a constant value. Also, in different glasses undergoing APS, the constant nucleation rates approached at longer times were identical since the compositions of the baria-rich phases in the glasses were identical and determined by the immiscibility boundary. In contrast, in the glasses not showing APS the N_v against time plots were linear, i.e. the nucleation rate was constant with time at a given temperature.

Some results illustrating these conclusions are shown in Figs 3.14 and 3.15. Figure 3.14 compares the N_v against time plots at 718°C for two glasses containing 28.3 and 29.7 mol % BaO. In glass 29.7, which did not show

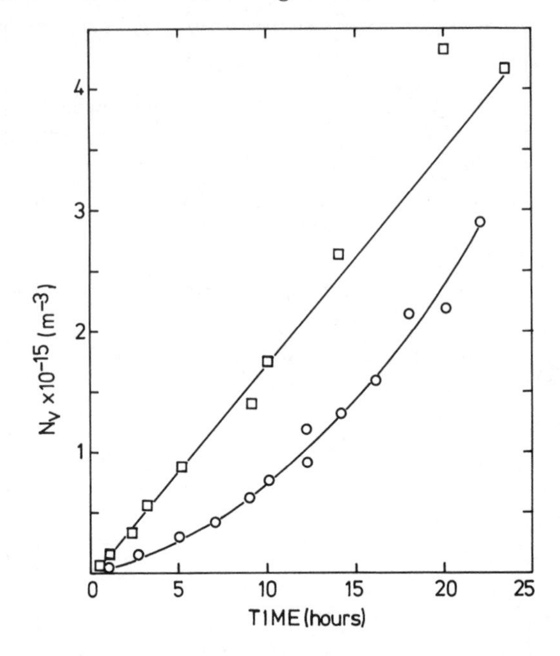

Fig. 3.14 N_v against time plots for $BaO–SiO_2$ glasses containing 28.3 mol % BaO (points ○) and 29.7 mol % BaO (points □) at nucleation temperature of 718°C. (N_v is the number of internally nucleated crystal spherulites of barium disilicate per unit volume of glass). Samples given short (10 to 30 min) growth treatment at 810 to 830°C. After Zanotto *et al.* (1986).

APS, the nucleation rate is constant. In glass 28.3, in which phase separation occurred simultaneously with crystal nucleation, the nucleation rate increased continuously with time. Figure 3.15 compares results at 743°C for glasses containing 28.3 and 27 mol % BaO. Both glasses showed APS at this temperature. The nucleation rate of glass 27 increases up to a time of 2 h, compared with the 7 h period required for glass 28.3 to reach a constant rate. After these periods the rates are constant and equal in both glasses. For glasses 28.3 and 27 a striking correlation was found at 743°C and 760°C between the times required for the amorphous (baria-rich) matrix to reach the equilibrium composition, revealed by a constant value of the integrated SAXS intensity Q, and the period of increasing crystal nucleation rates. From Fig. 3.16 the constant value of Q is approached after 3 to 4 h at 760°C and after 7 h at 743°C for glass 28.3. For glass 27 at 743°C, Q is constant from a time of less than 2.5 h heat treatment, indicating very rapid phase separation in this glass.

Thus for the glasses studied the crystal nucleation rate depended primarily on the composition of the baria-rich phase. Also in the phase separated

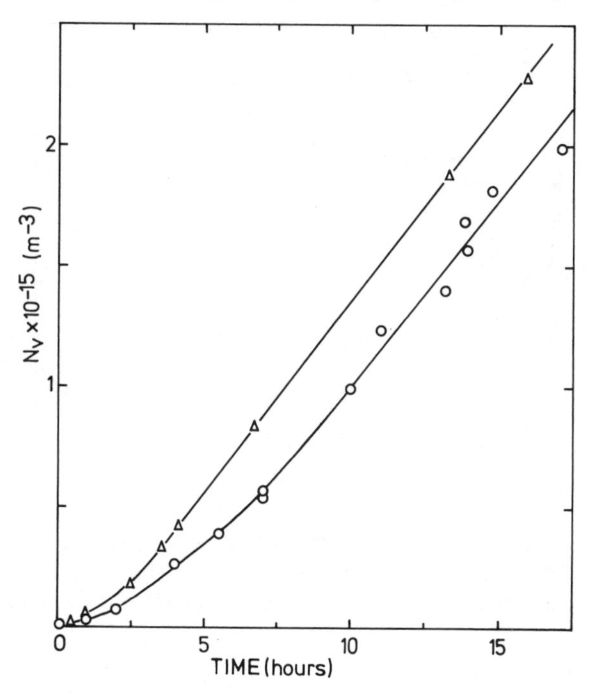

Fig. 3.15 N_v against time plots for $BaO–SiO_2$ glasses containing 28.3 mol % BaO (points ○) and 27.0 mol % BaO (points △) at nucleation temperature of 743°C. Samples given short growth treatment (as Fig. 3.14). After Zanotto *et al.* (1986).

glasses no direct correlation was found between the crystal nucleation rates and the interfacial area and number of the amorphous droplets determined by SAXS or TEM. However, an additional, much smaller effect to that of the compositional changes was observed at higher temperatures such as 760°C. Slight inflexions in the N_v against time plots occurred when the phase separation was in the early stages. These were possibly caused by some preferential (heterogeneous) nucleation within the diffusion zones (silica depleted regions) around the amorphous droplets. This was a minor effect which appeared to be masked and undetectable at temperatures where the homogeneous nucleation rate was high (i.e. in the region of 700°C). Finally, the effect of composition (BaO content) on the nucleation rates was analysed further by means of viscosity and crystal growth rate measurements. It was concluded that changes in both ΔG_D and W^* with BaO content were important, although changes in W^* had the larger effect.

The effects of phase separation on crystal nucleation in the $Li_2O–SiO_2$ system, which has similarities to the $BaO–SiO_2$ system, have also been recently studied (Zanotto and James, 1983). The high silica part of the phase

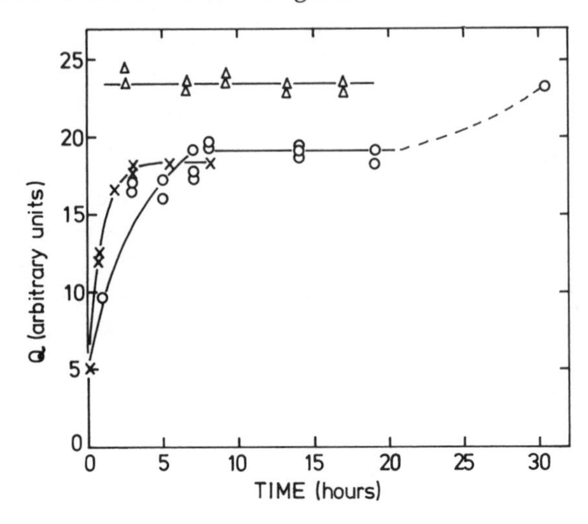

Fig. 3.16 Integrated SAXS intensity Q for BaO–SiO_2 glasses heat treated at 743 and 760°C against time of heat treatment. (△) glass 27.0 at 743°C; (○) glass 28.3 at 743°C; (×) glass 28.3 at 760°C. After Zanotto *et al.* (1982) and (1986).

diagram showing the miscibility gap is given in Fig. 3.17. The crystal nucleation kinetics of lithium disilicate were determined for three glasses; glasses containing 17.7 and 31 mol % Li_2O, which showed phase separation, and a glass containing 33.2 mol % Li_2O (close to $Li_2O{\cdot}2SiO_2$) which was just outside the immiscibility region and did not phase separate. Glass 17.7 was in the spinodal region of the miscibility gap for the temperatures studied in the work and showed a fine interconnected phase separation after heat treatment. Glass 31 was in the 'nucleation and growth' region, and after heat treatment showed silica rich droplets in a lithia rich matrix. Figures 3.18 and 3.19 show the nucleation results (number of crystal spherulites per unit volume, N_v, plotted against time) for heat treatments at 481°C. This temperature was chosen because APS was known to reach completion (equilibrium compositions of the phases) within a reasonable period of a few hours, and non-steady state crystal nucleation was negligible. After preparation, samples of glasses 17.7 and 31 were quenched to suppress APS, prior to the crystal nucleation heat treatment at 481°C. The samples were then given a short growth treatment to develop the nuclei at 570 or 600°C for 30 to 70 min. Finally N_v was determined by optical microscopy (section 3.3.1).

As in the BaO–SiO_2 system, the observations could be explained in terms of the composition of one of the amorphous phases. Thus, the crystal nucleation rates in glasses 17.7 and 31 undergoing phase separation increased with time at 481°C because of the shift in composition of the lithia

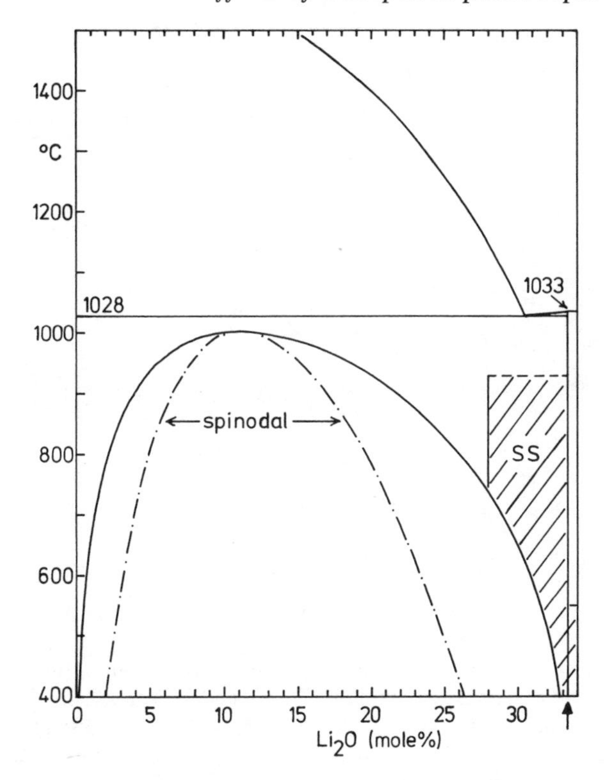

Fig. 3.17 Part of the Li_2O–SiO_2 phase diagram. Sub-liquidus liquid–liquid immiscibility boundary (binodal) and spinodal curve are from Haller *et al.* (1974). The shaded area represents the range of solid–solution formation (West and Glasser, 1971). $Li_2O \cdot 2SiO_2$ composition indicated by ↑ . After Zanotto and James (1983).

rich amorphous phase in which crystal nucleation occurred (Fig. 3.18). Moreover, the crystal nucleation rates in glasses 17.7 and 31 were equal for longer times (slopes of plots in Fig. 3.18), clearly as a result of the identical compositions of the lithia rich phases in the two glasses after the completion of phase separation.

The effect of a heat treatment at 497°C for 5 h to produce APS, prior to the nucleation treatment at 481°C, is shown in Fig. 3.19. In this case the crystal nucleation rates in the prior phase separated glasses were equal and constant from the beginning, being lower than the rates in the as-quenched glasses (broken lines). The lower rates observed can be attributed to the slightly lower Li_2O contents in the lithia-rich phases for the glasses heat treated at 497°C compared with the glasses only heated at 481°C (as defined by the miscibility boundary in Fig. 3.17). In principle, further secondary phase separation could occur at 481°C for the glasses prior heat treated at

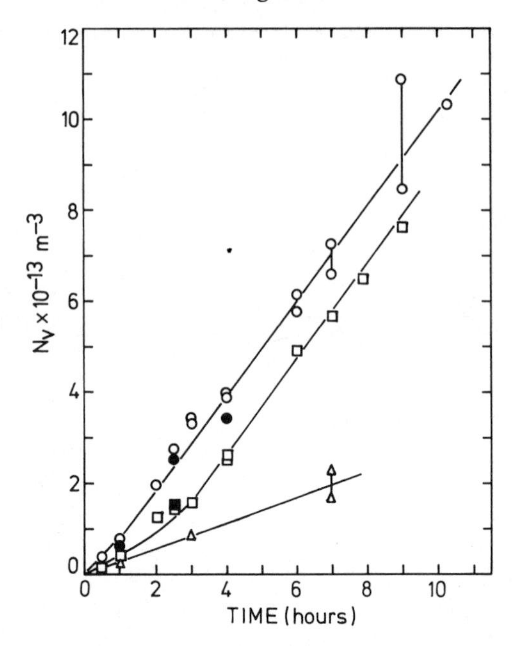

Fig. 3.18 Number of crystal particles ('spherulites') per unit volume (N_v) against time for Li_2O–SiO_2 glasses containing 17.7, 31.0 and 33.2 mol % Li_2O at nucleation temperature of 481°C. Glass 17.7 (□, ■); glass 31.0 (○, ●); glass 33.2 (△). Development (growth) temperatures: 570°C (○, □, △), 600°C (●, ■). After Zanotto and James (1983).

497°C. However, since the matrix compositions after treatment at 497°C would be only just inside the miscibility boundary corresponding to 481°C the kinetics of such secondary phase separation would be very slow. Therefore the composition of the lithia rich phase did not reach equilibrium at 481°C even after 7 h (Fig. 3.19) whereas equilibrium was attained much more quickly in the as-quenched glasses at 481°C (Fig. 3.18). It should be noted that the intercepts on the N_v axis in Fig. 3.19 were caused by some crystal nucleation during the initial treatment at 497°C, although the crystal nucleation rate at 497°C was much smaller than at 481°C (see, for example, Fig. 3.1).

Thus the 'composition' theory gave a clear explanation of the results in the Li_2O–SiO_2 system. There was, however, one striking difference from the BaO–SiO_2 system. The 33.2 mol % Li_2O glass (nearly LS_2), outside the immiscibility region had a lower crystal nucleation rate than in the fully phase separated glasses 17.7 and 31, in marked contrast to the BaO–SiO_2 system where the BaO·2SiO_2 glass had the highest rates. This behaviour was explained as follows. The crystalline phase precipitated in the phase

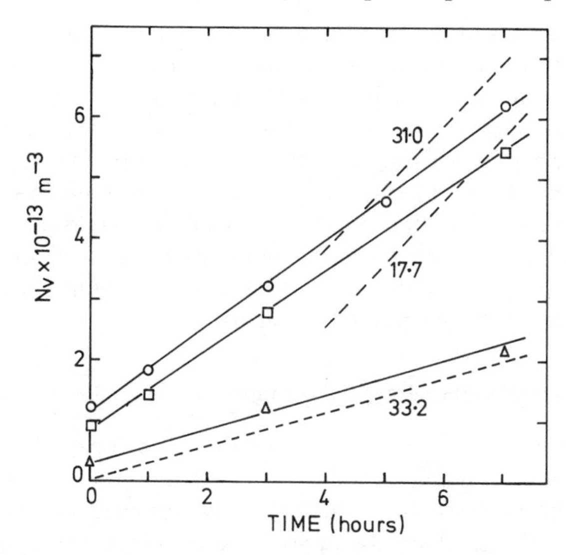

Fig. 3.19 N_v against time plots for Li_2O–SiO_2 glasses containing 17.7, 31.0 and 33.2 mol % Li_2O at nucleation temperature of 481°C; ---------- as-quenched glasses (see Fig. 3.18); ——— glasses given heat treatment at 497°C for 5 h to produce amorphous phase separation prior to crystal nucleation treatment at 481°C; glass 17.7 (□); glass 31.0 (○); glass 33.2 (△); Growth temperature 570°C. After Zanotto and James (1983).

separated glasses was a solid solution phase which differed significantly in composition from the stoichiometric lithium disilicate phase precipitated in the $Li_2O \cdot 2SiO_2$ glass (see Fig. 3.17). It was therefore likely that the thermodynamic driving force (ΔG) for crystallization was higher for the phase separated glasses than for the LS_2 glass, giving higher nucleation rates for the former glasses. It is easily shown that such an effect could occur from consideration of the probable shape of the free energy against composition curve for the solid solution phase, as discussed in section 3.4. Another possibility is that the interfacial energy σ is lower for the phase separated glasses. It would appear that in the BaO–SiO_2 system such effects either do not occur or are of minor importance, and that the simpler explanation, given previously, applies.

Finally, in the lithia–silica system, as in the baria–silica system, there was some evidence for enhanced crystal nucleation within the diffusion zones around the amorphous droplets in the early stages of phase separation (see Zanotto and James, 1983). However, it must be emphasized that this was a minor effect compared with the predominant influence of compositional changes.

In conclusion, the studies described demonstrate that APS has an important role in the formation of glass ceramics. Nevertheless even in these relatively 'simple' systems the effects are quite complex and caution must be exercised before generalizing the results to more complex multi-component systems. Thus the increases in crystal nucleation rates resulting from APS are usually relatively 'small' (an order of magnitude or less) when compared with the effect of nucleating agents, such as TiO_2 or P_2O_5, which may increase the nucleation rates by many orders of magnitude in some systems. As already seen in section 3.6 such nucleating agents frequently have complex roles and often promote immiscibility as well as high crystal nucleation, but it should not be assumed that amorphous phase separation necessarily plays a major part in their action.

3.8 SUMMARY AND CONCLUSIONS

In recent years progress has been made towards a better understanding of crystal nucleation in glass-forming systems. Quantitative studies of 'simple' or 'model' silicate glasses have aided this advance by identifying the factors affecting nucleation, which also apply in more complex compositions. The relatively low nucleation rates coupled with high viscosities of silicate (and other) glasses make them ideal for model nucleation studies. Thus nucleation rates can be conveniently and accurately measured in contrast to non-glass-forming systems, including most metallic and ceramic systems and both steady state and non-steady state nucleation can be investigated.

Studies of stoichiometric compositions such as LS_2, NC_2S_3, BS_2 and others, all of which exhibit volume nucleation of a crystal phase with the same composition as the parent glass, have revealed striking similarities. The values of T_{max}/T_m, where T_{max} is the temperature of maximum nucleation and T_m the melting point (or effective melting point) were in the range 0.54 to 0.59 for seven compositions. Values of T_d/T_m, where T_d is the 'just detectable' nucleation temperature ($I = 10^6\ m^{-3}\ s^{-1}$) were 0.62, 0.64 and 0.66 for LS_2, NC_2S_3 and BS_2 respectively. These represented somewhat higher undercoolings for the onset of homogeneous nucleation than observed in 'droplet' studies of non-silicate systems. Values of W^*/kT, where W^* is the thermodynamic barrier to nucleation, in the various systems also showed close similarities, being ~ 30 at T_{max} and ~ 50 at T_d. The remarkably consistent pattern of these results, and other evidence, indicates that the nucleation is predominantly homogeneous in these compositions. However, there are many other stoichiometric compositions, including NS_2, KS_2, anorthite and various geologically important systems which apparently exhibit only surface nucleation. Reasons for this were considered.

$Li_2O \cdot 2SiO_2$ has been studied intensively since it is one of the few systems where detailed thermodynamic data exist. Experimental results for non-steady state nucleation were in good general agreement with the theory of Kashchiev (1969) both with respect to the variation of the number density of crystals with time at constant temperature and the variation of induction time with temperature. For the steady state nucleation rates I, plots of $\ln(I\eta/T)$ against $1/\Delta G^2 T$ for LS_2 and other systems (NC_2S_3 and BS_2) gave reasonable straight lines in accordance with classical nucleation theory. However, the experimental pre-exponential factors were much higher than predicted by theory. It was shown that this discrepancy could be explained by postulating a temperature dependent interfacial energy (σ). Unfortunately, at present there appears to be no means of checking this possibility, independently of nucleation studies.

The effects of glass composition on volume crystal nucleation rates, i.e. when the crystallizing phase has a different composition from that of the parent glass, have been studied systematically in only a few systems, including lithia–baria–silica and soda–lime–silica. The results could be explained in terms of variations in both the thermodynamic (W^*) and kinetic (ΔG_D) barriers in the classical theory with composition and temperature. For compositions in the vicinity of NC_2S_3 in the soda–lime–silica system viscosity measurements were used to assess changes in ΔG_D. An interesting result in this system was that the highest nucleation rates did not occur at the stoichiometric NC_2S_3 glass composition, as might be expected from thermodynamic (ΔG) considerations. This was attributed in part to the effect of the variation of viscosity with glass composition.

Recent work on heterogeneous nucleation on metallic particles in glasses was discussed. One recent development worthy of particular note was the direct measurement, using TEM, of the contact angle θ between NC_2S_3 crystals and platinum substrate particles, and also the direct determination of the heterogeneous nucleation rates. Using the homogeneous nucleation rates for the same system the thermodynamic barriers for homogeneous and heterogeneous nucleation (W^*_{hom} and W^*_{het}) were obtained at a given temperature without introducing viscosity into the calculation. However, an alternative calculation using viscosity but not θ gave values for W^*_{hom} and W^*_{het} in good agreement with the first method. These results are of great interest since they provide some vindication of the classical theories of homogeneous and heterogeneous nucleation.

The roles of non-metallic nucleating agents are often complex and frequently not well understood. A brief review of these roles was given and two nucleating agents, P_2O_5 and H_2O, were discussed in detail. The large increase in crystal nucleation rates in Li_2O–SiO_2 glasses on adding P_2O_5 may be attributed to a reduction of W^*, probably by lowering the interfacial energy. However, heterogeneous nucleation on extremely small lithium phosphate crystals remains a possibility. Low levels of hydroxyl content were shown to produce large increases in crystal nucleation and growth rates in silicate glasses. One effect of the hydroxyl content was to reduce the viscosity and the kinetic barrier to nucleation (ΔG_D) but there was also evidence that W^* was altered. For LS_2 glass, W^* was reduced above 490°C, indicating that σ was lowered above this temperature.

Recent detailed investigations of the effects of amorphous phase separation (APS) on crystal nucleation kinetics in the 'model' systems $BaO–SiO_2$ and $Li_2O–SiO_2$ were discussed. In both systems APS had a marked but indirect effect on crystal nucleation because of the accompanying shift in composition of the baria-rich or lithia-rich amorphous phases in which crystal nucleation occurred. Thus in the glasses undergoing APS the crystal nucleation rate increased with time approaching a constant value at longer times when the amorphous phases had reached their equilibrium compositions. In both systems there was no apparent relationship between the crystal nucleation rate and the interfacial area or numbers of amorphous droplets determined by SAXS or TEM, but at higher temperatures there was evidence for some additional preferential (heterogeneous) nucleation within the diffusion zones around the droplets in the early stages of phase separation. However this was a minor effect compared with the predominant influence of compositional changes. In spite of these similarities between the nucleation behaviour of the $BaO–SiO_2$ and $Li_2O–SiO_2$ systems there was one notable difference. In the $BaO–SiO_2$ system, the crystal nucleation rate (at a given temperature) in the $BaO \cdot 2SiO_2$ glass (just outside the immiscibility region) was higher than that in the fully phase separated glasses with higher overall SiO_2 contents. In contrast, in the $Li_2O–SiO_2$ system, the crystal nucleation rate in the $Li_2O \cdot 2SiO_2$ glass was lower than in glasses that had undergone phase separation. This unexpected behaviour in the lithia–silica system was attributed to the precipitation of a solid solution crystal phase in the glasses exhibiting immiscibility.

Finally, it is appropriate briefly to consider likely areas for future studies of crystal nucleation in glasses. Studies of a number of 'simple' silicate systems in which the crystallizing phase has the same composition as the parent glass have revealed an interesting pattern in volume nucleation behaviour. The work should be extended to many more such systems and to other oxide or non-oxide compositions including phosphates and borates.

Further comparisons between classical nucleation theory and experiment should be made but, with a few exceptions accurate thermodynamic data are not presently available. There is an urgent need for such data in conjunction with accurate viscosities and nucleation rates. However, until the interfacial energy can be determined independently of nucleation studies no complete test of classical theory will be possible. The assumption that the kinetic barrier term ΔG_D or effective diffusion coefficient D can be calculated from viscosity data also requires further investigation. A different approach is to derive ΔG_D from measurements of crystal growth rates, in the same temperature range as the nucleation measurements, assuming that the kinetic barriers for growth and nucleation can be equated. Although such growth rates are very low they could be measured by means of long heat treatments, using electron microscopy. Another possibility is to determine

D or ΔG_D from the induction times for transient nucleation (see (3.9) and discussion of section 3.3.1).

Further studies of heterogeneous nucleation on metal particles using the approach discussed previously for NC_2S_3 would be well worthwhile and it is considered that much greater attention should be given to the kinetics of surface nucleation in view of its fundamental and technological interest (McMillan, 1979). There is much scope for work on the roles of nucleating agents in silicate and non-silicate systems and for nucleation studies of glasses prepared by unconventional methods including vapour deposition and sol-gel techniques.

In conclusion, it is hoped that the quantitative approach stressed in this chapter and the results described will stimulate many future studies of nucleation and growth in glasses.

ACKNOWLEDGEMENTS

I wish to acknowledge the contributions made by friends and colleagues, in particular those associated with the Universities of Sheffield and Warwick, past and present, to the work discussed in this chapter.

Various illustrations and micrographs are reproduced by kind permission of the Society of Glass Technology, *Glastechnische Berichte*, *Journal of Materials Science*, *Journal of Microscopy*, *Journal of Non-Crystalline Solids* and *Journal de Physique*.

Finally, I would like to record my special thanks to the late Professor Peter McMillan, who first introduced me to the subject of glass-ceramics, for his encouragement and inspiration and, above all, for his friendship.

REFERENCES

Beall, G. H. and Duke, D. A. (1983) in *Glass: Science and Technology*, Vol. 1 (eds D. R. Uhlmann and N. J. Kreidl), Academic Press, New York, pp. 403–45.

Berezhnoi, A. I. (1970) *Glass Ceramics and Photositalls*, Plenum Press, New York.

Berkebile, C. A. and Dowty, E. (1982) *Am. Mineral.* **67**, 886–99.

Burnett, D. G. and Douglas, R. W. (1971) *Phys. Chem. Glasses*, **12**, 117–24.

Craievich, A. F., Zanotto, E. D. and James, P. F. (1983) *Bull. Mineral.*, **106**, 169–84.

Christian, J. W. (1975) *The Theory of Transformations in Metals and Alloys*, 2nd edn, Pergamon Press, London.

Filipovich, V. N., Kalinina, A. M. and Sycheva, G. A. (1975) *Neorg. Mater.*, **11**, 1305–8.

Fletcher, N. H. (1958) *J. Chem. Phys.*, **29**, 572–6.

Fletcher, N. H. (1959) *J. Chem. Phys.*, **31**, 1136.

Fokin, V. M., Kalinina, A. M. and Filipovich, V. N. (1981) *J. Cryst. Growth*, **52**, 115–21.

Freiman, S. W., Onoda Jr., J. W. and Pincus, A. G. (1972) *J. Am. Ceram. Soc.*, **55**, 354–9.

Gonzalez-Oliver, C. J. R. (1979) *Crystal Nucleation and Growth in Soda–Lime–Silica Glasses*, PhD thesis, University of Sheffield.
Gonzalez-Oliver, C. J. R. and James, P. F. (1980a) *J. Microscopy*, **119**, 73–80.
Gonzalez-Oliver, C. J. R. and James, P. F. (1980b) *J. Non-Cryst. Solids*, **38/39**, 699–704.
Gonzalez-Oliver, C. J. R. and James, P. F. (1982) in *Advances in Ceramics*, Vol. 4 (eds J. H. Simmons, D. R. Uhlmann and G. H. Beall), American Ceramic Society, Columbus, Ohio, pp. 49–65.
Gonzalez-Oliver, C. J. R., Johnson, P. S. and James, P. F. (1979) *J. Mater. Sci.*, **14**, 1159–69.
Gutzow, I. (1980) *Contemp. Phys.*, **21**, 121–37; 243–63.
Gutzow, I. and Toschev, S. (1971) in *Advances in Nucleation and Crystallization in Glasses*, (eds L. L. Hench and S. W. Freiman) American Ceramic Society, Columbus, Ohio, pp. 10–23.
Gutzow, I., Toschev, S., Marinov, M. and Popov, E. (1968) *Krist. Tech.*, **3**, 337–54.
Gutzow, I., Zlateva, E., Alyatov, S. and Kovatscheva, T. (1977) *J. Mater. Sci.*, **12**, 1190–202.
Haller, W., Blackburn, D. H. and Simmons, J. H. (1974) *J. Am. Ceram. Soc.*, **57**, 120–6.
Headley, T. J. and Loehman, R. E. (1984) *J. Am. Ceram. Soc.*, **67**, 620–5.
Hillig, W. B. (1962) in *Symposium on Nucleation and Crystallization in Glasses and Melts* (eds M. K. Reser, G. Smith and H. Insley), American Ceramic Society, Columbus, Ohio, pp. 77–89.
Hoffman, J. D. (1958) *J. Chem. Phys.*, **29**, 1192–3.
Isard, J. O., James, P. F. and Ramsden, A. H. (1978) *Phys. Chem. Glasses*, **19**, 9–13.
Jackson, K. A. (1966) in *Nucleation Phenomena*, American Chemical Society, Washington, DC, pp. 35–40.
James, P. F. (1974) *Phys. Chem. Glasses*, **15**, 95–105.
James, P. F. (1982) in *Advances in Ceramics*, Vol. 4 (eds J. H. Simmons, D. R. Uhlmann and G. H. Beall), American Ceramic Society, Columbus, Ohio, pp. 1–48.
James, P. F. (1985) *J. Non-Cryst. Solids*, **73**, 517–40.
James, P. F. and Keown, S. R. (1974) *Philos. Mag.*, **30**, 789–802.
James, P. F. and McMillan, P. W. (1971) *J. Mater. Sci.*, **6**, 1345–9.
James, P. F. and Rowlands, E. G. (1979) in *Phase Transformations*, Vol. 2, The Institution of Metallurgists, Northway House, London, Section III, pp. 27–9.
James, P. F., Scott, B. and Armstrong, P. (1978) *Phys. Chem. Glasses*, **19**, 24–7.
Kalinina, A. M., Filipovich, V. N. and Fokin, V. M. (1980) *J. Non-Cryst. Solids*, **38/39**, 723–8.
Kashchiev, D. (1969) *Surf. Sci.*, **14**, 209–20.
Kelton, K. F. and Greer, A. L. (1986) *J. Non-Cryst. Solids*, **79**, 295–309.
Kelton, K. F., Greer, A. L. and Thompson, C. V. (1983) *J. Chem. Phys.*, **79**, 6261–76.
Klein, L. C., Handwerker, C. A. and Uhlmann, D. R. (1977) *J. Cryst. Growth*, **42**, 47–51.
Matusita, K. and Tashiro, M. (1973) *J. Non-Cryst. Solids*, **11**, 471–84.
Maurer, R. D. (1959) *J. Chem. Phys.*, **31**, 444–8.
McMillan, P. W. (1979) *Glass Ceramics*, 2nd edn, Academic Press, London.
Miyazawa, Y. and Pound, G. M. (1974) *J. Cryst. Growth*, **23**, 45–57.
Moir, G. K. and Glasser, F. P. (1974) *Phys. Chem. Glasses*, **15**, 6–11.
Neilson, G. F. and Weinberg, M. C. (1979) *J. Non-Cryst. Solids*, **34**, 137–47.

Oishi, Y., Terai, R. and Ueda, H. (1975) in *Mass Transport Phenomena in Ceramics*, Materials Science Research, Vol. 9 (eds A. R. Cooper and A. H. Heuer), Plenum Press, New York and London, pp. 297–310.
Ramsden, A. H. and James, P. F. (1984a) *J. Mater. Sci.*, **19**, 1406–19.
Ramsden, A. H. and James, P. F. (1984b) *J. Mater. Sci.*, **19**, 2894–908.
Rowlands, E. G. (1976) *Nucleation and Crystal Growth in the Lithia–Baria–Silica System*, PhD thesis, University of Sheffield.
Rowlands, E. G. and James. P. F. (1979) *Phys. Chem. Glasses*, **20**, 1–8 and 9–14.
Spaepen, F. (1975) *Acta Metall.*, **23**, 729–43.
Spaepen, F. and Turnbull, D. (1976) *Proc. 2nd Int. Conf. on Rapidly Quenched Metals* (eds N. J. Grant and B. C. Giessen), MIT Press, pp. 205–29.
Stookey, S. D. (1959) Fifth Int. Glass Congress, *Glastech. Ber.* **32K**, paper V/1.
Strnad, Z. and Douglas, R. W. (1973) *Phys. Chem. Glasses*, **14**, 33–6.
Takahashi, K. and Yoshio, T. (1973) *Yogyo Kyokai Shi*, **81**, 524–33.
Tashiro, M. (1985) *J. Non-Cryst. Solids*, **73**, 575–84.
Turnbull, D. (1950) *J. Appl. Phys.*, **21**, 1022–8.
Turnbull, D. (1952) *J. Chem. Phys.*, **20**, 411–24.
Turnbull D. (1956) in *Solid State Physics*, Vol. 3 (eds. F. Seitz and D. Turnbull), Academic Press, New York, pp. 225–306.
Uhlmann, D. R. (1971) in *Advances in Nucleation and Crystallization in Glasses* (eds L. L. Hench and S. W. Freiman), American Ceramic Society, Columbus, Ohio, pp. 91–115.
Uhlmann, D. R. (1985) *J. Non-Cryst. Solids*, **73**, 585–92.
Wang, Tian-He and James, P. F. (1984) unpublished work, University of Sheffield.
West, A. R. and Glasser, F. P. (1971) in *Advances in Nucleation and Crystallization in Glasses* (eds L. L. Hench and S. W. Freiman), American Ceramic Society, Columbus, Ohio, pp. 151–65.
Zanotto, E. D., Craievich, A. F. and James, P. F. (1982) *J. Phys. (Paris)*, **43**, C9, 107–10.
Zanotto, E. D. and James, P. F. (1983) Proc. 13th Int. Glass Congress, Hamburg, *Glastech. Ber.*, **56K**, 794–9.
Zanotto, E. D. and James, P. F. (1985) *J. Non-Cryst. Solids*, **74**, 373–94.
Zanotto, E. D., James, P. F. and Craievich, A. F. (1986) *J. Mater. Sci.*, **21**, 3050–64.
Zarzycki, J. (1982) in *Advances in Ceramics*, Vol. 4, (eds J. H. Simmons, D. R. Uhlmann and G. H. Beall), American Ceramic Society, Columbus, Ohio, pp. 204–17.

4

Oxynitride glasses and their glass-ceramic derivatives

G. Leng-Ward and M. H. Lewis

4.1 INTRODUCTION

Efforts to obtain silicate-based glasses with new or improved properties have been mainly directed towards replacing the network-forming silicon cation by elements such as aluminium, boron and phosphorous, or the network-modifying cations such as sodium and calcium by other alkalis, alkaline earths and a wide range of other cations. The oxygen atoms of the network tetrahedral $Si(O_4)$-groups may also be replaced to a limited degree by substituting for oxygen the most chemically and structurally similar elements such as sulphur, fluorine and nitrogen. For sulphur and fluorine substitution this has not led to the development of a range of glasses with inherently improved properties. The strength of the silicon–sulphur bond is an order of magnitude weaker than that of silicon–oxygen (Weyl, 1967) and the difficulty in preparing oxysulphide glasses because of the volatilizing and oxidizing nature of sulphides in glass melts means that few studies have been undertaken on such glasses. The substitution of small amounts of monovalent fluorine for divalent bridging oxygens led to the formation of $Si(O_3F)$ tetrahedra (Rabinovich, 1983), causing a decrease in melt viscosity and a weakening of the glass structure. Consequently fluoride additions are used as fining agents in the melting of many commercial glasses. However, the strength of the Si–N bond, the nitrogen atom's trivalency, its ability to substitute for oxygen in $Si(O_4)$ tetrahedral units as exemplified by the existence of crystalline oxynitride phases built up of $Si(O_{4-x}N_x)$ tetrahedra, has led to the development in recent years of a range of oxynitride glasses with significantly modified properties.

Oxynitride glasses were originally discovered in the 1960s. Mulfinger (1966) produced an oxynitride soda–lime glass by dissolving Si_3N_4 in the melt and proposed that nitrogen was substituting for oxygen in the network tetrahedra under reducing conditions. Elmer and Norberg (1967) found that oxide glasses after nitriding with nitrogen or ammonia contained small amounts of nitrogen and Davies and Meherahli (1971) reported that up to 3 at % nitrogen could be dissolved in aluminosilicate slag melts. The modification of some glass properties on substitution of nitrogen for oxygen were reported in these early reports on oxynitride glasses; that is, increased softening temperature, viscosity and resistance to devitrification.

The major impetus influencing the study of oxynitride glasses was the development of silicon nitride-based ceramics which contain various metal oxide additives to promote liquid phase sintering. The discovery that the residual liquid in these materials forms an oxynitride grain boundary glass prompted attempts to prepare homogeneous oxynitride glasses by melting mixtures of oxides and nitrides in a nitrogen atmosphere. Jack (1977) reported the accommodation of much greater concentrations of nitrogen in glasses within the Mg–Si–Al–O–N and Y–Si–Al–O–N systems (up to 10 at %), and Shillito *et al.* (1978) and Loehman (1979) reported that the glass properties in these systems, such as hardness and glass transition temperature, increased significantly with increasing nitrogen content. In the last ten years a wide range of oxynitride glasses have been prepared by melting mixtures of oxides and nitrides and by the ammonolysis of melts and of porous silicate and phosphorus gels.

Over the past twenty years the controlled crystallization of silicate glasses has generated glass-ceramic materials with special thermal, mechanical or electrical properties derived from a variety of silicate phase combinations in microcrystalline form. The properties of glass-ceramic materials obtained by the crystallization of oxynitride glasses have not been intensively studied, although the modification in the thermomechanical properties of Si_3N_4 related ceramics following the crystallization of the grain boundary oxynitride glass is well recognized.

Work on oxynitride glasses and glass-ceramics is still finding its way in terms of recognizing possible applications for this new class of materials. This chapter presents a survey of the published literature on oxynitride glasses – their preparation, properties and crystallization.

4.2 OXYNITRIDE GLASS FORMATION

4.2.1 Glass-forming systems

A listing of oxynitride glass-forming systems from the published literature is given in Table 4.1 (up to December 1987). References are listed for systems where glass compositions are given together with information relating to preparation, properties, characterization or crystallization. References with information on glass crystallization are marked with an asterisk.

The majority of oxynitride glasses studied have been formed from silicate based melts. The stability of silicate based oxynitride melts formed by melting mixtures of oxides and nitrides at typical processing temperatures of

Table 4.1 Oxynitride glass systems with associated references

System	*M (metal atoms)*	*Reference*
M–Si–O–N	Na	Coon *et al.*, 1983; Wusirika, 1985
	Na–Ca	Coon *et al.*, 1983; Frishat and Scrimpf, 1980, 1983; Frishat and Sebastian, 1985
	Al	Roebuck, 1978; Chyung and Wusirika, 1978*; Wusirika and Chyung, 1980*
	Ca	Drew *et al.*, 1983
	Mg	Baik and Raj, 1985, 1987; Drew *et al.*, 1983; Jack, 1977; Shaw *et al.*, 1984*; Tsai and Raj, 1981
	Mg–Sc	Tredway and Loehman, 1985
	La	Makishima *et al.*, 1983
	Y	Drew *et al.*, 1983
M–Si–Al–O–N	Li	Fischer *et al.*, 1985*; Luping *et al.*, 1984*; Wusirika and Chyung, 1980*; Wusirika, 1985*
	Mg	Homeny and McGarry, 1984; Kenmuir *et al.*, 1983; Pasturnak and Verdier, 1983; Thorp and Kenmuir, 1981; Tredway and Risbud, 1985a; Wild *et al.*, 1981*, 1984*; Wusirika and Chyung, 1980*; Wusirika, 1985
	Mg–Li	Wusirika and Chyung, 1980*; Wusirika, 1985
	Mg–Sc	Tredway and Loehman, 1985
	Be	Wusirika and Chyung, 1980*; Wusirika, 1985
	Cs	Wusirika and Chyung, 1980*
	Cs–Mg	Wusirika and Chyung, 1980*
	Ca	Drew *et al.*, 1983; Jankowski and Risbud, 1983; Kenmuir *et al.*, 1983; Pasturnak and Verdier, 1983; Sakka *et al.*, 1983; Thorp and Kenmuir, 1981
	Mn	Pasturnak and Verdier, 1983; Wusirika, 1985
	Ba	Herron and Risbud, 1986*; Pasturnak and Verdier, 1983; Tredway and Risbud, 1983a, 1983b*, 1985b; Wusirika, 1985
	Ba–Mg	Wusirika and Chyung, 1980*; Wusirika, 1985
	Y	Thomas *et al.*, 1982*; Aujla *et al.*, 1986; Hampshire *et al.*, 1984; Jack, 1977; Kenmuir *et al.*, 1983; Leedecke and Loehman, 1980; Leng-Ward and Lewis, 1985*; Loehman 1979; Messier and Broz, 1982; Messier, 1982*; Messier and Deguire, 1984; Messier, 1985*, 1987; Mittl *et al.*, 1985; Shillito *et al.*, 1978; Wusirika and Chyung, 1980*

Table 4.1 *Continued*

System	*M (metal atoms)*	*Reference*
	Sc	Ball *et al.*, 1985*
	Nd	Kenmuir *et al.*, 1983; Pasturnak and Verdier, 1983
M–Si–B–O–N	Ca–Al	Jankowski and Risbud, 1983
	Na	Coon *et al.*, 1983
	Na–Al–Ba	Brinker *et al.*, 1983
M–O–N	Ca–Al–Mg–Ba	Bagaasen and Risbud, 1983
	Na–B	Frishat *et al.*, 1984
M–P–O–N	H	Marchand, 1983
	Li	Larson and Day, 1986; Marchand, 1983
	Na	Marchand, 1983; Reidmeyer and Day, 1985; Reidmeyer *et al.*, 1986
	Na–Ba	Rajaram and Day, 1987
	Na–Mg	Rajaram and Day, 1987
	Na–Ca	Rajaram and Day, 1987
	Na–Sr	Bonnell *et al.*, 1987
	K	Marchand, 1983

* References with information on glass crystallization.

1500–1700°C is determined by melt composition, temperature and the furnace atmosphere. Considering the simple Si–O–N system at equilibrium, standard Gibbs energies of formation, ΔG^0, may be used to predict relative stabilities (from Loehman, 1983):*

$$<Si_3N_4> + 3/2(O_2) \rightarrow 3(SiO) + 2(N_2) \qquad \Delta G_f^0(1900\ K) = -652 \text{ kJ per mole } Si_3N_4 \quad (4.1)$$

$$<Si_3N_4> + 3(O_2) \rightarrow 3\{SiO_2\} + 2(N_2) \qquad \Delta G_f^0(1900\ K) = -1609 \text{ kJ per mole } Si_3N_4 \quad (4.2)$$

$$<Si_3N_4> + 3\{SiO_2\} \rightarrow 6(SiO) + 2(N_2) \qquad \Delta G_f^0(1900\ K) = +301 \text{ kJ per mole } Si_3N_4 \quad (4.3)$$

The large negative Gibbs energies for reactions (4.1) and (4.2) indicate that there is a strong driving force for reaction between Si_3N_4 and any small amount of oxygen in the furnace atmosphere. Consequently practical and fundamental limitations arise in the processing of oxynitride melts because of the necessity to work in highly reducing conditions. Reaction (4.3) shows that at 1900 K there is little tendency for Si_3N_4 to react with SiO_2 to give gaseous decomposition products as long as equilibrium is maintained. However, when the system is not at equilibrium, the loss of N_2 and/or SiO(g) through condensation on the cold furnace walls promotes decomposition of the melt.

* <solid>, {liquid}, (gas)

Oxidation–reduction reactions in the melt that produce gaseous products or precipitates limit the stability of oxynitride melts. Loehman (1983) used Ellingham diagram plots of ΔG^0 against temperature to compare the relative oxidation or reduction tendencies of the possible ingredients of oxynitride melts. The ΔG^0 values for a range of oxidation reactions at 1500°C are presented in Table 4.2.† Experimental observations of the stability of silicate and aluminosilicate oxynitride melts for a wide range of systems are consistent with this ranking of oxide compounds according to

† 1200°C for the formation of some alkali oxides.

Table 4.2 ΔG^0 values for a range of oxidation reactions

Reaction	ΔG^0 (kJ per mole O_2) 1500°C
$4\{K\} + (O_2) = 2\{K_2O\}$	−209*
$4\{Na\} + (O_2) = 2\{Na_2O\}$	−318*
$\{Sn\} + (O_2) = <SnO_2>$	−217
$2(Zn) + (O_2) = 2<ZnO>$	−217
$2<Ti_2O_5> + (O_2) = 6<TiO_2>$	−268
$<Mo> + (O_2) = <MoO_2>$	−276
$<P>$ white $+ (O_2) = 2/5(P_2O_5)$	−280
$2\{Mn\} + (O_2) = 2<MnO>$	−510
$<C> + (O_2) = 2(CO)$	−531
$4/3<AlN> + (O_2) = 2/3<Al_2O_3> + 2/3(N_2)$	−573
$\{Si\} + (O_2) = \{SiO_2\}$	−594
$<Ti> + (O_2) = <TiO_2>$	−627
$1/3<Si_3N_4> + (O_2) = <SiO_2> + 2/3(N_2)$	−640
$4<TiO> + (O_2) = 2<Ti_2O_3>$	−673
$4(Li) + (O_2) = 2<Li_2O>$	−677
$2<Ti> + (O_2) = 2<TiO>$	−706
$\{Ce\} + (O_2) = <CeO_2>$	−715
$2(Mg) + (O_2) = <MgO>$	−732
$4/3\{Al\} + (O_2) = 2/3<Al_2O_3>$	−736
$2\{Ba\} + (O_2) = 2<BaO>$	−744
$<Zr> + (O_2) = <ZrO_2>$	−769
$4/3\{La\} + (O_2) = 2/3<La_2O_3>$	−857
$4/3\{Nd\} + (O_2) = 2/3<Nd_2O_3>$	−899
$2(Ca) + (O_2) = 2<CaO>$	−903
$2\{Be\} + (O_2) = 2<BeO>$	−920
$4/3<Sc> + (O_2) = 2/3<Sc_2O_3>$	−928
$4/3<Y> + (O_2) = 2/3<Y_2O_3>$	−932

* ΔG^0 (kJ per mole O_2) at 1200°C.

their free energies of formation relative to the ease of oxidation of Si_3N_4 and AlN under highly reducing conditions. Metal oxides such as Li_2O, BaO and Y_2O_3 with ΔG^0 values more negative than that for the oxidation of Si_3N_4 and AlN are stable constituents of M–Si–Al–O–N melts under highly reducing conditions, while oxides such as B_2O_3, P_2O_5 and TiO_2 with ΔG^0 values more positive are reduced while Si_3N_4 and/or AlN is oxidized. The reduced species are lost from the melt as a liquid or gas, or precipitated within the melt as a solid. For example, globules of metallic tin formed on the outside surface of a Y–Si–Al–O–N glass containing 2 wt % SnO_2 melted at 1650°C (Leng-Ward, private communication); large weight losses of potassium vapour occurred in the preparation of K–Si–Al–O–N melts (Messier and Deguire, 1984); a gold coloured film attributed to TiO was deposited at the surface of a Mg–Si–Al–O–N glass melted at 1650°C (Tredway and Risbud, 1985a). In general oxides with the more negative Gibbs free energies of formation should be favoured in formulating oxynitride compositions. The redox stability criterion also accounts for the unreactivity of molybdenum and boron nitride (BN) in contact with oxynitride melts and their subsequent use as crucible materials, or as a powder bed in the case of BN.

The emphasis on references to Y and Mg-containing glass forming oxynitride systems is a reflection of the importance of MgO and Y_2O_3 as sintering additives in Si_3N_4 related ceramics. As Table 4.1 shows, the majority of oxynitride glasses successfully melted and studied have constituents which are stable according to the redox criteria. The main exceptions to this are Na and K-containing glasses where precautions need to be taken to minimize the loss of these volatile species.

The formation of oxynitride glasses that are unstable according to the thermodynamic limitations is possible where alternative low temperature techniques exist. For example M–B–O–N glasses can be formed by the ammonolysis of gels (Brinker and Haaland, 1983). Nitrogen containing phosphate glasses can be prepared by the ammonolysis of alkaline polyphosphate or phosphorous oxide melts as well as by doping a melt with metal nitrides (Marchand, 1983; Rajaram and Day, 1987; Reidmeyer and Day, 1985; Reidmeyer *et al.*, 1986).

4.2.2 Glass-forming compositions

Oxynitride glass-forming regions are essentially extensions of oxide glass-forming regions broadly centred around eutectic compositions, but modified by the particular characteristics that the trivalent nitrogen atom has on melt structure and properties. Figure 4.1 shows schematically the glass-forming volumes in the Mg–Si–Al–O–N and Y–Si–Al–O–N systems as determined by cooling melts at moderate cooling rates from 1700°C (from Hampshire *et al.*, 1985).

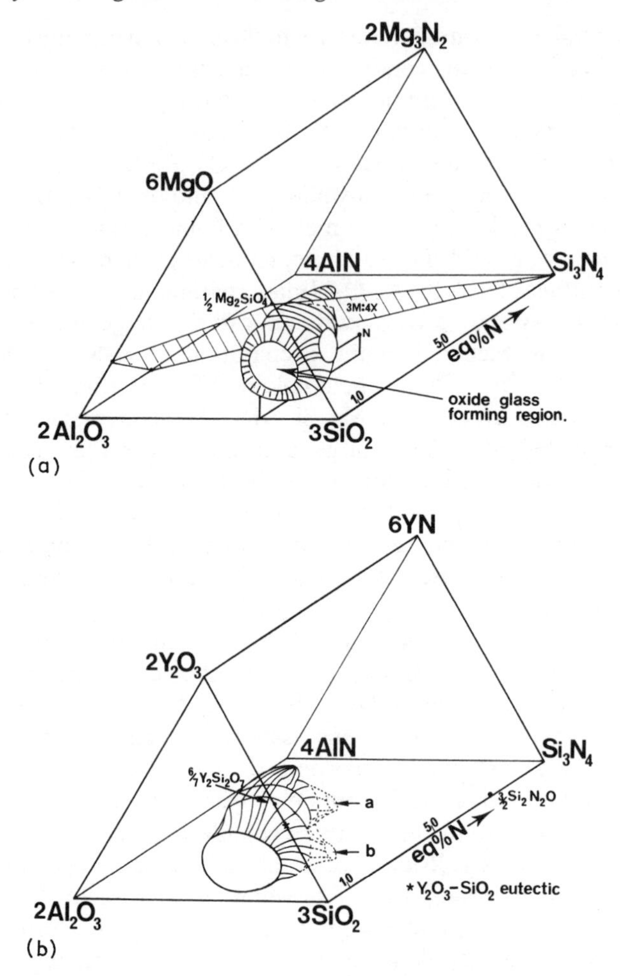

Fig. 4.1 Mg–Si–Al–O–N (a) and Y–Si–Al–O–N (b) glass-forming regions on cooling at moderate rates from 1700°C (from Jack, 1977; Hampshire *et al.*, 1985).

The substitution of small to moderate amounts of nitrogen for oxygen generally leads to a lowering in the melting temperature (T_m), increases in melt viscosity and the inhibition of the nucleation of pure silicate phases as $Si(O,N)_4$ tetrahedra are formed in the melt. This leads to increases in the extent of glass-forming regions as nitrogen is initially substituted for oxygen. This is illustrated in Fig. 4.1 where for both systems the extent of the glass-forming regions are at their greatest at ~ 10–15 eq % N.* The lowering in eutectic temperatures (T_e) in M–Si–O–N systems as compared with the

* In such Janecke prism composition diagrams the axes are in equivalent units; i.e. eq % N = 3[N]/(2[O] + 3[N]), eq % Mg = 2[Mg]/(3[Al] + 4[Si] + 2[Mg]).

Table 4.3 Eutectic temperatures (°C) for M–Si–O and M–Si–O–N systems (from Hampshire *et al.*, 1985)

Eutectic temperatures (°C)		
M (metal atom)	*M–Si–O (M_xO_y–SiO_2)*	*M–Si–O–N*
Mg	1540	1390
Ca	1485	1440
Al	1595	1470
Y	1640	1500

corresponding M–Si–O systems is shown in Table 4.3 (from Hampshire *et al.*, 1985).

The substitution of nitrogen for oxygen may enable oxynitride glasses to be formed where it is not possible or very difficult to form equivalent oxide glasses. For example, near-eutectic composition Y_2O_3–SiO_2 melts crystallize very rapidly into yttrium disilicate and silicon dioxide on cooling, while near-eutectic melts in the Si_3N_4–$Y_2Si_2O_7$–$Y_5(SiO_4)_3N$ and the Si_3N_4–$Y_2Si_2O_7$–SiO_2 regions of the Y–Si–O–N system are easily cooled through the $T_m \rightarrow T_g$ temperature range without crystallizing; hence the two distinct glass-forming regions marked 'a' and 'b' in the Y–Si–O–N square of the Y–Si–Al–O–N triangular prism in Fig. 4.1(b).

Oxynitride melt compositions well within a glass-forming region may be cooled at relatively slow rates involving switching the furnace off and allowing the melt to cool at typically ~50–250°C min^{-1} through the $T_m \rightarrow T_g$ range. However, melts near the limits of the glass-forming region may require rapid removal from the hot zone and quenching to avoid the onset of crystallization. For example, in the base Si–Al–O–N system there exists a quite extensive liquid-forming region at 1700°C as shown in Fig. 4.2(b); however, very rapid liquid-quenching is required to form Si–Al–O–N glasses (Roebuck, 1978; Loehman, 1980; Hampshire, 1985). Roebuck (1978) prepared a glass by rapidly quenching a melt of batched composition marked 'G' in Fig. 4.2(b), whereas on slow cooling this melt crystallized into X-phase, mullite and silicon oxynitride.

Figure 4.3 shows the microstructure of a high nitrogen-content (30 eq % N) Y–Si–Al–O–N melt after slow cooling in the furnace. During cooling β'-sialon crystals have precipitated from the melt, followed by yttrium aluminium garnet (YAG) crystals nucleating and growing sending the residual glass composition towards the two-liquid region extending out from between the Y_2O_3–SiO_2 eutectic and SiO_2 corner. Phase separated SiO_2 droplets are clearly visible as black spheres in the micrograph with the

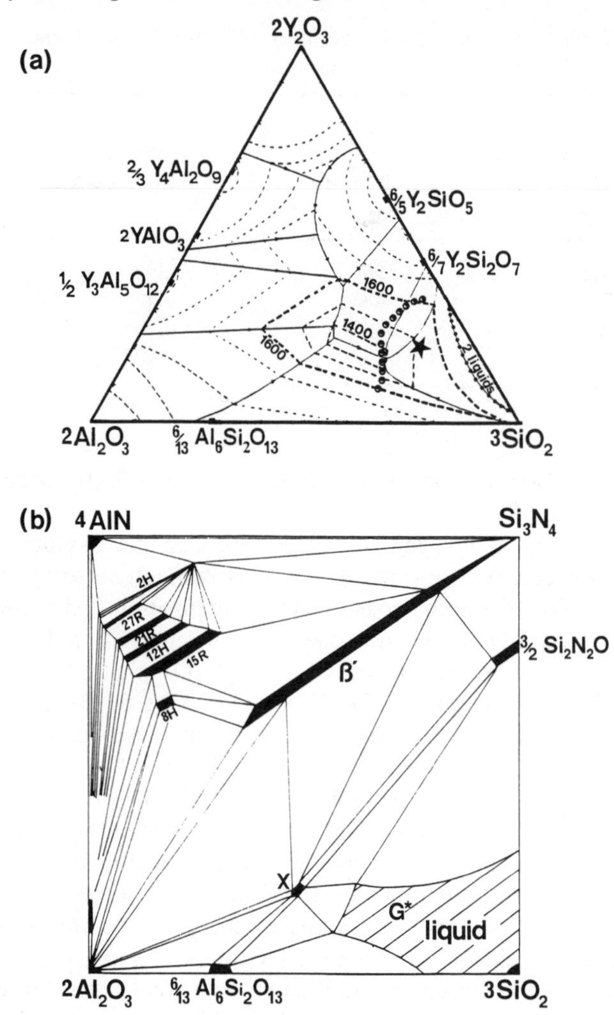

Fig. 4.2 (a) The Y_2O_3–SiO_2–Al_2O_3 phase diagram and (b) the Si–Al–O–N behaviour diagram at 1700°C (from Roebuck, 1978).

scale of phase separation increasing with Al/N deficiency; i.e. in regions of glass close to both YAG and β'-sialon crystals. However, by quickly removing this melt-containing crucible from the hot zone to room temperature in ~1 second the precipitation of β'-sialon and the crystallization of YAG is suppressed at the high cooling rate and a homogeneous glass is formed.

In forming silicate based oxynitride glasses by melting in a reducing atmosphere the glass-forming region is bounded by melt compositions with a

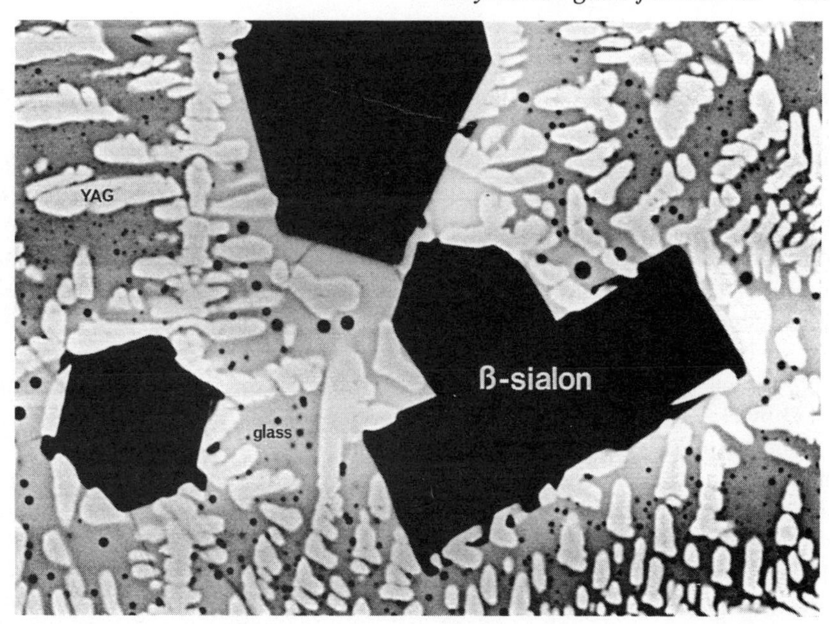

Fig. 4.3 SEM micrograph (backscatter mode (bs)) of a slow cooled high nitrogen content Y–Si–Al–O–N melt.

melting temperature at an upper limit determined by melt stability, and by the tendency for the melt to crystallize during cooling. For example, at temperatures above 1725°C significant weight losses due to decomposition reactions occur in Y–Si–Al–O–N melts. This together with the precipitation of high nitrogen content phases such as β'-Si_3N_4, β'-sialon, Si_2N_2O and the AlN based polytypes such as 15R from melts on cooling appears to limit the formation of oxynitride glasses in M–Si–Al–O–N systems to nitrogen contents no higher than ~30–35 eq % N. The extent of the glass-forming region tends to contract severely at such high nitrogen levels as shown in Fig. 4.1.

The miscibility gaps common in binary M_xY_y–SiO_2 systems extend into M–Si–O–N and M–Si–Al–O–N systems. Melt compositions within an immiscibility region form phase-separated glasses consisting of a SiO_2-rich droplet phase within a modifier-rich matrix. Glasses with compositions close by in sub-liquidus regions of immiscibility tend to phase separate during cooling through the T_m–T_g range or during annealing. The partial plot of the miscibility gap in the Mg–Si–O–N system reported by Shaw *et al.* (1984) shows how most of the glasses in this restricted glass-forming region phase separate on cooling and that the incorporation of nitrogen tends to restrict the extent of the miscibility gap. Energy loss spectra indicate some preferential partitioning of nitrogen into the Mg-rich matrix phase.

In Fig. 4.2(a) the extent of the sub-liquidus immiscibility region in the Y–Si–Al–O system is indicated. Phase separated glasses are formed for glass compositions prepared between the two-liquid region and the curved line of partially blacked circles (Leng-Ward, private communication). Figure 4.4(a) illustrates such a phase separated microstructure for a quenched oxide glass of composition $Y_{0.84}Si_{2.01}Al_{0.48}O_6$, marked by a star in Fig. 4.2(a). On replacing 10 eq % of the oxygen in this composition by nitrogen, the scale of the phase separation becomes much finer as shown in Fig. 4.4(b). At higher levels of nitrogen substitution above ~12 eq % N homogeneous glasses can be formed. The tendency for SiO_2-rich M–Si–Al–O–N glasses to phase separate into SiO_2 droplets during cooling is suppressed by both the substitution of Si by Al and by the replacement of O by N.

The incorporation of more than one modifier oxide into more complicated melt formulations may extend the range of glass formation by altering T_m, T_g, the ease of crystal nucleation, and by enabling the co-ordination requirements in the melt of the mixture of modifier ions of different radii to be more easily satisfied. For example, Tredway and Loehman (1985) formed Mg/Sc–Si–O glasses where equivalent Sc–Si–O and Mg–Si–O melts would not form glasses. In both the Y–Si–O–N and Nd–Si–O–N systems the rapid crystallization of respectively yttrium disilicate and neodymium disilicate prevents the formation of glasses in the vicinity of the Y_2O_3–SiO_2 and Nd_2O_3–SiO_2 eutectic compositions. However, by using 50:50 Y:Nd modifier ion levels instead of equivalent single Y or Nd levels the crystallization of the disilicate phases is suppressed and glasses may be easily formed (Leng-Ward, private communication). This extends the glass-forming region marked (a) in the Y–Si–O–N system of Fig. 4.1(b) back to the oxide face for the Y/Nd–Si–Al–O–N system. This is despite the existence of Y/Nd solid solubility in the disilicate phase.

To summarize, the most extensive oxynitride glass-forming regions occur in systems containing more than four elemental components, in particular aluminosilicate oxynitride glasses with the addition of one or more metallic elements. These additional elements are usually 'network modifiers' except in the case of Be and Mg which may take a network structural role. The position of the glass-forming region in the phase diagram for a particular system is centred about the eutectic temperature with the extent of this region determined by an upper melt temperature stability limit and the tendency for oxide and oxynitride phases to crystallize from the melt during cooling.

4.3 OXYNITRIDE GLASS PREPARATION

Oxynitride glasses have been prepared in a number of ways, the most common being the melting of mixtures of oxides and nitrides in a highly

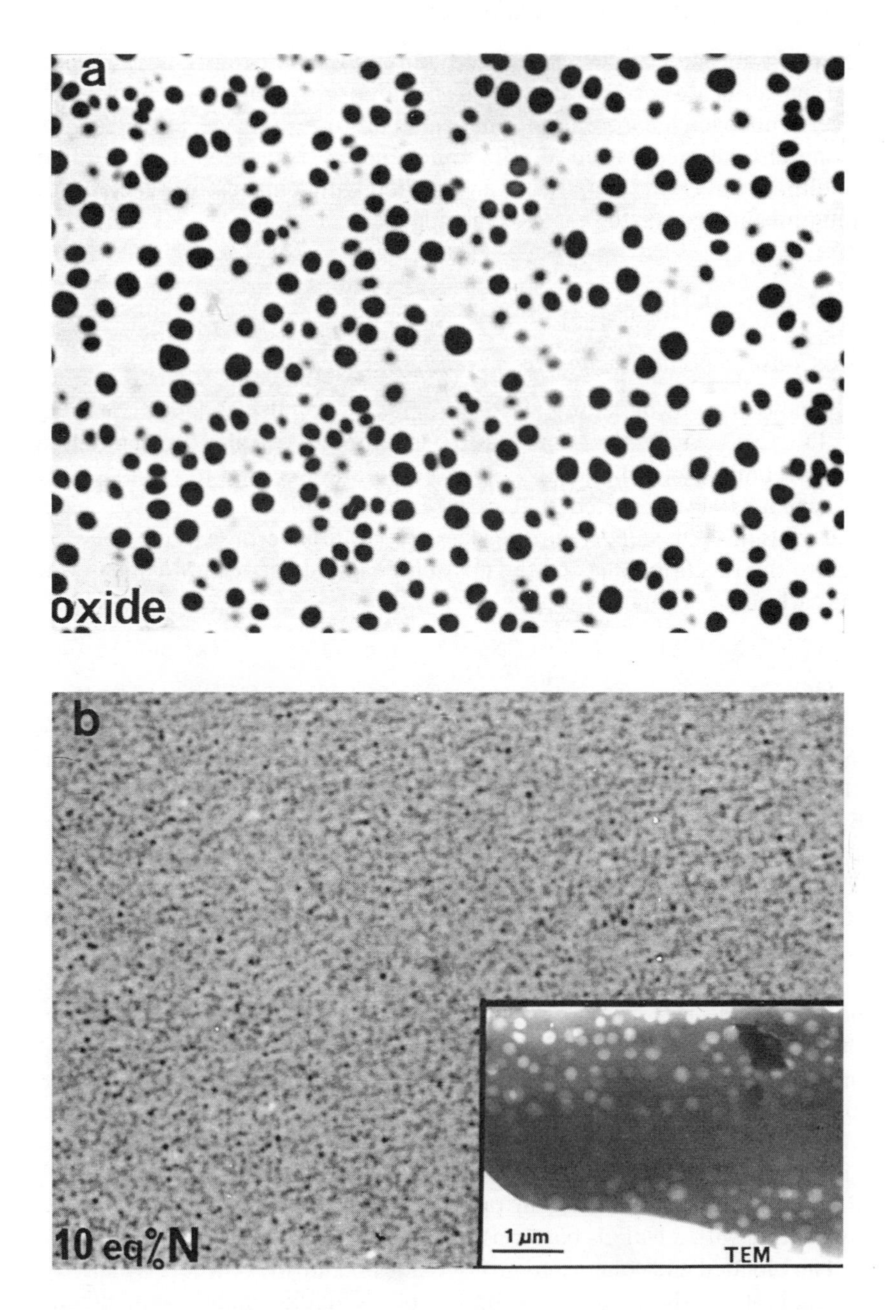

Fig. 4.4 SEM (bs) micrographs of (a) quenched oxide, and (b) quenched 10 eq % N $Y_{0.84}Si_{2.01}Al_{0.48}(O,N)$ glasses. This Y/Si/Al composition is marked as a star in Fig. 4.2(a).

reducing furnace atmosphere. The bubbling of nitrogen or ammonia through a silicate based oxide melt, by comparison, results in a relatively low level of nitrogen incorporation. The recent upsurge in sol–gel processing has prompted the preparation of ultrahomogeneous oxide gels which can then be either converted to a powder and melted with a nitride, or reacted with ammonia and converted to an oxynitride glass.

4.3.1 Direct melting

(a) Constituents and melting temperature

Typical oxide powders used as constituents of oxynitride glasses are SiO_2, Al_2O_3, Li_2O, MgO, Y_2O_3 and Nd_2O_3. Carbonates can also be used as long as a preliminary holding step at low temperature is used to remove CO_2. Wusirika (1984) reported that carbonates should not be used as batch materials in a one-step oxynitride glass preparation because of the large loss due to the reaction (4.4) and the possible reaction between Mo and C with the carbonate decomposition.

$$Si_3N_4 + CO_2 \rightarrow 3SiO_2 + 6CO + 2N_2 \quad \Delta G^0(1650°C) = -733\,\mathrm{kJ\,mol^{-1}} \quad (4.4)$$

The sol–gel processing route can be used to produce an amorphous, intimately mixed, pure ultrahomogeneous oxide powder which can then be melted with a nitride to form an oxynitride glass. Tredway and Risbud (1985b) used tetraethylorthosilicate (TEOS), aluminium sec-butoxide, and barium acetate as precursor materials for preparing Ba–Si–Al–O–N glasses. Unfortunately there is no method for incorporating nitrogen in the solution stage of this process. For preparing phosphorous oxynitride glasses Reidmeyer and Day (1985) prepared the $NaPO_3$ base oxide glass by melting reagent grade $NaHPO_4 \cdot H_2O$ in a platinum crucible at 750°C.

For incorporating nitrogen in melts Si_3N_4 and AlN are the two main nitrides used. Messier and Deguire (1984) concluded from their work on thermal decomposition in the Y–Si–Al–O–N system that the use of AlN instead of Si_3N_4, where possible, does reduce the number of spherical elemental Si precipitates remaining as a dispersion in the glass. Of the lower melting point nitrides, Ca_3N_2 has been used by Jankowski and Risbud (1983) to prepare Ca–Si–Al–O–N glasses and Mg_3N_2 was used by Reidmeyer and Day (1985) in preparing phosphorous oxynitride glasses by melting with a $NaPO_3$ base glass.

The blended and compacted mix of oxide and nitride powders is melted at a temperature where the melt is stable and its fluidity enables good mixing. At high nitrogen contents this usually involves operating within a rather narrow temperature range, and for a time period, that just gives enough melt fluidity but avoids decomposition reactions that occur at too high a temperature. For example, in the Y–Si–Al–O–N system melting temperatures in the 1650–1700°C range have been used by most workers for preparing high nitrogen content glasses. Nearer 1750°C weight losses become high (i.e. >10 wt %) and the glass becomes discoloured with decomposition products. It is possible to modify the fluidity of silicate-containing glasses with small fluoride additions to reduce the viscosity of the melt.

Tredway and Risbud (1985b) reported that a lower melting temperature could be used when melting Si_3N_4 with their 'higher reactivity' gel-derived powders than with a one-step melt from oxide powders. This gave a better quality glass free from metallic inclusions resulting from Si_3N_4 decomposition reactions. In contrast to the high melting temperatures needed in silicate-based glasses melting temperatures in the 580–650°C range were used by Reidmeyer and Day (1985) to melt $NaPO_3$ glass doped with 3 wt % AlN.

Although most oxynitride glasses are prepared with a single-step melt special requirements may necessitate a more complicated procedure. For example, in order to prepare a Na–Ca–Si–O–N glass with a set Na content, Frishat and Scrimpf (1980) used a three-step melting procedure involving a base oxide melt at 1350°C, a melt with Si_3N_4 at 1350°C, followed by a final 1150°C melt in a vacuum furnace under Ar reduced pressure with oxide additions to correct for losses of Na in the earlier steps.

(b) Containment

In a highly reducing atmosphere at high temperatures oxynitride melts may be contained within a molybdenum crucible or a graphite crucible slurry-coated with BN powder. The BN powder must be of very high quality as the presence of small quantities of B_2O_3 destablizes the melt. The use of a BN powder coating and/or powder bed within a graphite crucible helps maintain the local reducing conditions and facilitates removal of the glass pellet from the crucible after cooling. The use of a Mo crucible enables the melt to be poured. Boron nitride crucibles machined from hot pressed BN have been used but require cleaning involving baking at ~1900°C in nitrogen to eliminate contamination (Messier, 1987).

There are a few reported preparations of oxynitride glasses melted in graphite crucibles (Wusirika and Chyung, 1980). There were no visual signs of any reaction between a Y–Si–Al–O–N melt and graphite crucible although the melt did adhere to the graphite and consequently penetrated ~0.5 mm into the pores. Rajaram and Day (1987) used a graphite crucible to prepare a phosphorous oxynitride glass and reported no reaction with or adherence of the melt to the crucible.

Wusirika and Chyung (1980) prepared batches of ~500 g, pouring the melt in a nitrogen atmosphere into a preheated graphite mould. No other attempts to prepare large quantities of oxynitride glass by pouring into a mould have been documented as the majority of studies have been small-scale preparations of the order of 10–100 g batches.

(c) Atmosphere

High purity (oxygen-free) nitrogen is the most suitable atmosphere, normally maintained as a static atmosphere at slightly above atmospheric pressure. The use of a nitrogen atmosphere may result in some nitrogen dissolution in the melt and the inhibition of decomposition reactions producing nitrogen gas. However, at ~1 atm pressure and for relatively short melting periods, typically up to 1 or 2 h, both these potentially advantageous effects are probably minimal (for a discussion of nitrogen dissolution in Ca–Si–O melts see Dancy and Janssen (1976)). Argon atmospheres have been used by a number of workers. In principle maintaining a modest partial pressure of SiO in the furnace atmosphere should suppress thermal decomposition (Wusirika, 1984); however, this is too difficult experimentally.

The retardation of reactions producing gaseous SiO and N_2 decomposition products (see reactions (4.1)–(4.3)) using a high-pressure nitrogen atmosphere has been reported for several glass compositions. Makishima *et al.* (1983) reported the preparation of a transparent La–Si–O–N glass with the highest reported analysed nitrogen content of 36.8 eq % N, formed by reacting an 'as-batched' powder pellet of composition $La_2Si_2O_4N_2$ at 1650–1700°C for 35 min at 30 atm N_2. Loehman (1983) reported the preparation of a dense, pore-free, Ca–Si–Al–O–N glass melted under 300 MPa N_2 whereas when melted under 0.1 MPa N_2 a highly porous glass was formed.

When the preparation of an homogeneous high nitrogen content glass is prevented by the precipitation of a nitrogen-containing phase such as β'-sialon during cooling, the use of high nitrogen pressure is of no use. Mittl *et al.* (1985) used pressures of 2142 atm N_2 in hot isostatic pressing of Y–Si–Al–O–N compositions with increasing nitrogen content but were unable to prepare a glass of higher nitrogen content than the levels of ~30 eq % N reported by Hampshire *et al.* (1985), Messier and Broz (1982) and Leng-Ward and Lewis (1985).

The presence of water vapour in the furnace atmosphere where melting silicate glasses leads to the incorporation of Si–OH groups into the silicate tetrahedra. However, the incorporation of –OH groups into oxynitride melts is modified by dehydration reactions involving nitrogen, as typified by (4.4).

$$6H_2O + Si_3N_4 \rightarrow 2N_2 + 3H_2 + 3SiO_2 \qquad (4.5)$$

Mulfinger and Franz (1965) noted that a large addition of Si_3N_4 lowered the water content of oxynitride glasses to zero. This effect is illustrated by the near-infra-red spectra of a Ca–Si–Al–O and a 18 eq % N Ca–Si–Al–O–N glass in Fig. 4.7(b) (from Hampshire *et al.* 1985). Consequently it is preferable to use an ultra-dry furnace atmosphere to minimize any nitrogen loss through the reaction of Si_3N_4 or $Si(O,N)_4$ groups in the melt with water.

4.3.2 Thermochemical ammonia nitridation

(a) Initial oxide preparation

Oxynitride glasses may be prepared by the reaction of ammonia with silicate and phosphorus melts and with gels. Nitridation via ammonolysis is most effective when applied to microporous oxide gels of large surface area which can then be coverted into dense oxynitride glasses. The silicate oxide gel is formed by a solution technique based on the controlled hydrolysis/polymerization of tetraethoxysilane (TEOS). A pure silicon dioxide gel can be prepared, or gel compositions from many 'glass-forming' systems provided the proper organometallic precursor materials are available. An example of a flow chart for a sol–gel preparation process for the Ba–Si–Al–O system was published by Tredway and Risbud (1985b). This illustrates the carefully controlled and time consuming solution and drying procedures necessary to give a highly porous monolithic gel made up of glass-like macromolecules containing network-forming cations linked by bridging oxygens.

For the preparation of phosphorous oxynitride glasses, base oxide glasses can be prepared by melting reagent grade $NaH_2PO_4 \cdot H_2O$, $NH_4H_2PO_4$ and oxides or carbonates such as MgO, CaO, $SrCO_3$ and $BaCO_3$ in air at ~ 900–1000°C as reported by Rajaram and Day (1987).

(b) Ammonolysis and gel–glass conversion

Oxide gels can be converted into oxynitride gels by treatment with flowing anhydrous ammonia at temperatures in the upper part of the range 400–1100°C. The sequence of reactions involves the reaction of ammonia with ≡Si–OH and ≡Si–O–Si≡ groups to form amine groups at lower temperatures, followed by condensation reactions to form nitride species with the loss of oxygen as water at higher temperatures. For specific reaction mechanisms in a pure silica gel see Brow and Pantano (1987) and for multi-component silicate gels see Brinker and Haaland (1983). The use of flowing ammonia gives a continuous replenishment of gaseous reactant and also removes the water produced by the various condensation reactions, critical to maximizing the incorporation of nitrogen.

Brow and Pantano (1987) reported being able to replace up to 75 eq % of the oxygen in a very thin (~ 100 nm) silicon dioxide gel film by nitrogen, giving a limiting composition of Si_2N_2O. They verified that the silicon

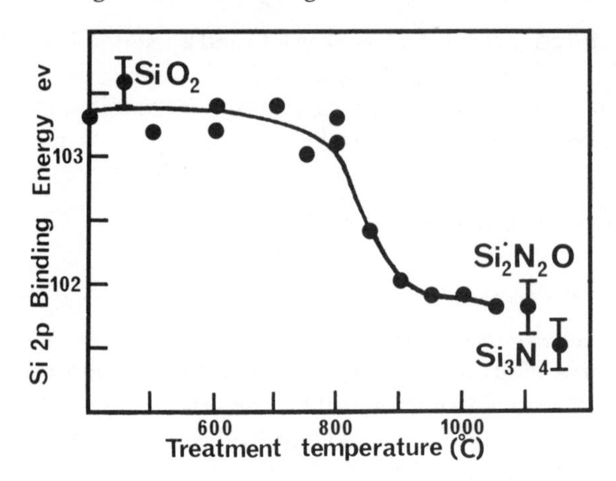

Fig. 4.5 X-ray photoelectron spectra (XPS) of thin (~ 100 nm) silica gels nitrided at various temperatures (from Brow and Pantano, 1987).

dioxide gel films were converted from oxide materials to oxynitride materials when the treatment temperature exceeded 800°C using X-ray photoelectron spectra (XPS) analysis. The effect of temperature on the Si 2p binding energy for such ammonia-treated sol–gel derived films is shown in Fig. 4.5. However, for the incorporation of nitrogen into bulk gels the degree of nitrogen substitution for oxygen may be very much lower.

Oxynitride gels may be converted to oxynitride glasses by heating in a vacuum at just above the glass transition temperature, typically ~900°C. A proportion of the nitrogen is lost during gel–glass conversion. Brinker and Haaland (1983) synthesized dense, colourless and homogeneous Na–Al–Ba–Si–B–O–N oxynitride glasses containing up to ~2 eq % N, reporting evidence for B–N bonding which is difficult to achieve by direct melting.

Phosphorous oxynitride glasses may be prepared by remelting the oxide glass frit in anhydrous ammonia (see Reidmeyer *et al.* (1986) for possible reaction mechanisms). Rajaram and Day (1987) achieved nitrogen levels of up to 25 eq % N in their M–Na–P–O–N glasses assuming that two nitrogens are replacing three oxygens on nitridation. Marchand (1983) characterized a 9.7 wt % N Na–P–O–N glass as having a formula of $NaPO_2N_{0.67}$ from weight loss considerations on nitridation, corresponding to a 33.3 eq % N glass. Na–P–O–N glasses of up to 12 wt % N have been prepared by Reidmeyer *et al.* (1986).

4.3.3 Glass characterization

Weight losses of Si, O, and N have been reported to varying degrees during the preparation of aluminosilicate glasses by direct melting. Messier and

Deguire's (1984) analysis of possible decomposition reactions occurring in silicate oxynitride melts and analyses by a number of workers is consistent with gaseous SiO being the main product lost from melts and condensing on cooler parts of the furnace.

Reports of nitrogen loss show a large variation, illustrated by a nitrogen retention between 38 and 88% reported by Homeny and McGarry (1984) in melting Mg–Si–Al–O–N glasses. In this case the higher nitrogen losses are associated with high melting temperatures (over 1700°C) used to melt high nitrogen content glasses. Where compositions melt easily below 1700°C, where the nitrogen atmosphere is free of oxygen or water impurities and is static, a number of reports show that the nitrogen retention of between 90 and 100% can be achieved (Messier, 1987; Homeny and McGarry, 1984).

A number of workers have suggested that it is the spherical Si precipitates that give silicate oxynitride glasses their grey coloration, the intensity of which increases with nitrogen content. This could be caused by the small amount of Si_3N_4 decomposition that occurs according to (4.6) at melting temperatures over 1600°C of

$$\langle Si_3N_4 \rangle \rightarrow 3\{Si\} + 2(N_2) \qquad (4.6)$$

To eliminate frothing in Mg–Si–O–N glasses Baik and Raj (1985) have in fact suggested doping Mg–Si–O–N glasses with elemental silicon to raise the activity of silicon in the melt and hence depress the vapour pressure of N_2.

Aluminosilicate oxynitride glasses made using AlN as the only nitrogen source, however, still contain Si precipitates and the associated grey colour. This Si must come from impurities in the AlN or result from some reduction of $Si(O,N)_4$ tetrahedra to Si in the melt. Significantly the Ba–Si–Al–O–N glasses prepared from gel-derived oxide powders and melted at relatively low temperatures between 1550 and 1590°C were essentially colourless (Tredway and Risbud, 1985b) as were the Si-free Ca–Al–Mg–Ba–O–N glasses of Bagaasen and Risbud (1983) and phosphorous oxynitride glasses in general.

4.4 OXYNITRIDE GLASS STRUCTURE

Discussions on the structure of oxynitride glasses have until recently been confined to inferences based on the progressive change of physical properties with composition and rather limited information from infrared spectra. Recently solid state NMR (MAS) has enabled the identification of discrete structural units within aluminosilicate oxynitride and phosphorous oxynitride glasses.

Descriptions of oxide glass structure are mainly based on these local 'molecular' structural units, similar to that found in crystalline phases, but bound together in a structure which lacks periodicity and a long range order.

Hence aluminosilicate glasses can be described as being made up of a network of corner sharing $Si(O)_4$ and $[Al(O)_4]^-$ tetrahedra, with any network-modifying ions present (e.g. Na^+, Ca^{2+}, Al^{3+} and Y^{3+}) bonded ionically to locales of negative charge such as non-bridging oxygens [$\equiv$Si–$O^- \cdots {}^+Na$] or $[Al(O)_4]^-$. The degree of polymerization of the tetrahedral structural units is determined by the balance of modifying/structural cations. Little concrete information exists in the literature about middle range order although there are several hypotheses such as random network, crystallite and paracrystalline arrangements. A more detailed discussion of glass structure and the technique of MAS NMR is presented in Chapter 1.

4.4.1 Solid state NMR (MAS)

Evidence of the existence of discrete tetrahedral $Si(O,N)_4$ units in Y–Si–Al–O–N glasses is illustrated by the ^{29}Si spectra of Aujla *et al.*, (1986). The ^{29}Si spectra of the oxide glass, $Y_{1.03}Si_{1.27}Al_{1.27}O_6$, consists of one broad $Si(O_4)$ peak with a chemical shift centred at -83 ppm. The various types of $Si(O_4)$ tetrahedra arising from differing degrees of polymerization (i.e. Q_4, Q_3 and Q_2 units) and the various combinations of next-near-neighbour co-ordination are unresolved and contribute to the widening of the peak and some peak asymmetry. On replacement of 15 eq % of the oxygen in this glass with an equivalent amount of nitrogen giving a glass of composition $Y_{1.03}Si_{1.27}Al_{1.27}O_{5.1}N_{0.6}$, the ^{29}Si spectra shown in Fig. 4.6 can be resolved into three contributing peaks centred at -83 ppm, -71 (2) ppm and -60 (3) ppm which is consistent with the coexistence of respectively $Si(O_4)$, $Si(O_3N)$ and $Si(O_2N_2)$ tetrahedral units in the glass. Such down-field shifts in the 9–15 ppm range per replacement of oxygen by nitrogen in $Si(O,N)_4$ tetrahedral groups are known for a range of crystalline oxynitride phases (Smith, 1987; Dupree *et al.*, 1985).

The distribution of $Si(O_4)$, $Si(O_3N)$ and $Si(O_2N_2)$ groups in this 15 eq % N glass is consistent with (i) a close to statistical distribution of nitrogen among the $Si(O,N)_4$ groups and (ii) a preference for N bonding to Si, with an approximate 2 : 1 ratio of Si–N bonds to Al–N bonds. This is illustrated by the close match of the 15 eq % N glass spectra and the simulated spectra based on these two factors as shown in Fig. 4.6. A simulated spectra based on nitrogen showing no preference between Si and Al does not match the glass spectra (Aujla *et al.*, 1986).

However, the existence of a small concentration of $Si(ON_3)$ groups predicted to exist in this Y–Si–Al–O–N 15 eq % N glass by a statistical distribution of N among the $Si(O,N)_4$ groups cannot be confirmed. In fact the ^{29}Si spectra of the 30 eq % N Y–Si–Al–O–N glass appears to consist of mainly $Si(O_3N)$ and $Si(O_2N_2)$ groups with $Si(ON_3)$ groups still not evident at this high N level, either because they do not exist to a significant degree or because the $Si(ON_3)$ component of the spectra cannot be resolved. This

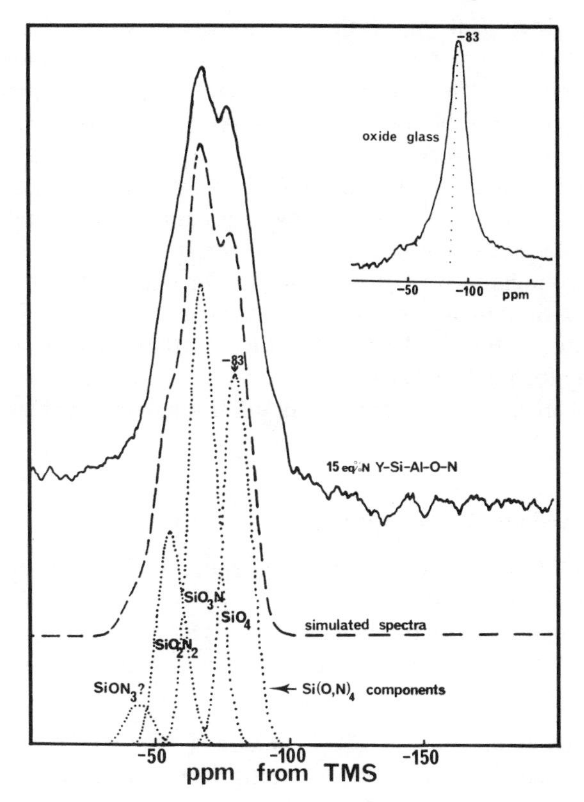

Fig. 4.6 ^{29}Si NMR (MAS) spectra of $Y_{1.04}Si_{1.27}Al_{1,27}O_{5.1}N_{0.6}$ glass (from Aujla *et al.*, 1986) together with simulated spectra and proposed spectral components. Inset shows the spectra of the equivalent oxide glass.

suggests that the co-ordination requirements of $Si(ON_3)$ groups are difficult to satisfy in the melt and that one finds a build-up of $Si(O_2N_2)$ groups at high nitrogen contents; that is the statistical distribution of nitrogen among $Si(O,N)_4$ groups breaks down at high nitrogen contents. As the 30 eq % N glass is very close to the nitrogen solubility limit it is probable that the eventual formation of $Si(ON_3)$ groups in the melt at still higher nitrogen contents results in the nucleation and precipitation of β'-sialon crystals on cooling, thus preventing the formation of a homogeneous glass.

The coexistence of three $Si(O,N)_4$ groups in Y–Si–Al–O–N glasses is paralleled by the distribution of $P(O,N)_4$ groups found in Na–P–O–N glasses by Bunker *et al.* (1987). The parent $NaPO_3$ glass spectrum showed a single $P(O_4)$ peak at a chemical shift of -20.3 ppm. As nitrogen substitutes for oxygen two new peaks are progressively formed at approximately -10 and 0 ppm, respectively assigned to $P(O_3N)$ and $P(O_2N_2)$ species. The changes in relative NMR peak intensities are consistent with a gradual random replacement of oxygen by nitrogen in the glass structure to form these two oxynitride species. The ^{15}N spectra of these glasses suggest that approximately three P–N=P bonds exist for every trico-ordinate >N– bonded to three different phosphorus atoms.

In both the Y–Si–Al–O–N and Na–P–O–N NMR (MAS) spectral studies, the respective ^{29}Si and ^{31}P peaks show a gradual positive chemical shift as nitrogen replaces oxygen in the glass network. This is consistent with the lower electronegativity of the nitrogen. The identification of $Al(O,N)_4$ groups in M–Si–Al–O–N glasses has not been possible so far due to the broad nature of the ^{27}Al spectra. Much further structural information should come from NMR (MAS) work on ^{29}Si, ^{27}Al and ^{31}P nuclei together with ^{15}N nuclei and from a range of modifier ion nuclei such as ^{89}Y.

4.4.2 Infra-red and Raman spectra

Chu *et al.* (1968) reported the infra-red spectra of a series of amorphous Si_3N_4–SiO_2 films at nitrogen contents of 6 eq % N, 25 eq % N and 35 eq % N together with those of amorphous SiO_2 and Si_3N_4. These spectra are reproduced in Fig. 4.7(a) with their two main features being, as nitrogen replaces oxygen:

1. The main band in the SiO_2 spectra (1080 cm^{-1}) attributable to asymmetrical Si–O–Si bond stretching vibrations shifts to lower wave numbers. This is consistent with the lower wave number of the Si–N–Si bond stretching vibration (850 cm^{-1});
2. The increased half-width of the Si–N band, approximately four times that of the Si–O band, and the merging of the Si–O and Si–N band components causes a widening of the spectra.

These two trends are also found in all the reported oxynitride glass infra-red spectra (e.g. Frishat and Schrimp, 1983; Brow and Pantano, 1984; Luping *et al.*, 1984). However, these trends are less easily observable in the infra-red spectra of oxynitride glasses because the modifier content generally needed to prepare high nitrogen content glasses results in the formation of non-bridging $\equiv$Si–O^- bonds in the glass. Such bonds have infra-red bands in the 950–1000 cm^{-1} wavenumber range which tends to mask spectra band shifts caused by replacement of O by N. There is no infra-red spectral evidence for non-bridging character in the nitrogen bonding in silicate oxynitride glasses and no evidence for Al–N bonding in the spectra of aluminosilicate oxynitride glasses.

Bunker *et al.* (1987) found that the Raman spectra of nitrided $NaPO_3$ glasses showed marked changes as nitrogen is incorporated into the glass. By comparing the spectra with those of phosphazene compounds and by

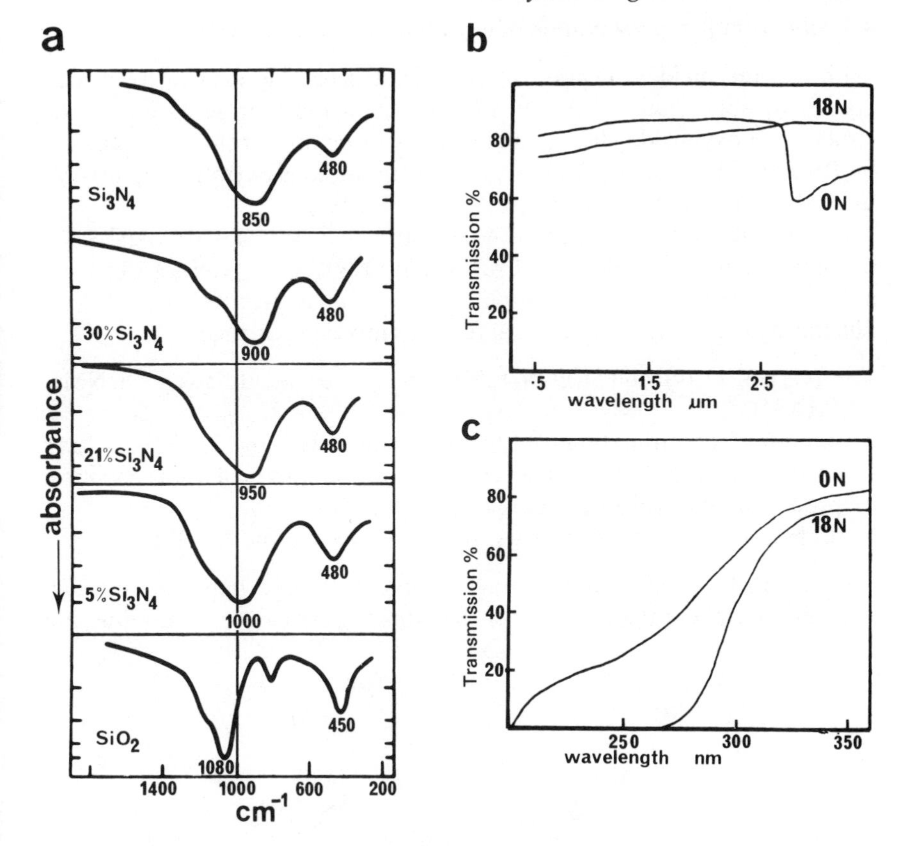

Fig. 4.7 (a) Infra-red spectra of amorphous Si_3N_4, SiO_2 and a range of nitrided SiO_2 films (from Chu *et al.*, 1968), (b) infra-red and (c) near-infra-red spectra of a Ca–Si–Al–O and an 18 eq % N Ca–Si–Al–O–N glass (from Hampshire *et al.*, 1985).

observing nitrogen isotope shifts a band at around 813 cm^{-1} was assigned to P–N=P bonds and that at 640 cm^{-1} to >N– bonds.

4.4.3 X-ray photoelectron spectra (XPS)

Brow and Pantano (1984) obtained a N 1s binding energy for Na–Si–O–N and Ca–Na–Si–O–N glasses of 398.0 ± 0.3 eV compared to 397.7 ± 0.5 eV for crystalline Si_3N_4. This indicates a similar three-fold co-ordination of nitrogen to silicon is probable in the glass. The possible existence of nitrogen atoms co-ordinated to only two silicons, that is, nitrogens with some non-bridging character ($Si–N^-–Si$), has not been confirmed although Brow and Pantano's spectra do have small but not well-defined low energy shoulders possibly due to this type of bonding.

4.4.4 Schematic representation of oxynitride glass structure

With the recent identification of specific structural groups in oxynitride glasses by solid state NMR (MAS) the schematic representation of the structure of oxynitride glasses can be taken a step further than was possible by Risbud in 1981 in his 'Analysis of bulk amorphous oxynitride structures using the network theory of glasses' (Risbud, 1981).

A schematic representation of the structural network of a M–Si–Al–O–N oxynitride glass containing a 1 : 1 ratio of $Si(O,N)_4$ to $Al(O,N)_4$ groups and a M^{3+} modifier ion content greater than Al/3 is given in Fig. 4.8(a), illustrating the following structural features involving nitrogen;

1. $Si(O_4)$, $Si(O_3N)$, and $Si(O_2N_2)$ structural groups (identified by ^{29}Si NMR (MAS))
2. three-fold co-ordination of nitrogen (suggested by XPS work and consistent with progressive changes in the physical and chemical properties of oxynitride glasses)
3. preference for Si–N bonding over Al–N (indicated by ^{29}Si NMR).

It is possible that nitrogen may be able to bridge between just two structural tetrahedral groups rather than three at high levels of modifier ion

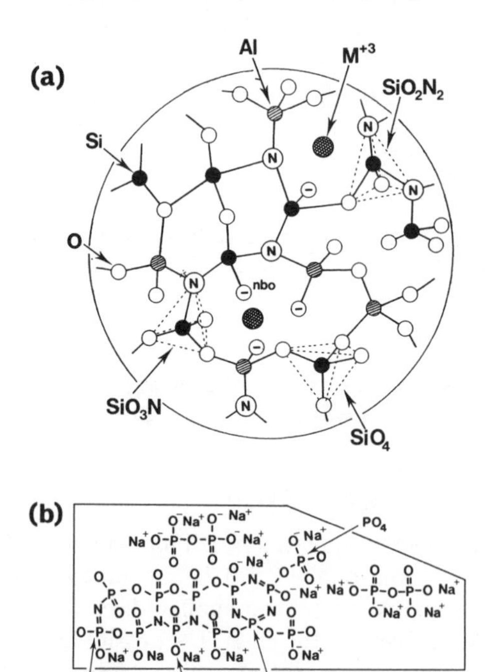

Fig. 4.8 Schematic structural representation of (a) a M–Si–Al–O–N glass and (b) a Na–P–O–N glass (from Bunker *et al.*, 1987).

content. This would parallel the trend of non-bridging oxygens replacing bridging oxygens at high modifier ion contents in silicate glasses. This type of co-ordination does exist in the nitrogen wollastonite phase, $YSiO_2N$, where each N atom bridges between two $Si(O_2N_2)$ groups and also bonds with a Y^{3+} (Morgan *et al.*, 1977).

In Fig. 4.8(b) a generalized model for nitrogen incorporation into $NaPO_3$ glass is presented (from Bunker *et al.*, 1987) illustrating the following structural features involving nitrogen:

1. $P(O_4)$, $P(O_3N)$ and $P(O_2N_2)$ structural groups (identified by ^{31}P NMR (MAS))
2. P–N=P and >N– bonding (suggested by both Raman and NMR)
3. cross-linking of phosphate chains by nitrogen (suggested by Raman, NMR and property changes).

The substitution of three-fold co-ordinate nitrogen for two-fold co-ordinate oxygen, and also the substitution of two-fold co-ordinate nitrogen for non-bridging oxygen, increases the cross-linking between the structural groups in silicate based oxynitride glasses and in phosphorous oxynitride glasses, creating a more tightly bonded glass structure. The degree of tightness in the glass structure is also influenced by the field strength of the modifier cation. Progressive changes in glass properties which reflect these structural modifications on substituting N for O are reviewed in the following section.

4.5 PROPERTIES OF OXYNITRIDE GLASSES

The properties of a glass are directly related to the glass structure, and, as noted in the previous section, it was the property changes that led to the generally held view that nitrogen increases the cross-linking of the network structure in oxynitride glasses. Examples of changes in measured properties on substituting nitrogen for oxygen in a range of oxynitride glasses are presented in Figs 4.9 to 4.12. In all cases the cation ratios are held constant as nitrogen replaces oxygen in the glass composition.

4.5.1 Physical properties

The density, refractive index and microhardness of a series of M–Si–Al–O–N glasses of constant M/Si/Al ratio and increasing nitrogen content are shown in Fig. 4.9(a),(b) and (c). The data are from Sakka *et al.* (1983) and Hampshire *et al.* (1984). The steady increase in these three properties with nitrogen content reflects an increased 'compactness' of the glass structure as nitrogens capable of bridging three network tetrahedral groups replace oxygens capable of bridging only two.

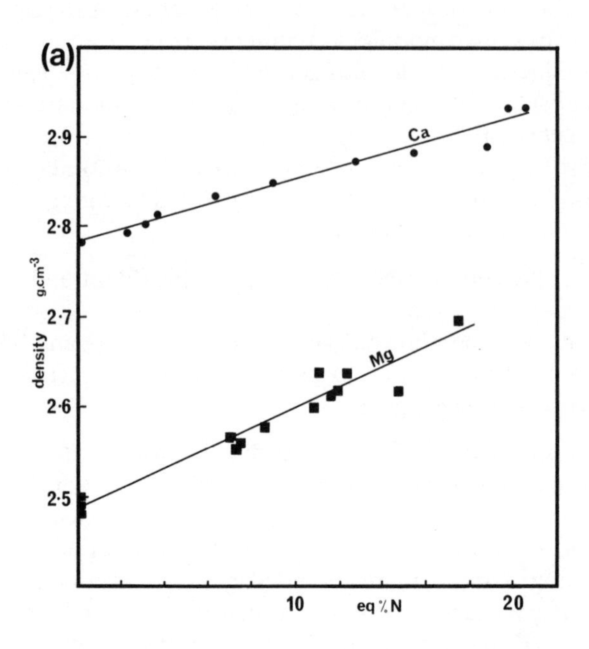
(a)
2·9
2·8
2·7
2·6
2·5
density g.cm^{-3}
Ca
Mg
10
eq % N
20

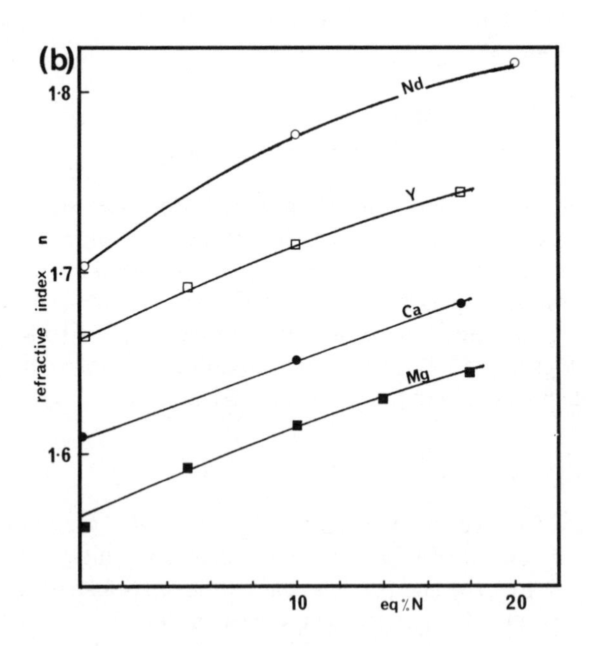
(b)
1·8
1·7
1·6
refractive index n
Nd
Y
Ca
Mg
10
eq % N
20

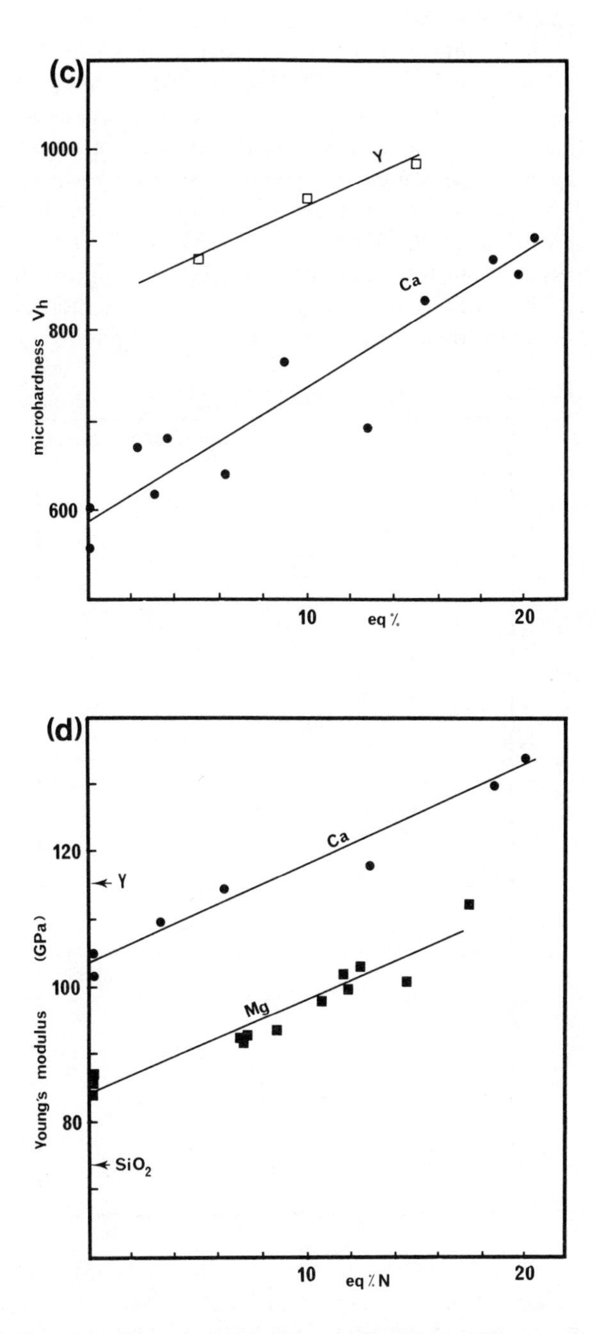

Fig. 4.9 (a) Density, (b) refractive index, (c) microhardness and (d) Young's modulus of M–Si–Al–O–N glasses (M = Ca,Mg,Nd,Y) (from Sakka *et al.*, 1983; Homeny and McGarry, 1984; Hampshire *et al.*, 1984).

The Young's modulus and the coefficient of thermal expansion depend on the strength of the chemical bonds in the short range structure of the glass. The Young's moduli of Ca–Si–Al–O–N glasses (from Sakka *et al.*, 1983), and Mg–Si–Al–O–N glasses (from Homeny and McGarry, 1984) show linear increases with nitrogen content (Fig. 4.9(d)). Decreasing linear coefficients of thermal expansion with nitrogen substitution are shown for a series of Na–Ca–Si–O–N glasses (from Frishat and Scrimpf, 1983), Li–Si–Al–O–N glasses (from Luping *et al.*, 1984), and for M–Na–P–O–N phosphorous oxynitride glasses (from Rajaram and Day, 1987) in Fig. 4.10. However, Sakka *et al.* (1983) found no significant decrease in the coefficients of thermal expansion for their series of Ca–Si–Al–O–N glasses.

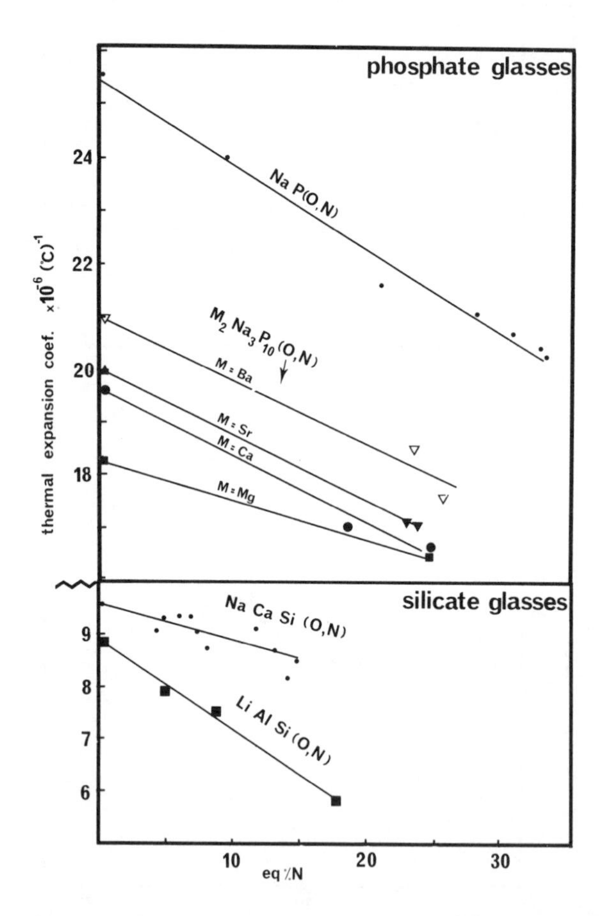

Fig. 4.10 Coefficients of thermal expansion for NaP(O,N) and $M_2Na_3P_{10}(O,N)$ glasses (averaged between 25 and 300°C from Rajaram and Day, 1987), Na–Ca–Si–(O,N) glasses (from Frishat and Scrimpf, 1983) and Li–Si–Al–O–N glasses (from Luping *et al.*, 1984).

The viscosity and glass transition temperature of a series of M–Si–Al–O–N glasses of increasing nitrogen content are shown in Fig. 4.11 (from Hampshire *et al.* 1984). The generally observable increases in these two properties is consistent with an increasing degree of connectivity as nitrogen replaces oxygen.

Messier (1987) reported high transverse rupture strengths for three Y–Si–Al–O–N glasses with the 30 eq % N glass averaging 186 MPa for 10 specimens tested in four point bending compared with strengths of around 115 MPa for Y–Si–Al–O glasses. The strength-limiting defects in these nitrogen glasses were observed to be the silicon spheres of typically ~ 100 μm diameter that result from thermal decomposition during melting and also small pores and devitrified regions. K_{1c} values of between 0.95 and 1.44 MN $m^{3/2}$ reported by Messier (1987) for Y–Si–Al–O–N glasses are higher than for typical oxide glasses (0.7–0.8 MN $m^{3/2}$). Homeny and McGarry (1984) found that the fracture toughness of a series of Mg–Si–Al–O–N glasses of constant cation content increased linearly with nitrogen content from 1.16 (oxide) to 1.42 MN $m^{3/2}$ for the 17.5 eq % N glass with a correlation coefficient of 0.90.

Electrical conductivity and dielectric loss properties depend mainly on the mobility of the network modifier ions. The work of Kenmuir *et al.* (1983) showed that while the dielectric constant always increased with nitrogen

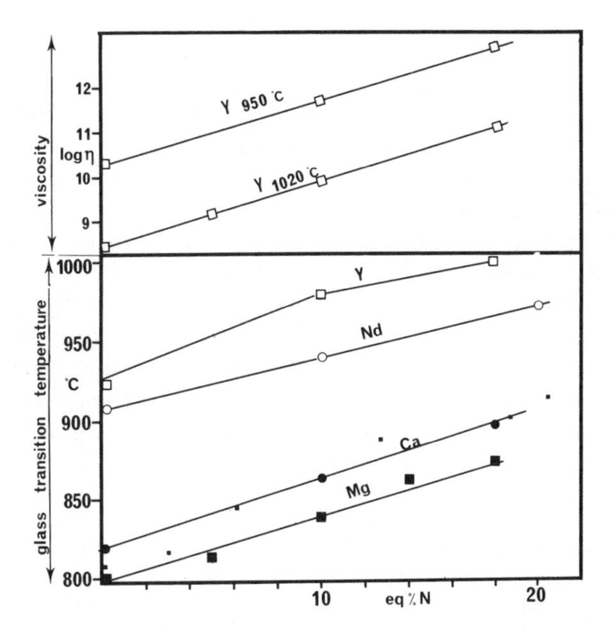

Fig. 4.11 Viscosity and glass transition temperature of M–Si–Al–O–N glasses (M = Mg,Ca,Y,Nd) (from Hampshire *et al.*, 1984).

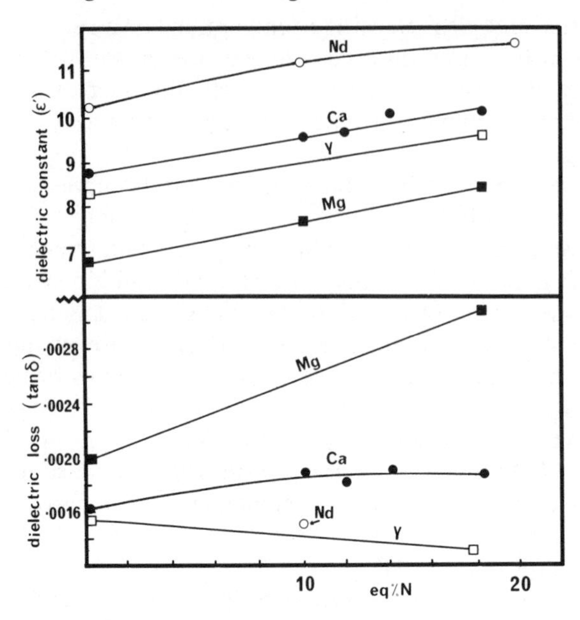

Fig. 4.12 Dielectric constant and dielectric loss of M–Si–Al–O–N glasses (M = Mg,Ca,Y,Nd) (from Kenmuir *et al.*, 1983).

content, the dielectric loss could either increase, as in Mg–Si–Al–O–N glasses, or decrease slightly in the case of Y–Si–Al–O–N glasses as evident in Fig. 4.12. At 1600 Hz values of electrical conductivity for Ca–Si–Al–O–N glasses rose from 12.8 for the oxide glass to 16.6×10^{-12} ohm^{-1} cm^{-1} for a 15 eq % N glass. Conductivity in Mg–Si–Al–O–N glasses also rose with nitrogen content but decreased slightly in Y–Si–Al–O–N glasses. Kenmuir *et al.*'s (1983) results together with earlier work by Leedecke and Loehman (1980), Thorp and Kenmuir (1981) and Drew *et al.* (1983) indicate that at room temperature and in the frequency range used (e.g. 500–10 000 Hz) dielectric polarization and a.c. conductivity results from a 'hopping' mechanism and that this is not changed by the substitution of nitrogen. No specific role has been attributed to nitrogen in determining the electrical properties and, as Thorp *et al.* (1984) point out, impurity effects cannot be ruled out as a factor in determining these properties.

4.5.2 Chemical properties

(a) Solubility in water

The increased cross-linking in the structure of an oxynitride glass would be expected to lead to improvements in chemical durability over the parent

Table 4.4 Dissolution rates of $Ca_2Na_3P_{10}(O,N)$ glasses in deionized water (30°C) (from Rajaram and Day, 1987)

Nitrogen content (eq % N)	*Dissolution rate ($g\,cm^{-2}\,min^{-1}$)*
0	1.3×10^{-4}
19	1.7×10^{-7}
25	9.2×10^{-8}

oxide glass. However this could only be expected at temperatures below the glass transition temperature where no catastrophic oxidation can take place.

This is illustrated by the improvements in the susceptibility of glasses to attack by water achieved by substituting nitrogen for water. Frishat and Sebastian (1985) found that the initial leach rate of a 4 eq % N Na_2O–CaO–SiO_2 glass in water at 60°C was 2/3 that of the equivalent oxide glass. Luping *et al.* (1985) reported that the leach rate of a Li–Si–Al–O glass in a 20% HF solution was halved by substituting 18 eq % of the oxygen with nitrogen. Improvements of over 1000× in the dissolution rates of $Ca_2Na_3P_{10}(O,N)$ and $Ba_2Na_3P_{10}(O,N)$ glasses in water with nitrogen substitution as determined by Rajaram and Day (1987) can be seen in Table 4.4.

Bunker *et al.* (1987) suggest that phosphorous oxynitride glasses dissolve via network hydrolysis where the cross-linking of the phosphate chains prevents these chains being released directly into solution as intact units. Approximately 60% of the phosphate groups may be cross-linked by nitrogen in sodium alkaline-earth phosphorous oxynitride glasses. Work on the hydrolysis of these phosphorous oxynitride glasses is prompted by the need to improve their chemical durability if practical advantage is to be taken of their high thermal expansion coefficients (refer to Fig. 4.10) and low melting temperatures (<1000°C).

It should be noted that the solubility of glasses in water can vary markedly with the modifier ion. For example pure silica is essentially insoluble in water (pH = 7) at temperatures of less than 250°C, and only very slightly soluble above this temperature. However, simple alkali oxide glasses such as $Na_2O \cdot SiO_2$ dissolve quickly at room temperature; the addition of Al_2O_3 enhances solubility while the addition of CaO can decrease it. Likewise in the phosporous oxynitride glasses of Rajaram and Day (1987) considerable variation in the water solubility with modifier ion is found.

(b) Susceptibility to oxidation

The use of oxynitride glasses in air at high temperatures is limited by their susceptibility to oxidation as illustrated in Fig. 4.13. At temperatures

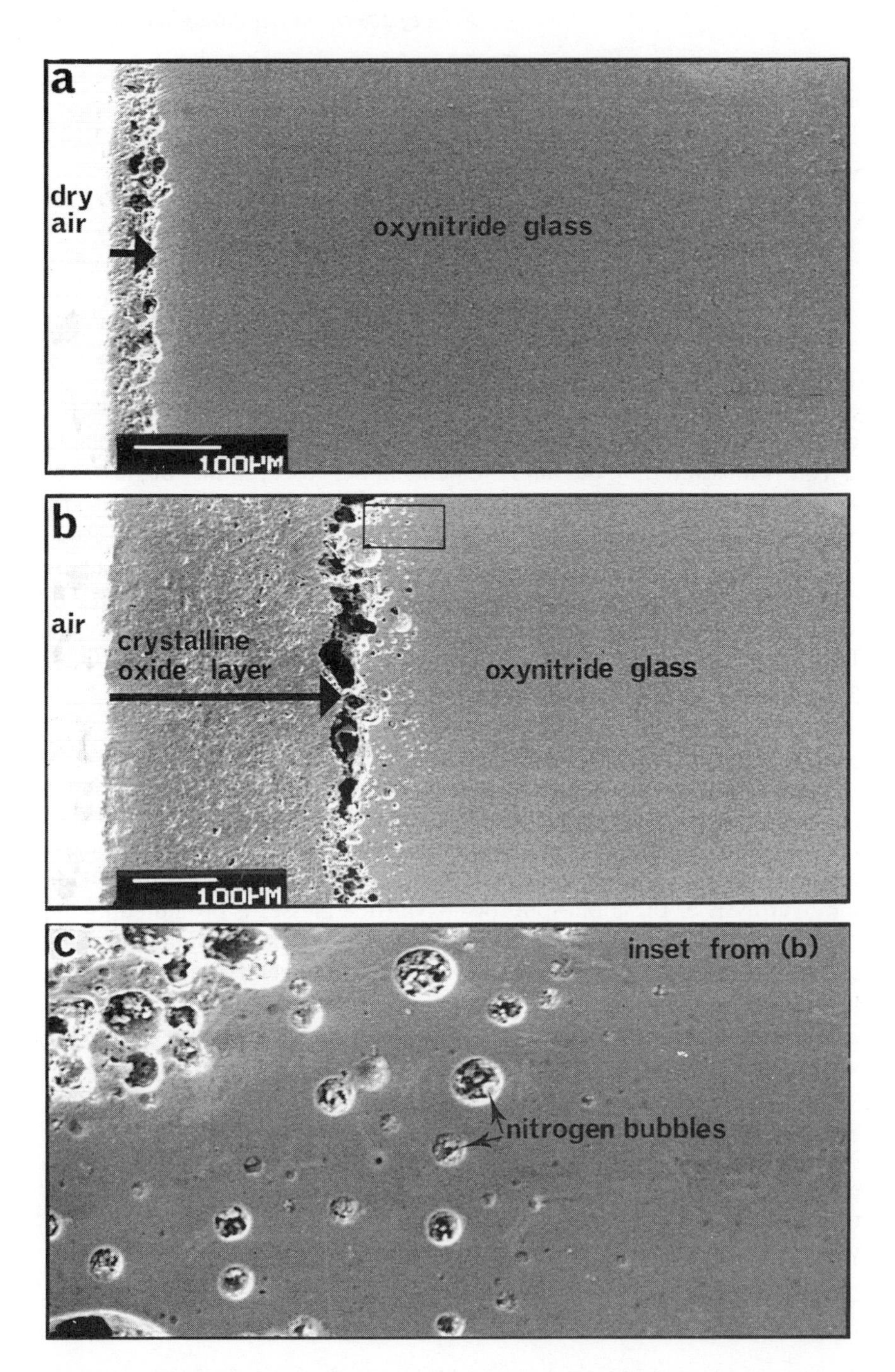

Fig. 4.13 SEM micrographs of 25 eq % N Y–Si–Al–O–N glass oxidized in air at 1100°C for 3 h: (a) dry air flowing through silica gel; (b) and (c) still atmospheric air.

appreciably above the glass transition temperature the oxidation process involves:

1. diffusion of oxygen into the glass
2. evolution of nitrogen gas, often creating enough pressure to cause foaming if the glass viscosity is low enough
3. usually crystallization of the growing oxide layer.

Wusirika (1985) surveyed the oxidation of a range of oxynitride glasses finding that aluminosilicate glasses containing modifier ions Li^+, Mg^{2+} and Mn^{2+} formed coherent, crystalline oxide layers of ~0.06–0.1 mm thickness after 24 h at 1000°C. Consequently they had a better oxide resistance than glasses containing Na and Ba which foamed on oxidizing. A Be–Si–Al–O–N glass showed no evidence of oxidation at 1000°C, a ~0.01 mm oxide layer formed after 4 h at 1300°C and a ~1.5 mm foamed layer after 2 h at 1400°C. This high stability in air is because Be acts as a glass network former tetrahedrally co-ordinated to O and N producing a tightly knit structural network with few non-bridging oxygens.

The susceptibility of silicate oxynitride glasses to oxidation is markedly affected by the presence of water vapour in the air as illustrated by the micrographs in Fig. 4.13; SEM micrographs show a Y–Si–Al–O–N glass oxidized for 3 h at 1100°C in dry air (a) and in the atmosphere (b). Water in the atmosphere may react with bridging Si–O–Si bonds at the glass surface producing ≡Si–OH groups, thus lowering the glass viscosity and facilitating the diffusion of oxygen into the glass. At 1200°C this Y–Si–Al–O–N glass oxidizes extremely rapidly leaving a bloated powdery crystalline specimen after 20 min.

To summarize, the generally linear changes in properties with nitrogen substitution for oxygen is consistent with the progressive formation of a more compact and interconnected glass network structure. The average interatomic distance of the Si–N bond of 1.74 Å in Si_3N_4 and 1.72 Å in Si_2N_2O is actually longer than that of the 1.62 Å Si–O bond. However, the local region around the nitrogen is more compact with three adjoining $(Si,Al)(O,N)_4$ groups present rather than the maximum of two such groups in the case of oxygen. The changes in physical and chemical properties on substitution of nitrogen for the oxygen is largely regarded as being beneficial, with the obvious exception to this being the susceptibility to oxidation at temperatures above the glass transition temperature.

4.6 CRYSTALLIZATION

Oxynitride glasses may be crystallized to form oxynitride glass-ceramic materials and grain-boundary oxynitride glasses in Si_3N_4-based ceramics may be crystallized to form multiphase nitride ceramics. An understanding of the compositional and heat treatment requirements which dictate the nucleation, growth kinetics and morphology of specific crystalline phases in these glasses is necessary to produce useful materials. This section surveys the current published data on oxynitride glass crystallization for a variety of systems.

4.6.1 Glass systems

(a) Y–Si–O–N and Y–Si–Al–O–N glasses

The main phases identified in the crystallization of these glasses are listed in Table 4.5. A distinguishing feature of the ternary oxide system (Y_2O_3–SiO_2–Al_2O_3) is that there is no crystalline phase with a composition within the bounds of the glass-forming region, a feature maintained on extending the composition into the Y–Si–Al–O–N system. Even the two intermediate phases of low thermal stability have compositions outside the region where glasses are easily formed.

Table 4.5 Main crystallization products of M–Si–O–N and M–Si–Al–O–N glasses (M = Y, Mg, Nd, La)

M (metal atom)	*Category*	*Crystalline phases*
Y	1	$Y_2Si_2O_7$, $Y_3Al_5O_{12}^{s}$, $Y_5Si_3O_{12}N^{s}$, $YSiO_2N^{s}$
	2	$Al_6Si_2O_{13}$, $Al_{18}Si_{12}O_{39}N_8$, $Si_2N_2O^{s}$
	3	Y_2SiAlO_5N, I_w
Mg	1	Mg_2SiO_4, $MgSiO_3$, $MgAl_2O_4^{s}$, [α-$Mg_2Si_5Al_4O_{18}$] Mg–N–petalite
	2	$Al_6Si_2O_{13}$, $Al_{18}Si_{12}O_{39}N_8$, $Si_2N_2O^{s}$
	3	[β''], [μ-$Mg_2Si_5Al_4O_{18}$]
Ln (Nd or La)	1	$Ln_2Si_2O_7$, $LnAlO_3^{s}$, $Ln_5Si_3O_{12}N^{s}$ ~[$Ln_4Si_4Al_3(O,N^{0-10})$]
	2	$LnAl_{11}O_{18}$, $Al_6Si_2O_{13}$, $Al_{18}Si_{12}O_{39}N_8$, $Si_2N_2O^{s}$
	3	~[$Ln_4Si_3Al_3(O,N^{20})$]

NB: The following points apply to Table 4.5:

(i) Category (1) phases are 'high M atom' content phases, (2) are 'low or zero M' content phases and (3) are intermediate phases of limited thermal stability.

(ii) The binary oxide phases Al_2O_3 and SiO_2 are not included in the table.

(iii) Superscript 's' indicates that a small amount of substitution of N for O (or vice versa) and Si for Al (or vice versa) may occur in this formula.

(iv) '[ ]' bracketing indicates phases with compositions within, or very close to, glass-forming regions.

(v) '( )' bracketing indicates that the oxygen/nitrogen composition has not been determined accurately; however, the N superscript is an estimate in units of 'eq % N' of the nitrogen content or composition range.

(vi) '~' indicates approximate composition based on TEM EDX analysis only.

The crystallization of Y–Si–Al–O–N glasses is illustrated by Leng-Ward and Lewis's (1985) work on a series of glasses of constant cation ratio, $Y_{1.04}Si_{1.27}Al_{1.27}$, and of increasing substitution of N for O at nitrogen levels of 0, 10, 20 and 30 eq % N. At an annealing temperature of 1250°C, high enough to nucleate the stable phases in categories 1 and 2 of Table 4.5, the base oxide glass crystallized into yttrium disilicate ($Y_2Si_2O_7$), Al_2O_3 and

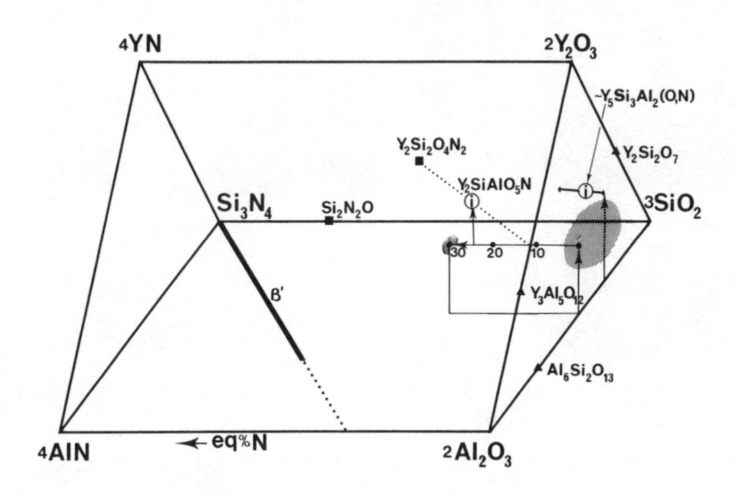

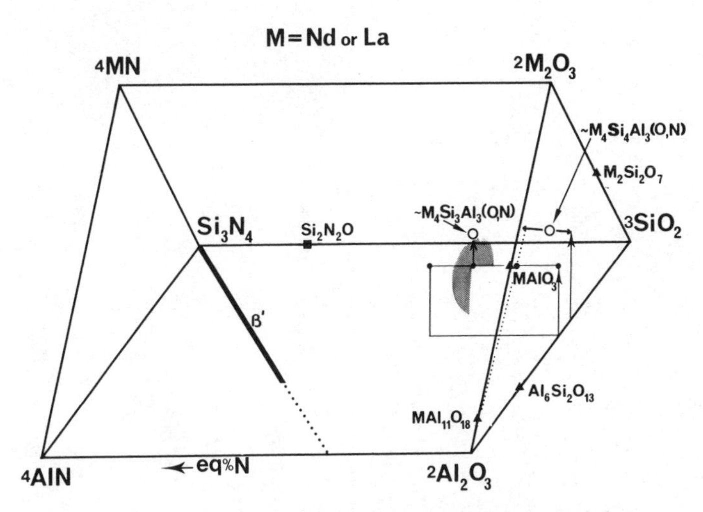

Fig. 4.14 (a) Y–Si–Al–O–N and (b) M–Si–Al–O–N (M = Nd,La) Janecke prisms showing main crystallizing phases. The compositions marked 0, 10, 20 and 30 along the top of the rectangle are glass-forming compositions in units of eq % N. Shading schematically indicates glass-forming regions which are broadly similar for both systems.

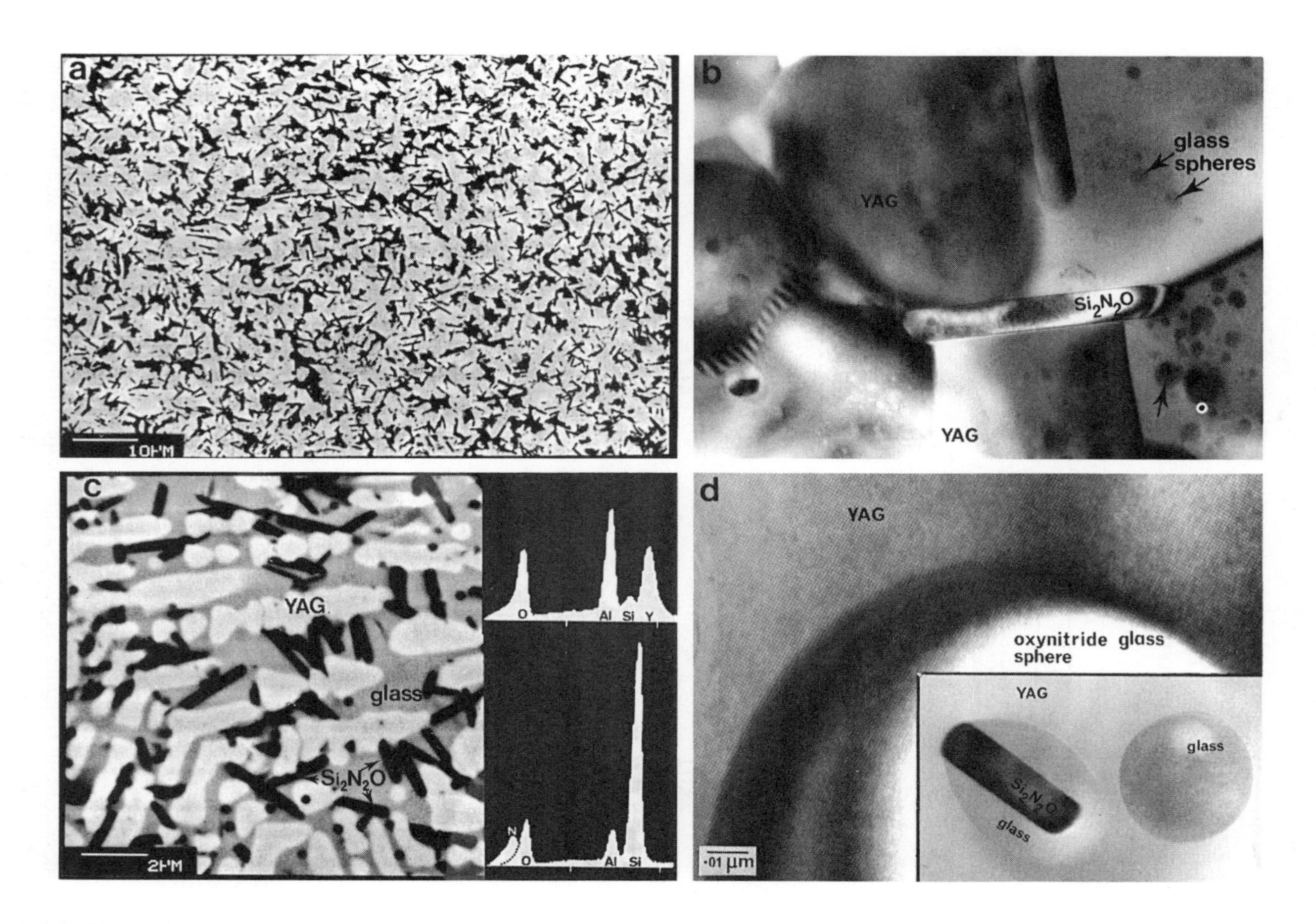

a
10μM
b
YAG
glass
spheres
Si_2N_2O
YAG
c
YAG
glass
Si_2N_2O
2μM
O
Al
Si
Y
N
O
Al
Si
d
YAG
oxynitride glass
sphere
YAG
glass
Si_2N_2O
glass
·01 μm

mullite ($Al_6Si_2O_{13}$). On substituting 30 eq % of the oxygen by nitrogen the resulting oxynitride glass crystallized at 1250°C into yttrium aluminium garnet ($Y_3Al_5O_{12}$) and Si_2N_2O with minor amounts of $Y_2Si_2O_7$, $Al_6Si_2O_{13}$ and $YSiO_2N$. This is consistent with this high nitrogen glass composition lying close to the $Y_3Al_5O_{12}$–Si_2N_2O tie line as shown in Fig. 4.14.

Melt compositions exactly on the $Y_3Al_5O_{12}$–Si_2N_2O tie line are very difficult to prepare because they are on the limit of the glass-forming region. The microstructure of a close to two phase $Y_3Al_5O_{12}$–Si_2N_2O glass-ceramic is shown in Fig. 4.15. The growth morphology of the $Y_3Al_5O_{12}$ results in the entrapment of glass between adjacent cellular projections of growing $Y_3Al_5O_{12}$ crystals. This glass becomes redistributed throughout the $Y_3Al_5O_{12}$ in the form of a multitude of small glass spheres. Figure 4.15(c) shows a lattice fringe image of cubic $Y_3Al_5O_{12}$ near a $Y_3Al_5O_{12}$–glassy sphere interface with the inset illustrating how occasionally a Si_2N_2O crystal nucleates within one of these spheres at the sphere–$Y_3Al_5O_{12}$ interface.

At relatively low annealing temperatures of ~950–1100°C the nucleation and growth of N-wollastonite ($YSiO_2N$) and the intermediate phases Y_2SiAlO_5N and I_w is kinetically favoured over that of the stable equilibrium phases $Y_3Al_5O_{12}$ and Si_2N_2O. This is illustrated in Fig. 4.16(a) by the crystallization of I_w and Al_2O_3 from a 10 eq % N glass annealed at 1100°C. TEM 'energy dispersive X-ray analysis' (EDX) of the phase I_w gave an average composition close to $Y_5Si_3Al_2[O,N^{0-10}]$ which appears consistent with Thomas *et al.*'s (1982) 'energy dispersive X-ray' (EDX) and 'energy loss spectral' (ELS) analyses of a similar phase. The small amount of nitrogen in this phase is difficult to quantify analytically by ELS analysis. However, a ^{29}Si MAS NMR spectrum (Fig. 4.16(b)) of phase I_w shows specifically three distinct $Si(O_4)$ peaks together with two smaller peaks which are consistent with being $Si(O_3N)$ and $Si(O_2N_2)$ peaks (Leng-Ward and Smith, private communication).

The simultaneous crystallization of $YSiO_2N$, Y_2SiAlO_5N and I_w in the same glass annealed at 1100°C and at 1000°C illustrates the proliferation of intermediate phases of varying stability that can occur in Al/Si/O/N systems (see XRD trace in Fig. 4.17). Although all three phases can be indexed on very similar hexagonal unit cells there is no solid solution between them. The Y_2SiAlO_5N phase is isostructural with $YSiO_2N$ involving the substitution of one Si by Al and one N by O resulting in a lowering of the

---

Figure 4.15 (a) SEM and (b) TEM micrographs of $Y_3Al_5O_{12}$–Si_2N_2O glass-ceramic, (c) SEM (bs) micrograph showing early stages of crystallization and EDX (windowless) analyses of $Y_3Al_5O_{12}$ and Si_2N_2O and (d) TEM micrograph of the $Y_3Al_5O_{12}$–glass interface of one of the oxynitride glass spheres trapped inside a $Y_3Al_5O_{12}$ crystal showing cubic $Y_3Al_5O_{12}$ lattice fringes and inset the growth of a Si_2N_2O crystal within one of these spheres.

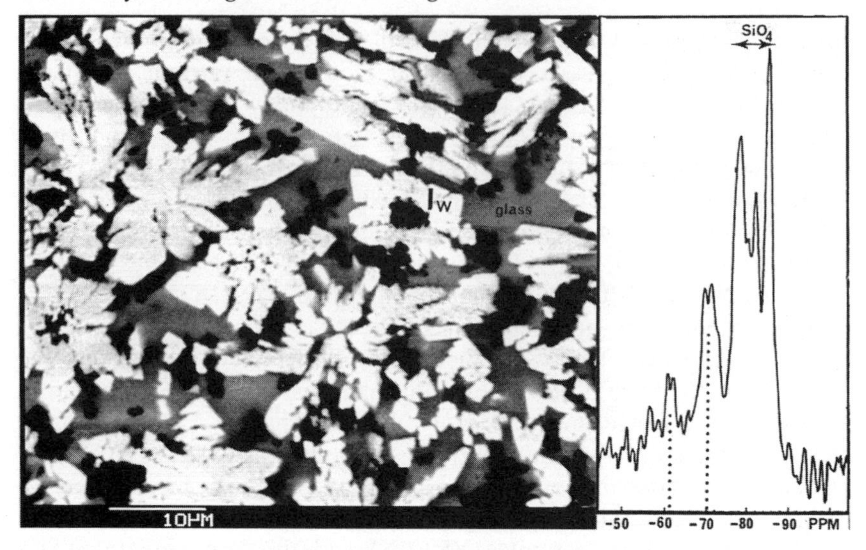

Fig. 4.16 (a) SEM micrograph of intermediate phase I_w and Al_2O_3 crystals growing in a 10 eq % N Y–Si–Al–O–N glass after annealing at 1100°C and (b) ^{29}Si NMR of phase I_w.

hexagonal *a*-dimension, an increase in the *c*-dimension and a decrease in thermal stability. Phase I_w exists over a range of composition from a pure oxide through to a nitrogen content of ~10 eq % N with its thermal stability at a maximum in the central part of this N composition range.

At much lower yttrium contents than used by Lewis and Leng-Ward and at moderate nitrogen contents X-phase ($Si_{12}Al_{18}O_{33}N_9$) crystallized as the main phase in Y–Si–Al–O–N glasses prepared by Wusirika and Chyung (1980) in their survey of a range of oxynitride glass-ceramic systems. At these low modifier ion contents the melts are very viscous and Wusirika and Chyung used AlF_3 additions as a fining agent.

Summarizing $Y_3Al_5O_{12}$, $Y_2Si_2O_7$, $Y_5Si_3O_{12}N$ and Si_2N_2O are the dominant stable phases reported crystallizing from Y–Si–Al–O–N glasses together with X-phase and mullite in low Y glasses.

(b) Mg–Si–O–N and Mg–Si–Al–O–N glasses

The main phases identified in the crystallization of these glasses are shown in Table 4.5. In contrast to the Y_2O_3–SiO_2–Al_2O_3 system the ternary oxide MgO–SiO_2–Al_2O_3 system does contain a stable equilibrium phase with a composition within the glass-forming region, cordierite ($Mg_2Al_5Si_5O_{18}$), the basis of many commercial glass-ceramics.

Tredway and Risbud (1985a) investigated the crystallization of a Mg–Si–Al–O–N glass into cordierite with the nitrogen being concentrated in the

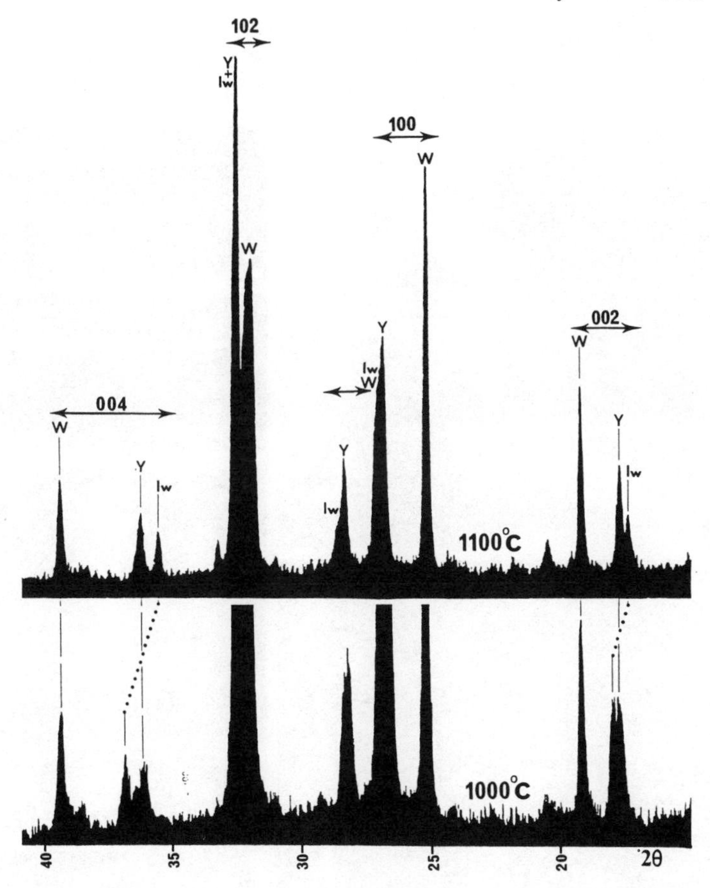

Fig. 4.17 X-ray diffraction traces of a Y–Si–Al–O–N glass after annealing at 1100°C and at 1000°C showing the simultaneous crystallization of $YSiO_2N$(W), Y_2SiAlO_5N(Y) and I_w.

residual glass. They concluded that a nucleating agent (e.g. ZrO_2) was necessary for bulk nucleation of cordierite as it is in silicate ceramics where TiO_2 is usually used. Differential thermal analysis by Jameel and Thompson (1986) showed that the resistance of cordierite glasses towards crystallization is greatly increased by the presence of nitrogen, with a ~13 eq % N 'cordierite-Si_3N_4' glass showing a very broad weak crystallization exotherm centred about 1090°C whereas the pure cordierite glass gave a distinct exotherm at 1037°C. This 13 eq % N glass crystallized into mixtures of cordierite, X-phase and Mg–N-petalite on prolonged annealing at 1050°C. Cordierite and TiN were the crystalline products when such a glass containing 7 wt % TiO_2 was annealed under the same

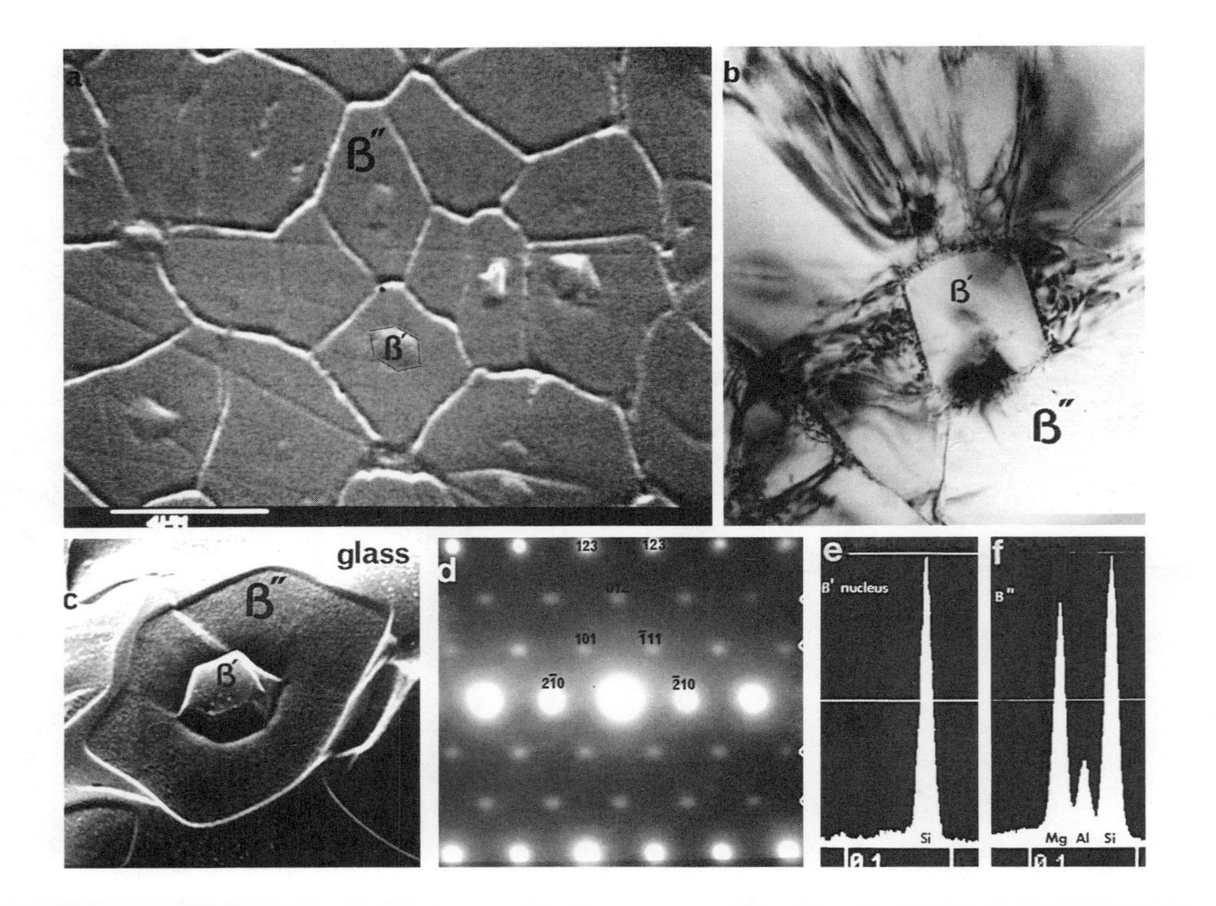
a
β″
β′
b
β′
β″
c
glass
β″
β′
d
101
e
β′ nucleus
Si
f
β″
Mg Al Si

conditions with there being no evidence that the TiN assisted the nucleation of the cordierite.

The magnesium analogue of petalite ($LiAlSi_4O_{10}$) sometimes occurs in microcrystalline form as a metastable crystallization product together with cordierite (Shreyer and Schairer, 1962). A Mg–N-petalite phase termed N-phase has been reported by Hampshire *et al.* (1985) as a crystallization product in Mg–Si–Al–O–N glasses. Leng-Ward *et al.* (1986) found tabular crystals of such a phase together with forsterite (Mg_2SiO_4) in a slow cooled Mg–Si–Al–O–N melt. EDX analyses of these Mg–N-petalite crystals gave an analysed composition of $Mg_{0.6}AlSi_2O_{3.4}N_{1.7}$ indicating a composition outside the nitrogen-rich end of the glass-forming region marked by 'N' in Fig. 4.1(a). Although there is a range in the unit cell dimensions of N-phase, indicating a range of composition, there is no evidence for a range of composition extending from a magnesium oxide petalite through the glass-forming region to the oxynitride petalite.

Although there are no stable oxynitride phases with compositions within the glass-forming region, Mg–Si–Al–O–N glasses with a cation/anion ratio of 3/4 may be crystallized at <1150°C into a high Mg content metastable β'' phase, a solid solution of Mg in β'-sialon (Wild *et al.*, 1981, 1984). β'' crystals nucleate epitaxially on β-Si_3N_4 or β'-sialon crystals that have precipitated from the melt on cooling as shown in the micrographs of Fig. 4.18. At temperatures above 1150°C the β'' phase transforms into mainly forsterite and Mg–N-petalite. Low Mg content β'' crystals with compositions outside the glass-forming region may form together with forsterite in a duplex microstructure during the slow cooling of Mg–Si–Al–O–N 3M/4X melts (Leng-Ward *et al.*, 1986); however, with too slow a cooling rate the more stable N-petalite phase forms in preference to β''.

Summarizing Mg_2SiO_4, $MgAl_2O_4$, $Mg_2Si_5Al_4O_{18}$ and the oxynitride phases Si_2N_2O and Mg–N-petalite are the dominant phases reported crystallizing from Mg–Si–Al–O–N glasses. The lack of crystalline quaternary Mg–Si–O–N phases means that Mg–Si–O–N glasses crystallize into enstatite and/or forsterite together with Si_2N_2O.

(c) Ln–Si–Al–O–N glasses (Ln = Nd or La)

Crystallization in these two systems has not been extensively reported. The glass-forming regions in these two systems appear broadly similar

---

Fig. 4.18 (a) SEM (dilute HCl etch) and (b) TEM micrographs of a 3M/4X Mg–Si–Al–O–N glass crystallized into β'' on annealing at 980°C; (c) SEM micrograph of the fracture surface of such a partially crystallized glass (dilute HF etch). The β' nuclei are clearly visible in (a), (b) and (c). The β'' $(12\bar{1})$ electron diffraction pattern is shown in (d) and the EDX (TEM) analyses of a β' nuclei and of a β'' crystal shown respectively in (e) and (f).

to that for the Y–Si–Al–O–N system and the Y–Si–O–N system. The disilicate, metasilicate, N-apatite and N-wollastonite phases also exist in the Ln–Si–O–N systems. There are considerable differences in the nature of the phases crystallizing in these Ln–Si–Al–O–N glasses due to the existence of two Ln-oxynitride phases stable up to 1300°C + and the perovskite ($LnAlO_3$) and β-alumina ($LnAl_{11}O_{18}$) phases of the Ln_2O_3–Al_2O_3 systems.

In Ln–Si–Al–O–N glasses of around ~20 eq % N a phase of analysed composition $Ln_4Si_3Al_3[O,N^{20}]$ (Ln = Nd or La) with hexagonal symmetry crystallizes (Lewis and Leng-Ward, 1988). This is an intermediate phase which is stable to relatively high temperatures of ~1350°C, above which transformation into stable equilibrium phases occurs, one of which is known to be $LnAlO_3$. This cubic phase composition is centrally situated in the glass-forming region with regard to Al/Si content and close to the upper glass-forming boundary with regard to modifier ion content (i.e. Nd^{3+} or La^{3+}). Near monophase glass-ceramics of this phase can be fabricated and its composition makes it well suited to forming multiphase glass-ceramics in combination with phases of 'low or zero' modifier content such as X-phase, $LaAl_{11}O_{18}$, Si_2N_2O, $BaSi_2Al_2O_8$ and also to form an intergranular phase in β'-sialon composite ceramics. Another Ln–Si–Al–O–N phase of approximate composition $Ln_4Si_4Al_3[O,N^{0-10}]$ crystallizes at lower nitrogen levels, as do the respective Ln–N apatite and $Ln_2Si_2O_7$ phases in Ln–Si–O–N glasses.

Figure 4.19 illustrates the microstructure in (a) of a Nd–Si–Al–O–N glass that has crystallized into $Nd_4Si_4Al_3[O,N^{0-10}]$, $NdAl_{11}O_{18}$ and Si_2N_2O and in (b) a La–Ba–Si–Al–O–N glass that has crystallized into a mixture of $La_4Si_3Al_3[O,N^{20}]$ and $BaSi_2Al_2O_8$.

(d) Other systems

In the Ca–Si–Al–O–N system Hampshire *et al.* (1985) have reported that there is a substantial range of nitrogen solubility in the gehlenite structure according to the formula $Ca_2(SiAl)_3(O,N)_7$. Consequently nitrogen–gehlenite tends to dominate as the main crystalline product in Ca–Si–Al–O–N glasses.

Reports of the crystallization of glasses in a number of other oxynitride systems has been published for 'occasional' glass compositions but not systematic studies exploring a range of possible glass-forming compositions. Most of these reports are of oxide phases such as barium feldspar ($BaSi_2Al_2O_8$), lithium disilicate ($Li_2Si_2O_5$) and β-eucryptite ($LiSi$–AlO_4). Fischer *et al.* (1985) have reported X-ray diffraction evidence for some substition of nitrogen for oxygen in a β-eucryptite ($LiSiAlO_4$) phase of an oxynitride glass-ceramic.

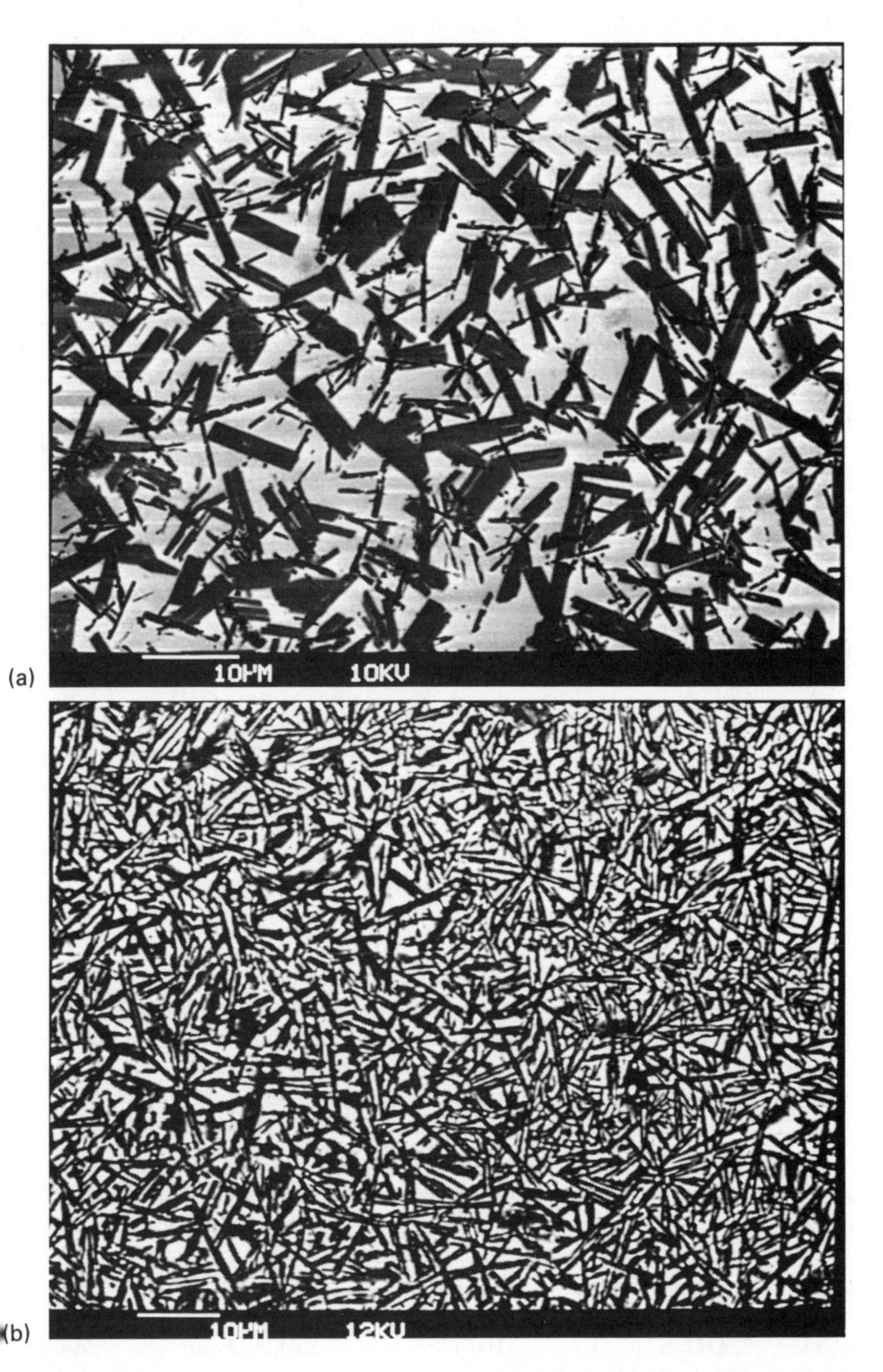

Fig. 4.19 SEM micrographs of (a) a $Nd_4Si_4Al_3(O,N)$–$NdAl_{11}O_{18}$–Si_2N_2O glass-ceramic and (b) a La–Ba–Si–Al–O–N glass-ceramic, both after crystallization at 1250°C.

The oxide phase barium feldspar ($BaSi_2Al_2O_8$) melts at 1750°C and is the most refractory compound that readily forms a glass. Ba–Si–Al–O–N glasses based on this composition have been prepared and crystallized by Tredway and Risbud (1983b). Annealing over a range of temperatures between 1000 and 1400°C these glasses were crystallized into the hexacelsian form of barium feldspar (60–70 vol %) with the nitrogen being concentrated into the residual glass. The crystal morphology varied from interlocking lath-like formations on annealing at 1020°C to blocky rectangular crystals on annealing at 1300°C and above. A study of SiC fibre reinforced glass-ceramics using Ba–Si–Al–O–N glass as the matrix led to the crystallization of a solid solution of cymrite ($BaSi_2Al_2O_8(OH)$) and hexagonal celsian ($BaSi_2Al_2O_8$) (Herron and Risbud, 1986).

Phases reported by Luping *et al.* (1984) crystallizing from Li–Si–Al–O–N glasses are confined to oxide phases which are found in $Li_2O–Al_2O_3–SiO_2$ glass-ceramics. In neither the Li–Si–Al–O–N nor Ba–Si–Al–O–N systems have new oxynitride phases been reported crystallizing from glasses.

4.6.2 Microstructural control

(a) Glass-ceramic materials

The requirement for high mechanical strength in glass-ceramics is favoured by a fine grained microstructure of small interlocking crystals. Annealing schedules used for silicate glass-ceramics (McMillan, 1975, Chapter 4) which involve a nucleation step in the temperature range of maximum nucleation density followed by a holding at higher temperatures for increased crystal growth rate generally apply to oxynitride glass ceramics. Many of the nucleating agents found necessary to crystallize silicate glass-ceramics such as TiO_2, Pt, SnO_2 and P_2O_5 are not stable in oxynitride melts under highly reducing conditions as discussed in section 4.2.1. However, the majority of oxynitride glasses have so far been found to be self-nucleating. Exceptions to this are the nucleation of some oxide phases such as cordierite and for the special nucleation requirements of the metastable β'' phase in the Mg–Si–Al–O–N system.

Some silicate glass ceramics are initially crystallized into metastable phases at relatively low temperatures and then transformed into more stable phases at higher temperatures in order to generate a particularly fine microstructure. There are no documented examples of this as yet for oxynitride glasses. In the Y–Si–Al–O–N system $Y_3Al_5O_{12}–Si_2N_2O$ glass-ceramics can only be initially partially crystallized into the intermediate phase $Y_2SiAlO_5N + Al_2O_3$ at 1100°C and there are no advantages in doing this prior to nucleating $Y_3Al_5O_{12}$ crystals at above ~1175°C. Fluoride additions to silicate glasses lower the viscosity and increase diffusion rates.

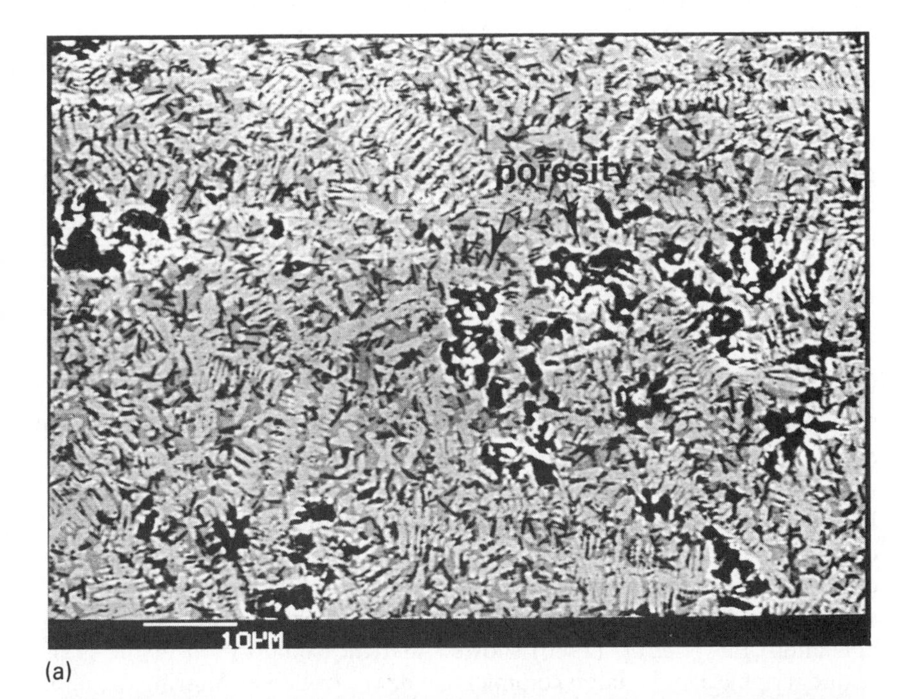

(a)

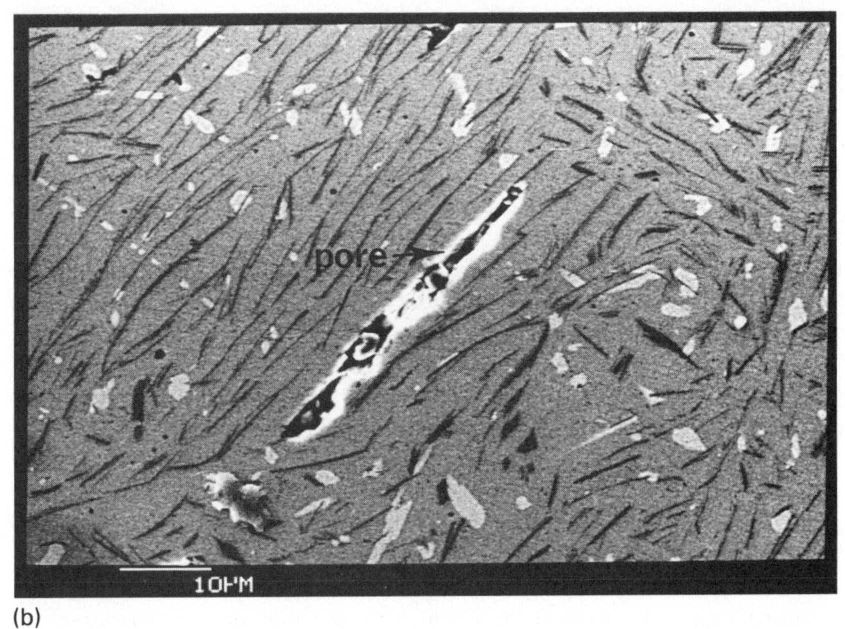

(b)

Fig. 4.20 SEM micrographs illustrating the development of porosity during annealing of (a) $Y_3Al_5O_{12}$–Si_2N_2O and (b) $La_{1.5}Si_2Al_1[O,N^{0-10}]$–$LaAl_{11}O_{18}$ glass ceramic materials.

This is liable to lead to lower nucleation densities and lessens the likelihood of diffusion controlled preference for intermediate phases.

Too-rapid crystal growth rates may give rise to high stresses in the material as phases of differing density and phases with high aspect ratio morphologies grow together from the glass. Examples of this giving rise to porosity in the microstructure are shown for the $Y_3Al_5O_{12}$–Si_2N_2O and $La_4Si_4Al_3$-$[O,N^{0-10}]$–$LaAl_{11}O_{33}$ glass-ceramics in Fig. 4.20 (Leng-Ward and Lewis, 1988b).

(b) Oxynitride grain-boundary glass crystallization

The crystallization of residual intergranular oxynitride glass boundaries in β-Si_3N_4 and sialon ceramics is essential if good mechanical properties are to be maintained to high temperatures above the glass softening point. In systems where the appropriate phase relations are known the bulk composition may be chosen so as to control the amount and composition of the resultant residual glass, and hence, to control the nature of the oxide and oxynitride phases that will crystallize in this glass on post fabrication annealing. Lewis *et al.* (1980) showed that the composition of the glass boundary phase in β'-sialon ceramics can be controlled by altering the bulk composition through AlN polytype additions so that the residual glass can be crystallized into $Y_3Al_5O_{12}$. This correlates with the nucleation and growth of $Y_3Al_5O_{12}$ in bulk Al/N rich Y–Si–Al–O–N glasses as discussed in section 4.6.1(a).

Bonnel *et al.* (1987) demonstrated the level of strict compositional control needed to predetermine the nature of the grain boundary crystallization products in hot-pressed β'-sialon ceramics. The relative amounts of sintering additive, Al_2O_3, AlN and SiO_2 added to Si_3N_4, were carefully tailored to generate a glass boundary phase after hot pressing that could be 'completely' crystallized into a microcrystalline cordierite microstructure when an MgO sintering oxide was used or $Y_3Al_5O_{12}$ when Y_2O_3 was used. The 'completeness' of crystallization needs to be qualified by the still probable existence in these materials of very fine intergranular amorphous layers of ~2 nm usually observable by high resolution lattice imaging techniques.

The choice of crystalline grain boundary phases in a 'β' + secondary phase' ceramic is limited to category 1 and [category 3] phases in Table 4.5 for the Mg–Si–Al–O–N, Y–Si–Al–O–N, and Ln–Si–Al–O–N systems. Factors important in deciding which of these phases are better suited as intergranular phases in order to maximize the high temperature strength, creep and fracture toughness of silicon nitride ceramics are their oxidation behaviour, thermal expansion mismatch with the β' matrix, nucleation density and morphology. For example, oxynitride phases such as

N-wollastonite which exhibit a large volume change on oxidation must be avoided.

4.6.3 Material properties

Very little published information on the properties of oxynitride glass-ceramic materials exists other than some microhardness (Messier, 1982), MOR and coefficient of expansion values (Wusirika and Chyung, 1980) on assorted crystallized glasses.

A comparison of the room temperature flexural strength and microhardness of an essentially $Y_3Al_5O_{12}$–Si_2N_2O glass-ceramic compared with a range of other silicate glass-ceramics is listed in Table 4.6 (Leng-Ward and Lewis, 1988). The extreme hardness of the material is derived from the hardness of the main two component phases, in particular the garnet $Y_3Al_5O_{12}$ phase which makes up ~75–85 vol % of the material. The strength-limiting flaws in this high strength glass-ceramic are the pores that develop during crystallization (see Fig. 4.20), and the ~10–15 μm size of the larger $Y_3Al_5O_{12}$ crystals.

Oxidation studies in air on this $Y_3Al_5O_{12}$–Si_2N_2O glass-ceramic carried out for 500 h at 1100°C showed no evidence of any oxidized layer. By comparison the parent Y–Si–Al–O–N glass is shown rapidly oxidizing after 3 h under identical conditions in Fig. 4.13(b). While the parent glass oxidizes catastrophically at 1200°C, in the derived glass-ceramic an oxide layer very slowly and progressively grows inwards from the glass-ceramic surface resulting in a ~500 μm layer after 20 days as shown in Fig. 4.21(a) (Leng-Ward and Lewis, 1988). The volume expansion on oxidation has cracked the specimen bars in from the corner edges at ~45° after 40 days

Table 4.6 Room temperature strength and microhardness of glass-ceramic materials

Glass-ceramic system	*Nucleating agent*	*MOR 3 point bend* (MPa)	*Microhardness Knoop* (kg mm^{-2})
Li_2O–SiO_2	–	30–50	
Li_2O–SiO_2	P_2O_5	110–398	
Li_2O–ZnO–SiO_2	P_2O_5	176–340	
Li_2O–Al_2O_3–SiO_2	TiO_2	112–122	
MgO–Al_2O_3–SiO_2	TiO_2	119–259	
ZnO–MgO–Al_2O_3–SiO_2	ZrO_2	69–103	
MgO–Al_2O_3–SiO_2	TiO_2	250	~700
MgO–Al_2O_3–SiO_2/SiC*	–	600	
Y_2O_3–Al_2O_3–SiO_2 ($Y_3Al_5O_{12}$–Si_2N_2O)	–	350–430	~1900

* SiC fibre reinforced glass-ceramic (Herron and Risbud, 1986).

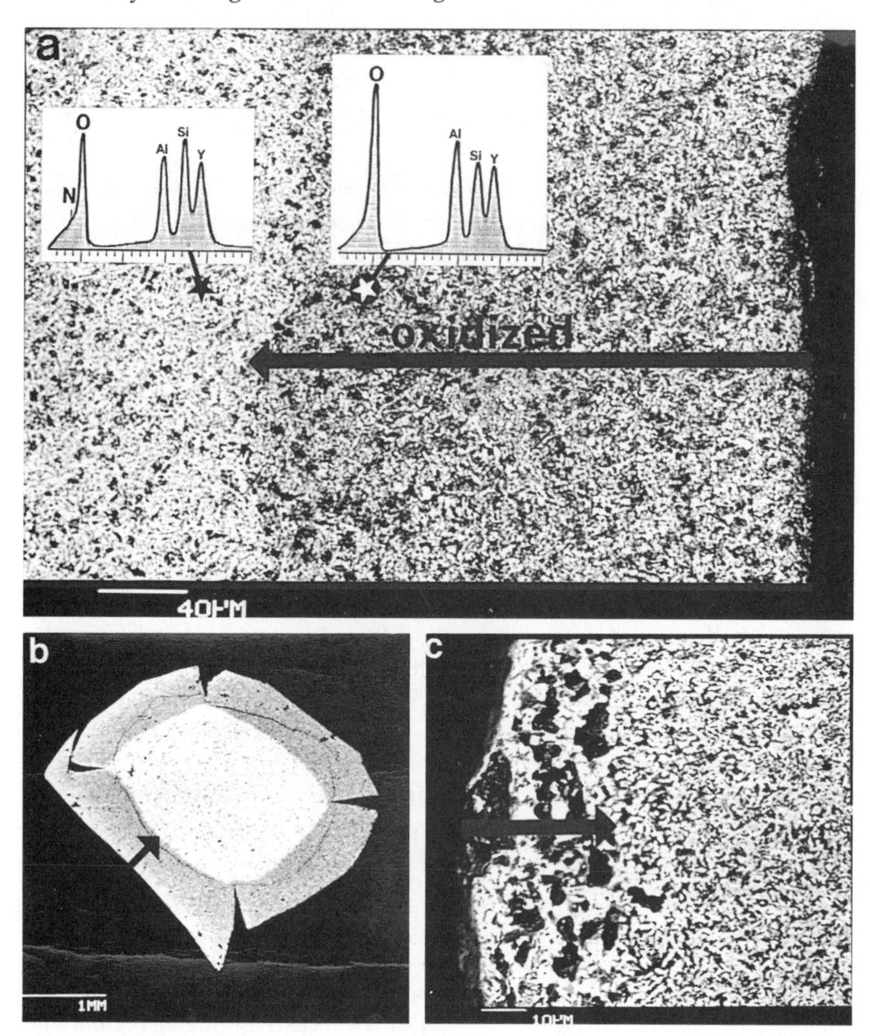

Fig. 4.21 SEM(bs) micrographs showing the oxidation of a $Y_3Al_5O_{12}$–Si_2N_2O glass-ceramic (a) after 20 days at 1200°C with EDX(SEM) windowless analyses either side of the oxidation front, (b) at low magnification after 40 days at 1200°C, and
(c) the oxidation fron after 3 h at 1300°C.

(Fig. 4.21(b)). Figure 4.21(c) shows the development of porosity with the much more rapid oxidation at 1300°C. Oxidation is accompanied by recrystallization into mainly $Y_2Si_2O_7$ and $Al_6Si_2O_{13}$. So despite the individual refractory nature of the main two phases $Y_3Al_5O_{12}$ and Si_2N_2O and the very low residual glass content this oxynitride glass-ceramic is not suitable for high temperature applications.

In formulating glass-ceramic materials the compositional requirements of the parent glass limit the number of available crystalline phases. However, the extension of glass-forming composition regions on substituting nitrogen for oxygen and the accessibility of crystalline phase assemblages not available in pure silicate glass-ceramic systems provides the incentive for investigating oxynitride glass-ceramics. The formation of an essentially $Y_3Al_5O_{12}$–Si_2N_2O glass-ceramic illustrates how the oxynitride glass-ceramic route can make available high proportions of an aluminate phase that is normally only present to a small degree in the oxide system. Crystalline phases with interesting electrical properties such as perovskite and β-alumina may be incorporated into oxynitride glass-ceramics as can be other as-yet undiscovered phases. However, to be useful, oxynitride glass-ceramics must have unique or markedly improved properties over silicate glass-ceramics to justify the environmental constraints imposed during preparation and high temperature application.

REFERENCES

Aujla, R. S., Leng-Ward, G., Lewis, M. H. *et al.* (1986) *Phil. Mag. B*, **54**, L51–L56.
Bagaasen, L. M. and Risbud, S. H. (1983) *J. Am. Ceram. Soc.*, **66**, C69–C71.
Baik, S. and Raj, R. (1985) *J. Am. Ceram. Soc.*, **68**, C168–C170.
Baik, S. and Raj, R. (1987) *J. Am. Ceram. Soc.*, **70**, C105–C107.
Ball, R., Lewis, M. H., Szweda, A. and Butler, E. (1985) *Mater. Sci. Eng.*, **71**, 137–45.
Bonnell, D. A., Tein, T. Y. and Ruhle, M. (1987) *J. Am. Ceram. Soc.*, **70**, 460–5.
Brinker, C. J. and Haaland, D. M. (1983) *J. Am. Ceram. Soc.*, **66**, 758–65.
Brinker, C. J., Haaland, D. and Loehman, R. J. (1983) *J. Non-Cryst. Solids*, **56**, 179–83.
Brow, R. K. and Pantano, C. G. (1984) *J. Am. Ceram. Soc.*, **67**, C72–C74.
Brow, R. K. and Pantano, C. G. (1987) *J. Am. Ceram. Soc.*, **70**, 9–14.
Bunker, B. C., Arnold, G. W., Rajaram, M. and Day, D. E. (1987) *J. Am. Ceram. Soc.*, **70**, 425–30.
Bunker, B. C., Tallant, D. R., Balfe, C. A. *et al.* (1987) *J. Am. Ceram. Soc.*, **70**, 675–81.
Chu, T. L., Szedon, J. R. and Lee, C. H. (1968) *J. Electrochem. Soc.*, **115**, 318–22.
Chyung, K. and Wusirika, R. R. (1978) *SiO_2=Al_2O_3–N Glass For Production of Oxynitride Glass-Ceramics*, US patent 4 070 198, January 24.
Coon, D. N., Rapp, J. C., Bradt, R. C. and Pantano, C. G. (1983) *J. Non-Cryst. Solids*, **56**, 161–6.
Dancy, E. A. and Janssen, D. (1976) *Can. Metall. Q.*, **15**, 103–10.
Davies, M. W. and Meherahli, S. G. (1971) *Metall. Trans.*, **2**, 2729–33.
Drew, R. A. L., Hampshire, S. and Jack, K. H. (1983) in *Progress in Nitrogen Glasses* (ed. F. L. Riley), Noordhoff International Publishers, pp. 323–6.
Dupree, R., Lewis, M. H., Leng-Ward, G. and Williams, D. S. (1985) *J. Mater. Sci. Lett.*, **4**, 393.
Elcolivet, C. and Verdier, P. (1984) *Mater. Res. Bull.*, **19**, 227–31.
Elmer, T. H. and Norberg, M. E. (1967) *J. Am. Ceram. Soc.*, **50**, 275–9.
Fischer, G. R., Wusirika, R. R. and Geiger, J. E. (1985) *J. Mater Sci.*, **20**, 4117–22.

Frishat, G. H., Krause, W. and Heubenthal, H. (1984) *J. Am. Ceram. Soc.*, **67**, C10–C12.
Frishat, G. H. and Scrimpf, C. (1980) *J. Am. Ceram. Soc.*, **63**, 714–5.
Frishat, G. H. and Scrimpf, C. (1983) *J. Non-Cryst. Solids*, **56**, 153–9.
Frishat, G. H. and Sebastian, K. (1985) *J. Am. Ceram. Soc.*, **68**, C305–C307.
Hampshire, S., Drew, R. A. L. and Jack, K. H. (1984) *J. Am. Ceram. Soc.*, **67**, C46–C47.
Hampshire, S., Drew, R. A. L. and Jack, K. H. (1985) *Phys. Chem. Glasses*, **26**, 182.
Harding, F. L. and Ryder, R. J. (1970) *Glass Technol.*, **11**, 54.
Herron, M. A. and Risbud, S. H. (1986) *Ceram. Bull.* **65**, 342–6.
Homeny, J. and McGarry, D. L. (1984) *J. Am. Ceram. Soc.*, **67**, C225.
Jack, K. H. (1977) in *Nitrogen Ceramics* (ed. F. L. Riley), Noordhoff International Publishers, pp. 257–62.
Jameel, N. S. and Thompson, D. P. (1986) *Special Ceramics*, **8**, Proc. Brit. Ceram. Soc., pp. 95–108.
Jankowski, P. E. and Risbud, S. H. (1983) *J. Mater. Sci.*, **18**, 2087–94.
Kenmuir, S. V. J., Thorpe, J. S. and Kulesza, B. L. (1983) *J. Mater. Sci.*, **18**, 1725.
Lang, J., Marchand, R., Verdier, P. and Pasturnak, R. (1982) *Mater. Sci. Monogr.*, **10**, [React. Solids, Vol.2], 506–11.
Larson, R. W. and Day, D. E. (1986) *J. Non-Cryst. Solids*, **88**, 97–113.
Leedecke, C. J. and Loehman, R. H. (1980) *J. Am. Ceram. Soc.*, **63**, 190.
Leng-Ward, G. and Lewis, M. H. (1985) *Mater. Sci. Eng.*, **71**, 101–11.
Leng-Ward, G. and Lewis, M. H. (1988) Oxidation resistance of a $Y_5Al_3O_{12}$–Si_2N_2O based Y–Si–Al–O–N glass ceramic, to be published.
Leng-Ward, G., Lewis, M. H. and Wild, S. (1986) *J. Mater. Sci.*, **21**, 1647–53.
Lewis, M. H., Bhatti, A. R., Lumby, R. J. and North, B. (1980) *J. Mater. Sci.*, **15**, 103–13.
Lewis, M. H. and Leng-Ward, G. (1988) Crystallization of Nd–Si–Al–O–N and La–Si–Al–O–N glasses, to be published.
Loehman, R. E. (1979) *J. Am. Ceram. Soc.*, **62**, 491–4.
Loehman, R. E. (1980) *J. Non-Cryst. Solids*, **42**, 433–45.
Loehman, R. E. (1982) *Proc. Ceram. Eng. Sci.*, **3**, 35–49.
Loehman, R. E. (1983) *J. Non-Cryst. Solids*, **56**, 123–34.
Luping, Y., Guanquing, H., Quanquing, F. and Jiazhi, H. (1984) *J. Chinese Silicate Soc.*, **12**, 387–95.
Makishima, A., Mitomo, M., Ii, N. and Tsutsumi, M. (1983) *J. Am. Ceram. Soc.*, **66**, C55–C56.
Marchand, R. (1983) *J. Non-Cryst. Solids*, **56**, 173–8.
McMillan, P. W. (1975) *Glass Ceramics*, Academic Press, London and New York.
Messier, D. R. (1982) *Proc. Ceram. Eng. Sci.*, **3**, 293.
Messier, D. R. (1985) *Rev. Chim. Miner.*, **22**, 518–33.
Messier, D. R. (1987) *Int. J. High Techn. Ceram.*, **3**, 33–41.
Messier, D. R. and Broz, A. (1982) *J. Am. Ceram. Soc.*, **65**, C123.
Messier, D. R. and Deguire, E. J. (1984) *J. Am. Ceram. Soc.*, **67**, 602.
Mittl, J. C., Tallman, R. L., Kelsey, Jr. P. V. and Jolley, J. G. (1985) *J. Non-Cryst. Solids*, **71**, 287–94.
Morgan, P. E. D., Carroll, P. J. and Lange, F. F. (1977) *Mater. Res. Bull.*, **12**, 251–60.
Mulfinger, H. O. (1966) *J. Am. Ceram. Soc.*, **49**, 462–7.
Mulfinger, H. O. and Franz, H. (1965) *Glastech. Ber.*, **38**, 235.
Pasturnak, R. and Verdier, P. (1983) *J. Non-Cryst. Solids*, **56**, 141–6.
Peng, Y. B. and Day, D. E. (1987) *J. Am. Ceram. Soc.* **70**, 232–6.
Rabinovich, E. M. (1983) *Phys. Chem. Glasses*, **24**, 54–6.

Rajaram, M. and Day, D. E. (1987) *J. Am. Ceram. Soc.*, **70**, 203–7.
Reidmeyer, M. R. and Day, D. E. (1985) *J. Am. Ceram. Soc.*, **68**, C188–C190.
Reidmeyer, M. R., Rajaram, M. and Day, D. E. (1986) *J. Non-Cryst. Solids*, **85**, 186–203.
Risbud, S. H. (1981) *Phys. Chem. Glasses*, **22**, 168–70.
Roebuck, P. H. A. (1978) PhD thesis, University of Newcastle upon Tyne, UK.
Sakka, S., Kamiya, K. and Yoko, T. (1983) *J. Non-Cryst. Solids*, **56**, 147–55.
Shaw, T. M., Thomas, G. and Loehman, R. E. (1984) *J. Am. Ceram. Soc.*, **67**, 643–7.
Shreyer, W. and Schairer, J. F. (1962) *Am. Mineral.*, **47**, 90–104.
Shillito, K. R., Wills, R. R. and Bennett, R. B. (1978) *J. Am. Ceram. Soc.*, **61**, 537.
Smith, M. (1987) *A high resolution multi-nuclear magnetic resonance study of ceramic phases*, PhD thesis, University of Warwick, UK.
Thomas, G., Ahn, C. and Weiss, J. (1982) *J. Am. Ceram. Soc.*, **65**, C185–C188.
Thorp, J. S., Ahmad, A. B., Kulesza, B. L. and Kenmuir, S. V. J. (1984) *J. Mater. Sci.*, **19**, 3211.
Thorp, J. S. and Kenmuir, S. V. J. (1981) *J. Mater. Sci.*, **16**, 1407–9.
Tredway, W. K. and Loehman, R. H. (1985) *J. Am. Ceram. Soc.*, **68**, C131–C133.
Tredway, W. K. and Risbud, S. H. (1983a) *J. Am. Ceram. Soc.*, **66**, 324–7.
Tredway, W. K. and Risbud, S. H. (1983b) *J. Non-Cryst. Solids*, **56**, 135–40.
Tredway, W. K. and Risbud, S. H. (1985a) *J. Mater. Sci. Lett.*, **4**, 31–3.
Tredway, W. K. and Risbud, S. H. (1985b) *Mater. Lett.*, **3**, 435–9.
Tsai, R. L. and Raj, R. (1981) *J. Am. Ceram. Soc.*, **65**, 270–4.
Weyl, W. A. (1967) *Coloured Glasses*, Society of Glass Technology, Sheffield, p. 45.
Wild, S., Leng-Ward, G. and Lewis, M. H. (1981) *J. Mater. Sci.*, **16**, 1815–28.
Wild, S., Leng-Ward, G. and Lewis, M. H. (1984) *J. Mater. Sci.*, **19**, 1726–36.
Winder, S. M. and Lewis, M. H. (1985) *J. Mater Sci. Lett.*, **4**, 241–3.
Wusirika, R. R. (1984) *J. Am. Ceram. Soc.*, **67**, C232–C233.
Wusirika, R. R. (1985) *J. Am. Ceram. Soc.*, **68**, C292–C297.
Wusirika, R. R. and Chyung, C. K. (1980) *J. Non-Cryst. Solids*, **38/39**, 39–44.

5

Optical properties of halide glasses

J. M. Parker and P. W. France

5.1 INTRODUCTION

Glasses based on halides have been known for a considerable time but because of specific disadvantages such as toxicity (BeF_2), hygroscopicity ($ZnCl_2$, BeF_2) and lack of stability (AlF_3) were not widely developed. However, in 1974 a new class of halide materials based principally on ZrF_4–BaF_2 glasses was discovered in France (Poulain *et al.*, 1975) with few of these difficulties. Their extended infra-red transparency compared with silica implied much lower minimum intrinisic absorption losses, because at longer wavelengths Rayleigh scattering is considerably reduced, and suggested the possibility that they might be used in long-haul telecommunications. A worldwide research programme was therefore initiated to develop a suitable melting and fabrication technology, and to optimize the glass compositions for stability and the required working properties, particular interest focusing on fibre production. At the same time many previously known glass systems have been re-examined and new compositions investigated.

Because of the wide interest in halide glasses several reviews have already been published (e.g. Trant *et al.*, 1984; Drexhage, 1985) and a number of conferences proceedings contain useful survey articles (e.g. Lucas and Moynihan, 1985; Almeida, 1986). El-Bayoumi *et al.* (1985) have listed many of the possible non-fibre applications of these glasses. This chapter concentrates on preparation techniques and those properties relevant to optical applications.

5.2 GLASS FORMATION AND STRUCTURE

On the basis of Zachariasen's rules, glass formation might only be expected in systems where the co-ordination number of the primary glass-forming component is low, normally four or less. Two halide glass-forming systems have been reported which satisfy this criterion. BeF_2 glasses were first discovered in 1926 (Goldschmidt, 1926). Their melts are viscous even at the liquidus and as a result form good glasses. The glasses have a network structure consisting of BeF_4 tetrahedra and are analogous to vitreous SiO_2 except that the lower ionic charges reduce the bond strengths and therefore T_l and T_g. $ZnCl_2$ is also isomorphous with SiO_2 and forms glasses based on $ZnCl_4$ units although the arrangement of tetrahedra is such that the Cl^- ions are approximately close packed (Sinclair *et al.*, 1980). It is more weakly bonded with a consequent lowering of melting temperature. T_g is only 103°C and T_l is 318°C. The viscosity at T_l has been reported to be 50 poise. The fluidity of its melts makes glass formation more difficult but the principal limitation to its wider application is its extreme hygroscopicity. Indeed study or use of the pure compound without water contamination is very difficult. While both of these halides form glasses in their own right a wide range of further glasses can be made by adding other halides. The fluoroberyllate glass-forming systems which have been studied are listed by Rawson (1967) and Baldwin *et al.* (1981). In particular AlF_3 and alkali fluorides can be used as modifiers. These compound glasses have lower melting temperatures, steeper viscosity–temperature curves and are usually more resistant to aqueous corrosion.

While no other pure halides form glasses many are conditional glass formers, i.e. require the presence of other additions although very few compositions are sufficiently stable to allow bulk glass production. Most are structurally interesting showing higher co-ordination numbers than for BeF_2 and $ZnCl_2$ glasses. Among these, halide glasses based on AlF_3 together with other components such as CaF_2, BaF_2 and PbF_2 have been known for some time (Sun, 1949) and are thought to be based on octahedrally co-ordinated networks. Recently quite stable fluoroaluminates have been synthesized using additives such as YF_3 (Kanamori *et al.*, 1981) and ThF_4 (Poulain *et al.*, (1981) allowing samples up to 6 mm thick to be manufactured.

The discovery of the heavy metal fluoride glasses (HMF) based on the ZrF_4–BaF_2 system extended the range of conditional halide glass formers and has given rise to considerable speculation as to their structure. Zr is regarded as the network-forming ion in conventional terminology but its co-ordination is variously thought to be 6 to 8, giving structural models ranging from chains to 3D networks (Parker, 1986). Because these glasses do not fit naturally into the normal concept of glasses an alternative approach has been to regard the glasses as based on random close packed arrangements of F^- and Ba^{2+} ions with the other cations occupying the various holes available in the structure (Poulain, 1981).

While a wide range of glass-forming systems have been discovered within this category the most stable compositions have been developed from the ZrF_4–BaF_2 binary. The T_c–T_g gap is frequently used as a measure of stability, sometimes normalized by dividing by the liquidus temperature, and on this basis the steady improvement in glasses achieved by the addition of network stabilizing fluorides LaF_3 (Poulain and Lucas, 1978) and AlF_3 (Lecoq and Poulain, 1980a) and network-modifying ions such as NaF or LiF

Table 5.1 Results are from DSC scans run at 10°C min^{-1} (taken from Bansal *et al.*, 1985)

Glass type	T_g (°C)	$T_c - T_g$ (°C)
ZrF_4–BaF_2	295	69
ZrF_4–BaF_2–LaF_3	299	92
ZrF_4–BaF_2–AlF_3	311	94
ZrF_4–BaF_2–AlF_3–LaF_3–NaF	267	124

(Lecoq and Poulain, 1980b) can be seen in Table 5.1. Other common additives are ThF_4 and HfF_4. The latter being almost chemically identical to ZrF_4 forms a very similar range of glasses but with a wider infra-red transmission. The role of different cations in HMF glasses has been categorized by Baldwin *et al.* (1981) as network formers, modifiers or intermediates on the basis of the cation–fluorine bond strength, an approach which has been useful in guiding compositional development.

In view of the potential of these fluorozirconate glasses their crystallization behaviour has been extensively studied. Low temperature crystallization behaviour is significant where the glasses are to be processed by reheating as for example in pulling fibres from preforms. Early experiments (Bansal *et al.*, 1983; MacFarlane *et al.*, 1984) used non-isothermal and isothermal DSC analysis to show that the process activation energy is high, typically 300–400 kJ mol^{-1}, although Moynihan (1986) has pointed out that more stable compositions give lower values. The crystalline phases forming particularly in the ZBLAN system have been studied by several authors (Bansal and Doremus, 1985; Parker *et al.*, 1986; Parker *et al.*, 1986); typically, metastable disordered barium and sodium fluorozirconates showing extensive solid solution are precipitated in the early stages which then subsequently transform to more stable forms. Also the glass which remains after this initial stage, being enriched in ZrF_4, LaF_3 and AlF_3 survives to higher temperatures but then crystallizes predominantly as $\beta BaF_2 \cdot 2ZrF_4$. At still higher temperatures a ternary sodium barium fluorozirconate is formed together with a lanthanum fluorozirconate. This complex sequence of events is mirrored in the multiple peaks normally present on DSC traces.

Interpretation of the early DSC experiments using the Johnson–Mehl–Avrami equation led to the conclusion that the mechanism of crystallization was predominantly three-dimensional growth from a constant number of nucleation sites (Bansal *et al.*, 1983; MacFarlane *et al.*, 1984). Parker *et al.* (1986) and Carter *et al.* (1986) have observed in bulk samples that crystallization commences from the surface and have measured crystal growth rates. Bansal and Doremus (1983) observed that crystallization of a

ZBL glass commenced with the initial formation of a surface crystalline layer. Ohishi *et al.* (1984) found that surface nucleation and growth was apparently enhanced in the presence of water vapour and proposed that dry atmospheres should be used during fibre pulling, although Carter *et al.* (1986) failed to substantiate this conclusion.

While considerable evidence points therefore towards heterogeneous surface nucleation as the initial step in bulk crystallization, many workers have found an increase in scattering loss from glasses heat treated below T_c, e.g. Tokiwa *et al.* (1985b and c) studied light scattering measurements from specimens heat treated at temperatures suitable for fabrication. There is no evidence that liquid–liquid phase separation occurs in ZBLAN based systems although it has been observed in several halide systems (Suscavage and El-Bayoumi, 1985; Simmons and Simmons, 1985; Cheng Ji-jian and Bao Shan-zhi, 1985). Homogeneous nucleation within the bulk of the glass for simple compositions has been proposed for example by Matecki (1983) and has also been reported for glasses melted in the presence of a reactive atmosphere containing chlorine (Neilson *et al.*, 1985), but recent experiments suggest that the exsolution of small oxide particles present as impurities in the glass is responsible for the increased loss (Lu and Bradley, 1985 and also see section 5.9.4).

While it is important to know the crystallization mechanisms on reheating, behaviour during cooling from the melt is a more useful indicator of basic glass-forming ability. Very little phase diagram information is available which might indicate crystallization routes or allow stability to be related to the positions of eutectics. Early attempts to predict the phase equilibria by computation using basic thermodynamic information, Kaufman *et al.* (1982) and Kaufman and Birnie (1983), were remarkably successful even though subsequent analysis (e.g. Parker *et al.*, 1986) has shown the existence of numerous binary and ternary compositions not included in the analysis. The main melting event on DSC traces of ZBLAN glasses typically is reported near 480°C. At these temperatures the phases present are sodium barium fluorozirconate and lanthanum difluorozirconate. There is some evidence from DSC studies that melting is not completed in these glasses, however, until somewhat higher temperatures and that the true primary phase in this system is LaF_3 (Carter *et al.*, 1986). During cooling melt temperatures often fall substantially below the liquidus before crystallization commences and a number of phases may appear metastably in addition to or instead of those predicted by the phase diagram.

An informative indicator of stability recently adopted is critical cooling rate which is a measure of the rate at which glasses can be cooled without crystallizing; measurements have largely been carried out without reference to the phases crystallizing. Busse *et al.* (1985) determined time–temperature–transformation curves for ZBLAN and ZBLALi glasses using isothermal DSC, and showed that the former were more stable. A similar study was carried out by Esnault-Grosdemouge *et al.* (1985) and Kanamori and Takahashi (1986) have made similar measurements but using a non-isothermal technique.

When fluoride ZBLAN or ZBLALi glasses are prepared by casting, a common feature is the presence of a few small crystals in the bulk of the glass at a typical level of 1 part in 10^{10}. Elegant work by Lu *et al.* (1984) and Lu and Bradley (1985) has identified many of these phases, typical crystals being AlF_3, LaF_3, ZrF_4 and barium fluorozirconates. Recognition of any of these phases in a particular melt can guide the direction of compositional modification. These authors also showed that crystallization can be initiated by heterogeneous nucleation on small Pt particles, whose origin was assumed to be the melt crucible and it is possible that oxide particles act in a similar way. Working methods therefore also play a role in minimizing this mode of crystallization.

Apart from the widely studied ZrF_4 based systems, glasses based on transition metal fluorides (e.g. ZnF_2, CdF_2, MnF_2 and FeF_3), actinide and lanthanide fluorides (ThF_4, UF_4, YbF_3) have been widely studied. Many are strongly coloured and therefore not particularly suitable for optical applications, except possibly as filters. Glasses with alkali fluorides (LiF, KF, RbF, CsF) as major components have also been reported. The review articles listed earlier provide extensive references to such work.

Glasses based on other halides tend to have been less well characterized partly because their physical properties are less favourable. Nevertheless they are likely to have extended infra-red transmissions and may also have other useful characteristics. $ZnCl_2$ is best used in combination with other halides such as KI. $BiCl_3$ and $BiBr_3$ act as glass progenitors and a range of glasses has been developed based on $ThCl_4$. A new development is glass formation amongst mixed alkali halides (e.g. AgX and CsX, X = Br or I, Nishii *et al.*, 1985).

The range of halide glass-forming systems have been extensively reviewed by Baldwin *et al.* (1981), Poulain (1983), Savage (1985) and Drexhage (1985). Recently Ma Fu Ding *et al.* (1984) have reviewed the potential for forming glasses in a wide range of chloride, bromide and iodide systems.

5.3 MELTING TECHNIQUES

The melting techniques developed for oxide glasses are generally unsuitable for halide systems. Firstly the materials are often toxic (especially BeF_2) and secondly hydrolysis can occur at high temperature according to:

$$F^-(m) + H_2O(g) \rightarrow OH^-(m) + HF(g)$$

so melting requires a controlled inert environment either in a specially designed glove box or in sealed liners flushed with dry gas. Since this reaction may occur slowly even at room temperature, all raw materials should be stored under inert conditions.

Molten halides are corrosive to many refractories, limiting the choice of crucibles to materials such as platinum, gold or vitreous carbon. Refractory furnace liners are often made of silica but many fluorides have high vapour pressures at melting temperatures, particularly ZrF_4, and can attack the liner according to:

$$SiO_2 + 4F^- \rightarrow SiF_4\uparrow + 2O^{2-}$$

SiC has been used as an alternative liner material. Volatilization is also troublesome with multicomponent systems such as the ZrF_4 based glasses where one component may be preferentially lost leading to an unspecified composition. Capping the crucible with a close fitting lid reduces loss as does minimizing gas flow through the liner.

A technique developed for halide glasses is reactive atmosphere processing (RAP). This is often desirable for several reasons. The dry atmospheres necessary to ensure melts free of OH, such as N_2 or Ar, are frequently able to reduce the melt. In particular dark deposits have been attributed to a reduced form of zirconium fluoride viz. ZrF_3 (France *et al.*, 1986a). Oxidizing atmospheres can redress the balance.

RAP can also reduce oxide and OH contamination according to:

$$O^{2-}(m) + F_2(g) \rightarrow 2F^-(m) + \tfrac{1}{2}O_2(g)\uparrow$$

$$2OH^-(m) + F_2(g) \rightarrow 2F^-(m) + H_2O(g)\uparrow + \tfrac{1}{2}O_2(g)\uparrow$$

The reactive species in these equations, viz. F_2 (or Cl_2) are either used directly or more often are produced *in situ* by thermal decomposition of another phase such as CCl_4, SF_6 or NF_3 (Nakai *et al.*, 1985a, b, 1986). However, heating the melt to a sufficiently high temperature under dry processing conditions (Poulain and Saad, 1984a) is sufficient to remove OH^- as follows:

$$OH^-(m) + F^-(m) \rightarrow O^{2-}(m) + HF(g)\uparrow$$

Besides improving the melt quality, oxidation can help to reduce TM absorption losses (see section 5.9.1). Oxygen has recently been proposed for this purpose by France *et al.* (1986b), who pointed out that oxygen would not replace fluoride ions in the melt and consequently little would be incorporated apart from the small amounts necessary to oxidize impurities.

An important consideration in using RAP is to ensure that attack of crucible material does not occur, producing for example, chloroplatinates in the melt, or that inert material such as carbon is not introduced either as particles which nucleate crystallization and increase scattering, or as complexes in solution which add to the infra-red absorption.

Another problem for many halide systems is the low viscosities at their liquidus. While this means that homogenizing and refining are rarely problems, there is a resulting tendency towards crystallization. To overcome it the melts must be rapidly quenched, usually by casting into preheated brass or gold moulds, at temperatures just below T_g. The more difficult glasses may require quenching between brass plates or even using rollers. Again dry, inert conditions are required. BeF_2 does not suffer from the same problem, being a viscous liquid even near T_l, and is therefore an attractive option despite its toxicity. After casting, annealing is typically carried out some 10°C below T_g.

5.3.1 ZrF_4 based glasses

This class of halide glasses has probably been the most extensively studied and several techniques have been developed for their synthesis. Most of the required fluorides are now commercially available at reasonable purities. These are mixed with a little ammonium hydrogen difluoride and melted according to a schedule which usually starts with a low temperature step to allow NH_4HF_2 to fluorinate any residual oxide, a melting step between 800 and 900°C and possibly low temperature conditioning before casting. Melting takes place in a sealed silica liner flushed with dry gases. For the more stable glasses the quenching rate required to suppress crystallization can be achieved simply by removing the liner from the furnace and cooling the glass within the melting crucible, or the molten glass may be cast.

More recently, carefully designed glove boxes fitted with melting and annealing ovens have been used. The ovens are generally installed in the base or top of the glove box allowing transfer of the batch or melt between the glove box and separate melting chambers. In this way the complete process can take place under inert conditions and reactive atmospheres can be introduced into the melting chamber without affecting the glove box interior.

Although resistance furnaces are normally used fluorides have also been melted by RF induction heating (Hutton *et al.*, 1984) in a vitreous carbon melting crucible inside a graphite susceptor. The technique, using a modified crystal-growing furnace, produced glass samples with extremely low scattering, which were therefore free of macroscopic inclusions.

Poulain and Lucas (1978) used oxides as the starting materials. These were converted to the fluorides using NH_4HF_2 according to:

$$2ZrO_2 + 7NH_4HF_2 \rightarrow 2(NH_4)_3ZrF_7 + NH_3 + 4H_2O$$
$$(NH_4)_3ZrF_7 \rightarrow ZrF_4 + 3NH_4F$$

A batch consisting of oxides or oxides and fluorides is mixed with an excess stoichiometric amount of NH_4HF_2. Fluorination takes place in two stages. Firstly a low temperature conversion to an ammonium complex occurs at 200°C, followed by decomposition of the complex to the metallic fluoride at temperatures >400°C. Volatile NH_4F evaporates from the batch. Similar equipment to that already described can be used except that care must be taken of the large volumes of HF and NH_3 liberated.

5.3.2 BeF_2 glasses

Two major problems with melting BeF_2 glasses are that it is extremely toxic and that high purity Be compounds are difficult to obtain. The most common method of preparation is to begin with technical grade $(NH_4)_2BeF_4$ and decompose it thermally to BeF_2 according to:

$$(NH_4)BeF_4 \rightarrow BeF_2 + 2NH_4F$$

Ammonium fluoride sublimes from the batch at 400°C. The actual melting equipment can be similar to that already described except that extra precautions are necessary to protect the operator.

Beryllium compounds are highly toxic causing respiratory failure as well as breakdown of the functions of the heart, liver and kidneys. The threshold limit value (TLV) is 0.002 mg m^{-3} which is one hundred times lower than arsine. Therefore apparatus should be contained in extracted enclosures at low pressures. The laboratory itself should also be a sealed room at low pressure with one entrance in regular use. Extracted air must be filtered before release to the environment and levels of Be must be below 0.01 μm m^{-3} at ground level.

BeF_2 forms a well behaved glass in contrast with fluorozirconates. Although its T_g is only 250°C, it still has a viscosity of 10^6 poise at its liquidus of 545°C. As a result melting temperatures must be of the order of 1000°C to ensure homogeneous glass. Crucibles can be platinum or vitreous carbon and control of the melting atmosphere is again advantageous. Because of the increased viscosity these glasses are stable to crystallization and do not have to be rapidly quenched.

Baldwin and Mackenzie (1979a) recently described a distillation technique used to purify BeF_2 prior to melting. Distillation was under an HF/H_2 atmosphere at 30 torr and gave significantly improved glasses. Sarhangi (1986) has reviewed potential vapour phase routes for making fluoroberyllates and Sarhangi and Thompson (1986) have recently described the successful manufacture of bulk BeF_2 and $BeF_2 \cdot AlF_3$ glasses by a chemical vapour deposition (CVD) route.

5.3.3 $ZnCl_2$ glasses

$ZnCl_2$ glasses are difficult to prepare and generally crystallize easily, supposedly due to residual water which promotes crystallization. Angell and Wong (1970) prepared the glass by passing HCl gas through molten $ZnCl_2$ and then dropping into liquid N_2. They were able to form small glass beads.

Robinson and Pastor (1982) reported glass making by reacting CCl_4 with Zn metal vapour at 650°C in a quartz reactor (nascent Cl^0 formed from CCl_4 reacts with Zn vapour to form the glass). On cooling to room temperature vitreous $ZnCl_2$ was formed but rapidly devitrified on exposure to air.

Binary glasses based on $ZnCl_2$ are more resistant to water and Savage (1982) reported combinations with $PdCl_2$, $CdCl_2$ and $CdBr_2$ as giving improved properties. These glasses were made by vacuum melting in silica ampoules at 500–700°C. The samples were cooled to 400°C before being quenched in liquid N_2.

5.4 OPTICAL TRANSMISSION CHARACTERISTICS

Of all halide glass properties, their optical characteristics are probably of most interest. Their extended transparency is illustrated by Fig. 5.1. The two major features of halide glasses are that firstly they have an extended infrared (IR) transmission, and secondly that, unlike chalcogenides, they have the advantage of transmitting through the visible into the ultraviolet (UV) region with a cut-off at about 250 nm. For BeF_2, reported spectra indicate a UV cut-off at about 150 nm, and an IR cut-off between 4 and 5 μm (Baldwin *et al.*, 1981).

One major application of halide glasses which takes advantage of their optical properties is that of IR optical fibres. These are being considered for IR imaging, low temperature thermometry, sensor applications and CO and CO_2 laser power transmission, uses which require only moderate losses but an extended IR transparency. By far the most important application, however, is ultra-low loss telecommunication fibres. Here the reduced

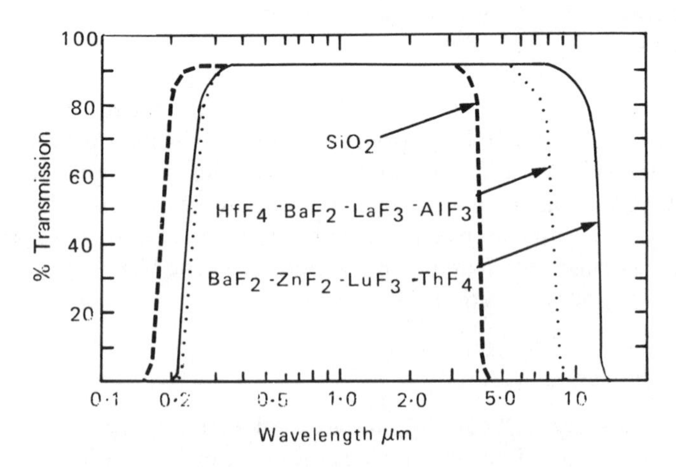

Fig. 5.1 Showing percentage transmission against wavelength for two HMF glasses and SiO_2. The samples were approximately 5 mm thick (after Drexhage, 1985).

scattering losses that are possible by operating further into the infra-red may ultimately allow high capacity transmission with repeater spacings of hundreds of kilometres. BeF_2 and ZrF_4 glasses offer most potential but other halide and mixed halide glasses are also being studied. To understand why extremely low losses may be possible the fundamental loss mechanisms will be considered next.

The total intrinsic losses are given by:

$$L = A \exp(a/\lambda) + B/\lambda^4 + C \exp(-c/\lambda) \tag{5.1}$$

where a, A, B, c and C are material constants. Units are normally expressed in dB km^{-1} for fibre systems although the conversion to absorbance can be made using 1 cm^{-1} = 10^6 dB km^{-1}.

The first term in (5.1) represents the ultraviolet absorption edge and is the tail of short wavelength absorptions due to electronic transitions within the material. This absorption is otherwise known as the Urbach edge and limits the transparency at short wavelengths. The second term represents Rayleigh scattering loss and arises from random composition and density fluctuations on a microscopic scale, which cause refractive index changes in the glass. The spatial period of these fluctuations is small compared to the wavelength of light and induces a scattering loss which decreases as λ^4. The last term represents the IR absorption edge and is a multiphonon absorption tail arising from overtones and combination bands of the fundamental material vibrations at longer wavelengths. The position of this absorption edge determines the suitability of the material as an IR transmitting medium. All three terms are discussed in more detail in subsequent sections.

For a silica based waveguide the minimum intrinsic loss is determined by the intersection of the Rayleigh scattering curve and the IR edge, and reaches a lowest value of about 0.15 dB km^{-1} at 1.6 μm (Miya *et al.*, 1979). Reported losses of 0.15 dB km^{-1} at 1.55 μm (Berkey and Sarkar, 1982) in silica fibres made by the MCVD process indicate that extrinsic losses have been virtually eliminated and the fundamental material limits reached. To prepare optical fibres with lower losses therefore required materials with IR edges at longer wavelengths.

5.4.3 Long wavelength materials

The long wavelength transmission limit is determined by IR absorption resulting from the multiphonon edge of the fibre materials. Since this edge is the tail of fundamental absorptions at longer wavelengths, the major criterion for the determination of suitable materials is that these fundamentals should be located at sufficiently long wavelengths.

The empirical Szigeti equation (Szigeti, 1950):

$$\omega = (k/\mu)^{1/2}/2\pi c \tag{5.2}$$

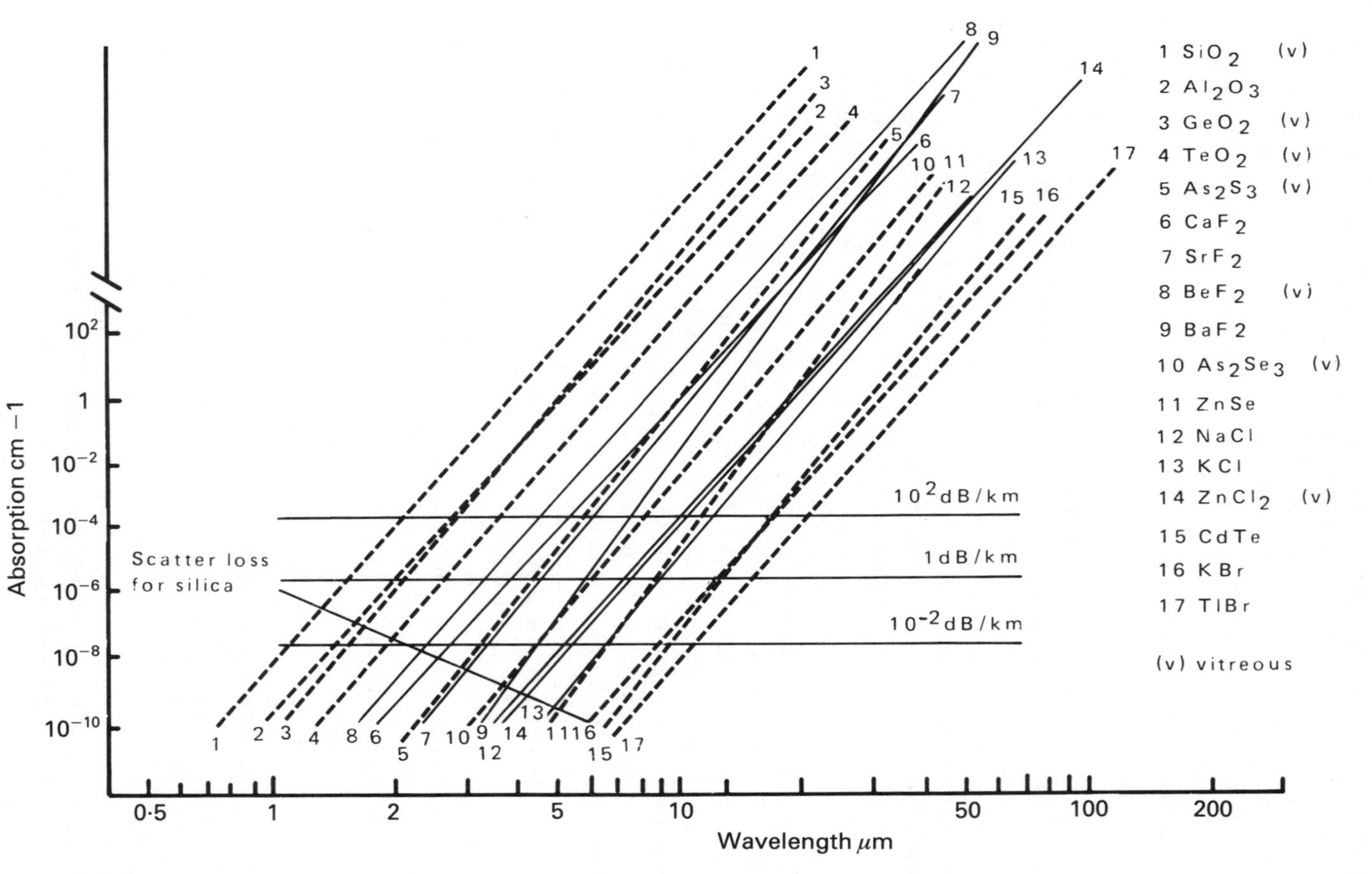

Fig. 5.2 The IR absorption of a range of materials transparent in the mid-IR (after Gannon, 1981).

stipulates that the position of the fundamental phonon frequency ω (in wavenumbers) is related to the force constant k and the reduced mass μ. Therefore heavier ions and weaker bonding are preferable for extended IR transmission.

The IR absorption of various candidate materials including vitreous and crystalline halide, chalcogenide and heavy metal oxide glasses is presented in Fig. 5.2. The increased mass of Ge over Si pushes the IR edge of GeO_2 out to longer wavelengths. Switching the anion from oxygen to fluorine reduces the force constant since fluorine is only singly charged and this also shifts out the IR edge of fluoride glasses, particularly the HMFs. Chalcogenide glasses have relatively weak covalent bonding and come next in the sequence of materials. Finally crystalline halides such as AgBr and KCl with their totally ionic lattices have the best IR transmission.

Of these materials probably the halide glasses, and in particular the fluorozirconates, offer the most potential for low loss fibres. These materials combine the attributes of reasonable IR transmission, glass stability, corrosion resistance and ease of fabrication. Chalcogenide glasses may well find applications at longer wavelengths but their opacity in the visible limits their use. Crystalline materials such as $ZnCl_2$ and TlBr have good IR transmission but poor mechanical properties and are difficult to convert into fibre at realistic rates.

5.5 INFRA-RED ABSORPTION

IR absorption is a dominant factor in determining intrinsic minimum losses. The IR spectrum can be subdivided into the fundamental and multiphonon region. At longer wavelengths the fundamental resonances provide intense absorptions. Here IR reflectivity and Raman spectra are normally used to predict edges and to obtain a deeper understanding of the glass structure, sufficiently thin specimens for analysis by transmission being difficult to obtain. At shorter wavelengths the multiphonon edge tails back towards the visible region and determines the minimum losses.

5.5.1 Far infra-red (FIR)

The fundamental modes of the halide glasses are in the 10–50 μm (FIR) region. Various materials are compared in Figs. 5.3 and 5.4.

(a) Reflection spectra

The main feature in the IR reflection spectra is two major broad peaks, whose positions depend on glass composition. SiO_2 and BeF_2 have remarkably similar spectra indicating a common underlying structure. The absorptions in silica, centred at 470 and 1120 cm^{-1} (21.3 and 8.9 μm)

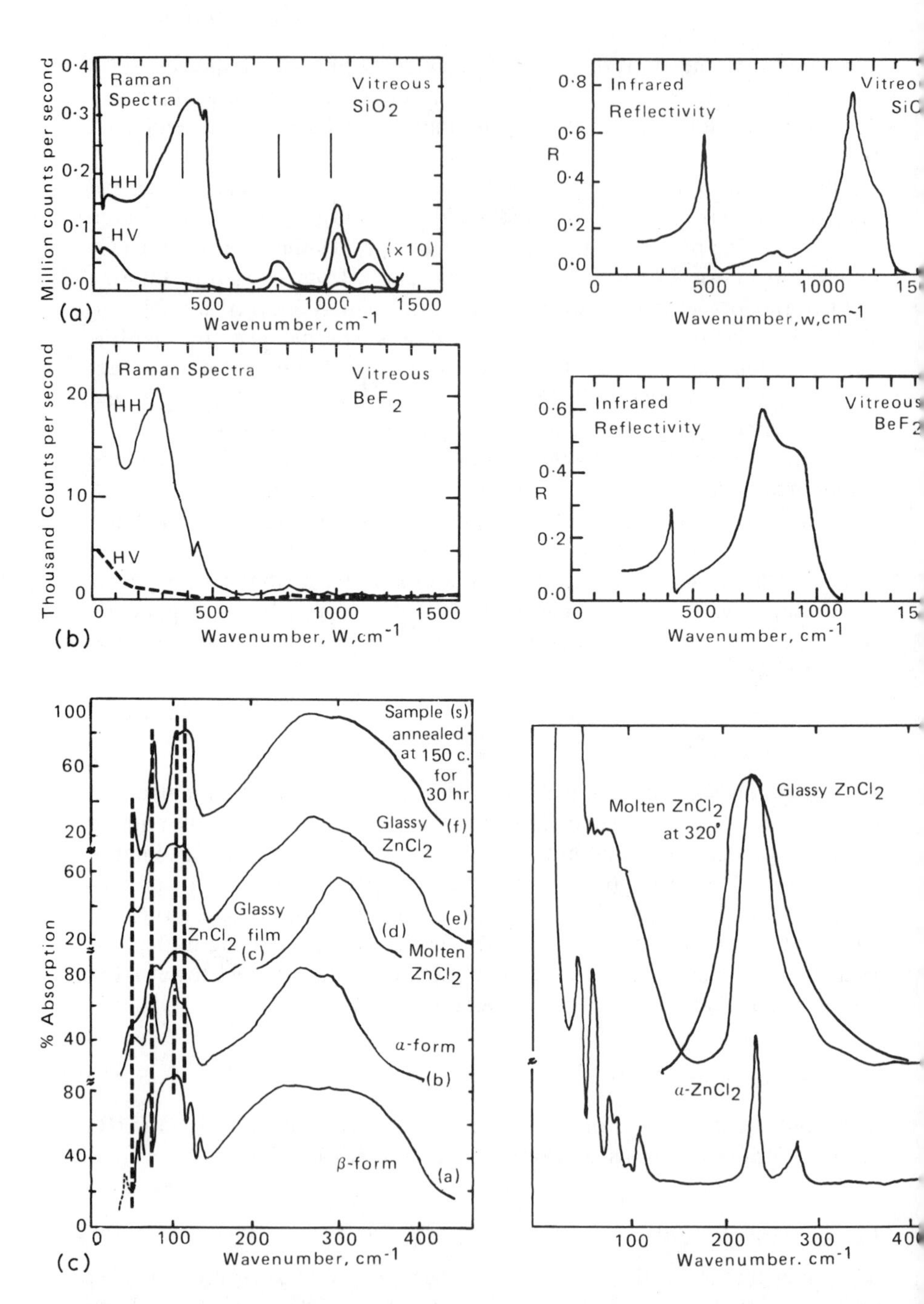

Fig. 5.3 IR reflectivity and Raman spectra for (a) vitreous SiO_2 (after Galeener and Lucovsky, 1976), (b) BeF_2 (after Galeener, 1978) and (c) $ZnCl_2$ (after Angell and Wong, 1970).

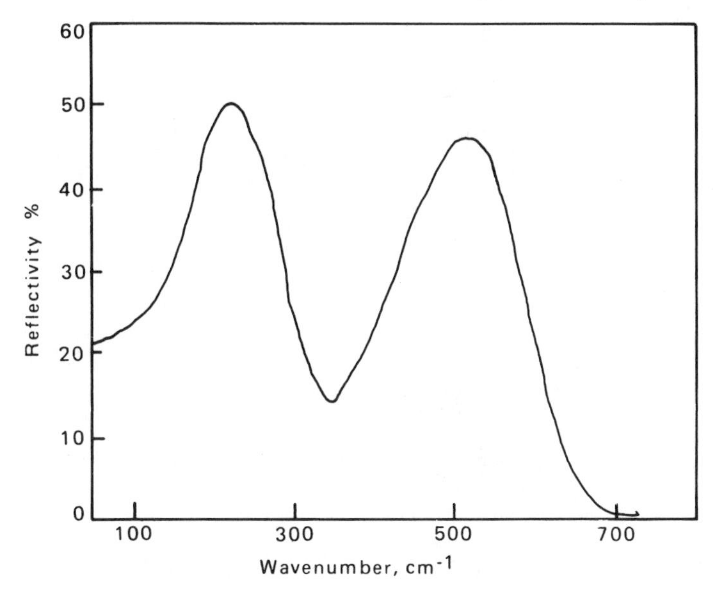

Fig. 5.4 IR reflectivity and Raman spectra for a HfF_4–BaF_2–ThF_4 glass (after Bendow *et al.*, 1982b).

correspond to ν_4 and ν_3 vibration modes (Fig. 5.5). A smaller absorption at 790 cm^{-1} (12.7 μm) is due to the ν_1 symmetric stretch. In GeO_2-doped silica fibres these absorptions are shifted to lower frequencies $\nu_3 = 1099\ cm^{-1}$ and $\nu_4 = 469\ cm^{-1}$ (Izawa *et al.*, 1977).

BeF_2 shows two major absorptions at 420 and 770 cm^{-1} (23.8 and 13.0 μm) again corresponding to ν_4 and ν_3 respectively (Baldwin *et al.*, 1981). The shift to lower frequencies results from the weaker electrostatic bonding in BeF_2 and the smaller reduced mass. Kondrat'ev *et al.* (1969) have studied the IR spectra of alkali fluoroberyllate glasses. Increased additions of a modifying fluoride such as KF reduced the intensity of the fundamental Be–F stretch at 770 cm^{-1} and also shifted it to lower frequencies. At 30 mol % KF it had shifted to 700 cm^{-1}. The magnitude of the shift increased according to reduced mass in the sequence Li > K > Rb > Cs.

The IR spectrum of vitreous $ZnCl_2$ has been measured by Angell and Wong (1970) and again has two major absorptions at 100 and 260 cm^{-1}. The spectrum is very similar to crystalline α-$ZnCl_2$, which is a 3D lattice of $ZnCl_4$ tetrahedra linked by Cl bridges. The results suggest a 3D network model analogous to vitreous SiO_2 and BeF_2.

The fundamental spectra of fluorozirconate and fluorohafnate glasses are composed of two well separated broad peaks near 250 and 525 cm^{-1} (40 and 19.0 μm) with the 525 cm^{-1} peak being considerably more intense. A

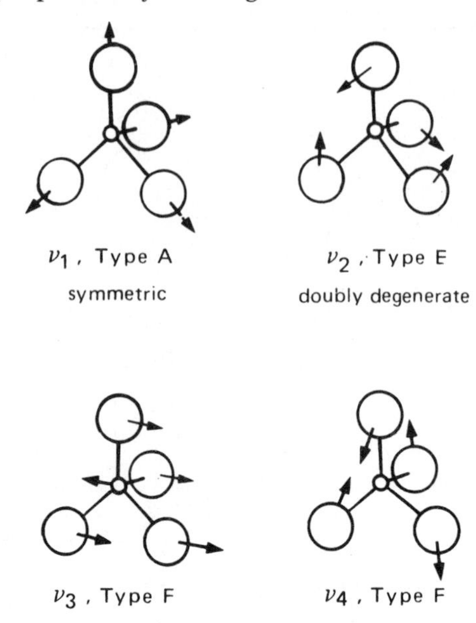

Fig. 5.5 Illustrating the vibration modes in SiO_2. ν_1 and ν_2 are only Raman active while modes ν_3 and ν_4 are both IR and Raman active (after Sigel, 1977).

detailed comparison with molten ZrF_4 salts suggested that the 525 cm^{-1} peak was due to Zr–F stretching vibrations while the 250 cm^{-1} peak was due to either a Ba–F or a Zr–F bend. In ternary glasses it appeared that the third component (LaF_3, GdF_3 or ThF_4) had a negligible effect (Bendow *et al.*, 1982b and 1985).

The FIR spectra of BaF_2–ZnF_2–ThF_4 with either YF_3 or YbF_3 added have been measured by Bendow *et al.* (1983) and are very similar to the fluorozirconates except that the peaks are shifted to lower frequencies. In the YF_3 glass peaks are observed at 400 and 250 cm^{-1} and in the YbF_3 glass they are at 435 and 250 cm^{-1}. On adding AlF_3 to this system an additional peak was observed at 625 cm^{-1} (16.0 μm). The height of this peak was proportional to the amount of AlF_3 added clearly identifying the origin of the absorption and correspondingly it has been shown that AlF_3 shifts the IR edge to shorter wavelengths. This would be expected from reduced mass considerations and suggests that AlF_3 should be avoided in the manufacture of low loss optical fibres. However, AlF_3 has other benefits in fluorozirconate glasses which at present require its inclusion.

Other glasses studied include a mixed halide glass containing ZrF_4–BaF_2–$BaCl_2$, with an extra peak at 470 cm^{-1} related to the presence of chloride in the glass (Almeida and Mackenzie, 1982).

(b) Raman spectra

In conjunction with IR reflection spectra, Raman scattering can yield useful data on glass structure. Here an incident photon is inelastically scattered by TO modes in the glass matrix. Phonons can be created or destroyed with a consequent shift in the frequency of the scattered light either to lower or higher energies (Stokes and anti-Stokes lines respectively). Although IR absorption and Raman scattering both involve the same set of energy levels, different selection rules allow different transitions. For example in the vibrational modes of SiO_2, illustrated in Fig. 5.5, ν_1 and ν_2 are only Raman active while ν_3 and ν_4 are both Raman active and IR active. Consequently the techniques are complementary.

A more detailed account of Raman spectra as applied to halide glasses is given by Bendow *et al.* (1982b). The technique can be used to define the number of different bonds in the glass network since it has been found that this correlates with the number of dominant polarized Raman lines. In general the normal selection rules are less well defined in a vitreous matrix when compared with a similar crystalline case and results in fewer, broader peaks with a higher background level. As shown in Fig. 5.3, the simple one component glasses SiO_2, BeF_2 and $ZnCl_2$ each exhibit one dominant line, at 450, 280 and 235 cm^{-1} respectively. Bendow *et al.* attributed this to symmetric stretch vibrations of the bridging atoms O, F or Cl, along the bisector of the angle that they make with their nearest neighbours.

Banerjee *et al.* (1981) reported the Raman spectra of fluorozirconate and fluorohafnate glasses. They exhibit a dominant peak in polarized VV spectra close to 580 cm^{-1} with a corresponding deep minimum in the depolarization ratio (Fig. 5.6). Despite the presence of multiple components in the glass the Raman spectrum is similar to the simple one-component systems and suggests a single bond. Ionic bonding was also thought to be retained in the vitreous state. Recently Almeida (1986) has reviewed more detailed interpretations of Raman spectra in HMF glasses and has concluded that the Zr co-ordination is 6. This conclusion has been contested by Kawamoto (1984) who suggested an alternative explanation based on 8-fold co-ordination. Saissy *et al.* (1985) showed that the Raman spectrum was unaltered in fibres compared with bulk glass. Walrafen *et al.* (1985) have interpreted the Raman spectra of fluorozirconate glasses containing PbF_2 and have also studied fluorozirconate melts at temperatures up to 800°C. In the latter case almost free ZrF_5, ZrF_6, ZrF_7 anions were indicated. Strom *et al.* (1984) investigated the effect of impurity ions in fluorozirconate glasses on their Raman and IR absorption.

Raman spectra of non-ZrF_4 containing glasses are scarce but information has been given on a BaF_2–ZnF_2–YbF_3–ThF_4 glass and related compositions by Bendow *et al.* (1985). The spectrum is much broader than those of fluorozirconates with a peak close to 400 cm^{-1}. The results suggest more disordering and a greater number of bonds.

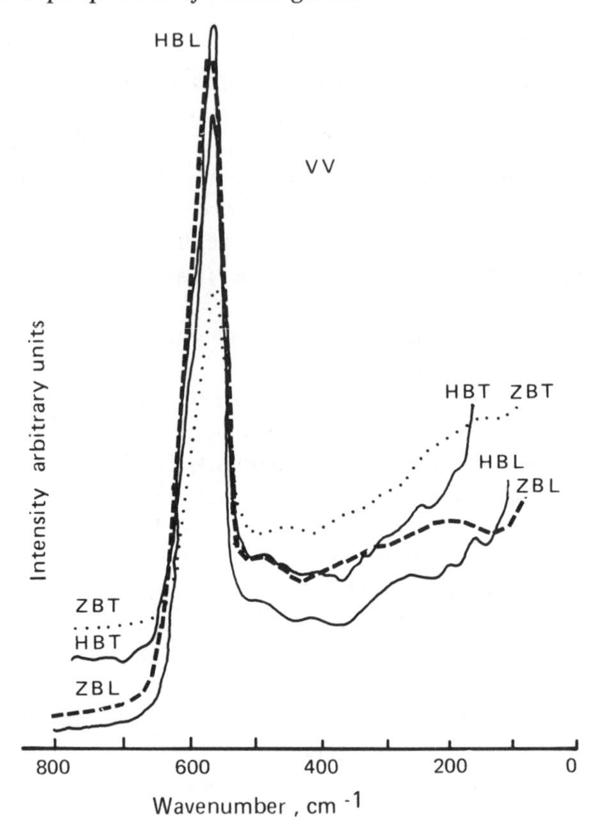

Fig. 5.6 The polarized Raman spectra of fluorozirconate and fluorohafnate glasses (after Banerjee *et al.*, 1981).

At high power inputs, non-linear interactions can occur giving rise to stimulated Raman scattering (SRS). While this can cause power loss in an incorrectly designed fibre optics system, it can also be put to good use to produce fibre amplifiers and tunable oscillators. Saissy *et al.* (1985) argued that SRS was expected to be enhanced in fluorides compared with silica because of the narrower width of the absorption band in the former case. They compared stimulated Raman gain in fluorozirconate and silica fibres pointing out the improvements needed in fluoride fibres before optimum results can be achieved. Stimulated Raman scattering and Raman amplification has also been reported by Durteste *et al.* (1985) who reported gains of 40 dB in 18 μm diameter, 10 m long fibres with a pump power of 1 kW.

Recently Monerie (1986) has measured the thresholds for stimulated Raman and Brillouin scattering in fluorides. He found a Brillouin gain factor of 3.5×10^{-11} mW^{-1} compared with 1.1×10^{-11} mW^{-1} in SiO_2 and therefore yielding thresholds about three times higher. On this basis for a fibre with a loss of 10^{-2} dB km^{-1}, a CW power level of 450 μW was shown to be feasible allowing a maximum distance between repeaters of 2000–3000 km. The gain factor for Raman scattering was found to be 1×10^{-13} mW^{-1}, comparable with silica. This would give a critical power level of 100 mW and would not give rise to any difficulties.

5.5.2 Multiphonon region

The fundamental absorption peaks mentioned above are very intense and although they lie in the FIR, they tail back to much shorter wavelengths. Essentially a fundamental absorption ω_0 has overtones (and combination bands) at $2\omega_0$, $3\omega_0$ etc. and although of decreasing intensity these overtones combine to give a higher frequency tail. This tail, known as the multiphonon absorption edge, limits the IR transparency of the material. More precisely multiphonon absorption occurs when incident photons couple with TO modes in the material. The TO mode can then decay into several lower energy phonons with frequencies related to the fundamental vibration in the material. The fundamental region is termed the 1-phonon region, and the first overtone is known as the 2-phonon region, the second overtone the 3-phonon region etc. Rupprecht (1964) first suggested an appropriate formula for the IR edge:

$$x_{\mathrm{IR}}(\omega) = C \exp(-c\omega) \quad (5.3)$$

where C and c are material constants, and this behaviour has been shown to be approximately correct in crystalline materials (Deutch, 1973; Sparks and DeShazer, 1981). This type of behaviour has also been observed in GeO_2 doped SiO_2 over the wavelength range 1.5 to 5 μm by a combination of measurements on optical fibres and bulk glasses (Izawa *et al.*, 1977). The fundamental ν_3 mode occurred at 1100 cm^{-1} (9.1 μm) and the 2-phonon and 3-phonon regions at 2220 cm^{-1} (4.5 μm) and 3330 cm^{-1} (3.0 μm) respectively. Other shoulders on this edge were caused by combination bands with the other fundamental mode ν_1 at 800 cm^{-1} (12.5 μm). Beyond the 3-phonon region the attenuation becomes logarithmic with frequency as predicted from (5.3).

Figure 5.7 shows the IR absorption spectrum of a ZrF_4–BaF_2–ThF_4 glass. The fundamental peak occurs at 500 cm^{-1} with a 2-phonon shoulder at 1000 cm^{-1} and the 3-phonon region at 1500 cm^{-1}. At shorter wavelengths higher order overtones begin to overlap and the edge becomes increasingly featureless. Bendow (1977) has shown that, in general, ionic materials have little structure in their IR edges when compared with semiconductor materials, and has attributed this to a larger anharmonicity and broader density of states in ionic materials.

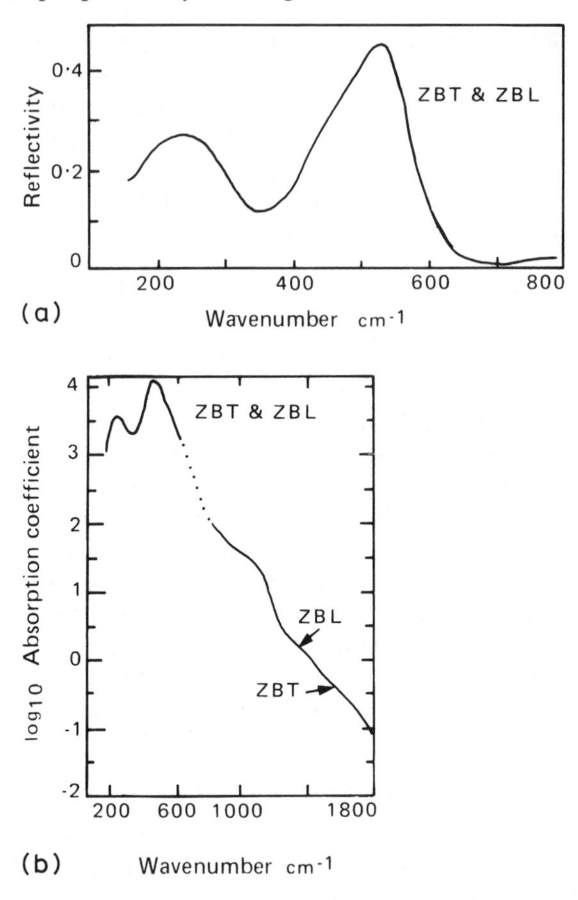

Fig. 5.7 IR absorption spectrum (a) in the fundamental and (b) in the multiphonon region in a ZrF_4–BaF_2–ThF_4 glass (after Bendow *et al.*, 1982b). The data combine FIR reflectivity results with IR transmission on thinned samples.

Figure 5.8 shows the IR edges in fluorozirconate and fluorohafnate glasses as given by Matecki *et al.* (1983) and compares them with a BaF_2–ZnF_2–YbF_3–ThF_4 glass and a mixed halide CdF_2–$CdCl_2$–$BaCl_2$ composition. The heavier anions shift the fundamental and hence give improved IR transmission. Moynihan *et al.* (1981) quoted a set of semi-empirical rules for predicting multiphonon behaviour in halide glasses:

1. For a pair of binary isostructural solid compounds containing a common atom (e.g. CaF_2 and BaF_2, KF and KCl) the member with the heavier atomic weight will exhibit a lower multiphonon absorption at a given frequency. (This results from the increased reduced mass and shift of the fundamental in the Szigeti equation);

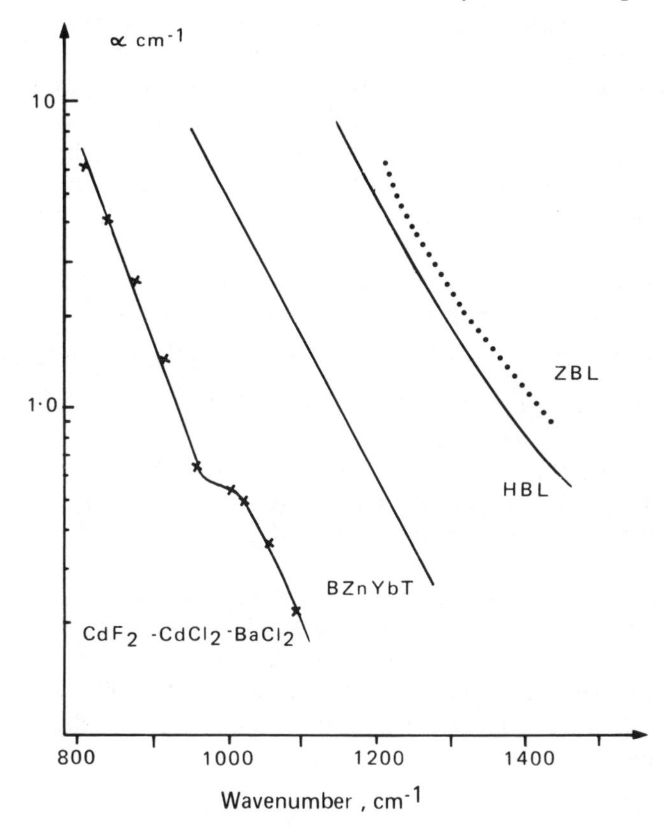

Fig. 5.8 IR edges for a range of halide glasses (after Matecki *et al.*, 1983).

2. For a pair of binary solid compounds with a common anion and with cations of similar atomic weight (e.g. KF and CaF_2) the lower charge cation will have the lower multiphonon absorption. (This results from the reduced force constant);
3. For ternary solids with a common anion the multiphonon absorption is given by a volume fraction weighted sum of attenuation coefficients of the end-member binary compounds.

Moynihan *et al.* came to the conclusion that monovalent cation fluorides no lighter than NaF, divalent fluorides no lighter than CaF_2, trivalent fluorides no lighter than LaF_3 and tetravalent fluorides no lighter than ThF_4 should contribute negligibly to the IR edge in fluoride glasses. The effect of AlF_3 on shifting the edge to shorter wavelengths was mentioned earlier.

An extrapolation of multiphonon edges back to 2.5 μm, on fluoride glasses that have been used to fabricate optical fibres, has been given by

Poignant (1982). He estimated an absorption near 10^{-2} dB km^{-1} depending on glass composition, but also commented that extrapolations from bulk measurements are unreliable since extrinsic impurity absorptions can modify the results. France *et al.* (1986c) have measured the position of the multiphonon edge in a ZBLAN glass on bulk specimens and fibres of lengths up to 100 m. As a result they have been able to obtain a more accurate estimate of 0.007 dB km^{-1} as the contribution of IR absorption to losses at 2.5 μm.

Temperature dependence of the multiphonon region has been evaluated by Bendow (1973) who extended (5.3) to the form:

$$x_{IR}(\omega,T) = x_0[N(\omega_0) + 1]^{\omega/\omega_0}[N(\omega) + 1]^{-1} \exp(-C\omega) \qquad (5.4)$$

where $N(\omega) = [\exp(\omega/kT) - 1]^{-1}$ is the Bose–Einstein function, ω_0 is an average optical phonon frequency, x_0 is a measure of the vibrational oscillator strength for coupling to light, and C is a measure of the vibrational anharmonicity. Drexhage *et al.* (1982) have shown that experimental data on the temperature dependence of the multiphonon edge in a BaZnYbTh glass agrees well with theory.

To extend transmissions to longer wavelengths other halides have been investigated. As an example Yamane *et al.* (1985) have examined the possibility of using $ZnCl_2$–KBr–$PbBr_2$ glasses for CO_2 laser power transmission and Angell *et al.* (1985) describe a number of lead and cadmium chlorides and iodides which form glasses in combination with alkali halides.

5.6 ULTRAVIOLET ABSORPTION

Insulating crystalline materials exhibit an energy gap (E_g) between their valence and conduction bands. Incident photons with energies greater than E_g will be absorbed and promote an electron from the valence into the conduction band. Although glasses lack long range periodicity, their short range order gives them similar absorption characteristics but usually with less well-defined structure, and consequently glasses also exhibit absorptions at high frequencies due to electronic transitions. This results in a high frequency cut-off given approximately by λ_0 (μm) $= 1.24/E_g$ (eV).

As with IR spectra, a technique frequently used to investigate the UV spectrum is reflectance. The UV reflectance spectra of SiO_2 (Phillip, 1966) and BeF_2 (Williams *et al.*, 1981) are again very similar, as for the IR, and confirm their structural relationship. The peaks observed correspond to different electronic transitions with the first absorption in SiO_2 at 10.3 eV (0.12 μm) well above the conduction band edge determined from photoconductivity (DiStefano and Eastman, 1971) to be at 8.9 eV (0.14 μm). The 10.3 eV peak has been attributed to an exciton resonance in the conduction band. Excitons are bound electron–hole pairs generated by photon absorption. Since the electron and hole have an attractive Coulomb interaction, stable bound states can be formed below the band gap. Transitions from the valence bands to exciton levels can therefore be observed at energies less than E_g. This has been observed in BeF_2 with the first 12.8 eV peak (0.08 μm) attributed to an exciton absorption, well below the direct allowed band edge estimated by Williams *et al.* to lie at 13.8 eV (0.07 μm).

UV reflection spectra for barium fluorozirconate glasses were obtained by Izumitani and Hirota (1985) who found a peak at 11 eV corresponding to bridging fluoride ions with weaker peaks at 9.5 and 7.4 eV assigned to non-bridging fluorides.

As the frequency of incident photons is increased the number absorbed will also increase. The absorption edge is not however a step function, but is a smoothly increasing monotonic function of frequency. An empirical rule characterizing this edge was first proposed by Urbach (1953). He reported that the absorption coefficient could be fitted to:

$$\alpha_{uv} = \alpha_0 \exp\left[(E - E_g)/\Delta E \right]$$

This empirical relationship has been found to hold for many materials, including both crystalline and amorphous solids over several orders of magnitude (Sigel, 1971). In particular Pinnow *et al.* (1983) showed that the UV edge in silica-based optical fibres closely followed the Urbach rule.

Practical cut-off wavelengths are often difficult to determine since impurities can mask their positions. Several 3d transition metal ions have charge transfer absorptions in the UV; Fe^{3+} and Cu^+ can all be a problem. The cut-off wavelength in SiO_2 has been reported to lie at 0.16 μm (Sigel, 1977). In BeF_2 it has been measured as close to 0.15 μm (Baldwin *et al.*, 1981). Fluorozirconates have cut-offs between 0.20 and 0.25 μm. Processing conditions are also reported to have a marked effect on the UV edge. For example Brown *et al.* (1982) showed that melting in Pt crucibles under CCl_4-containing atmospheres introduced an extra absorption at 350 nm and probably associated with Cl incorporation into the melt. Shelby *et al.* (1986) observed two peaks at 250 and 350 nm whose size apparently depended on the NaF concentration in the glass.

5.7 INTRINSIC SCATTERING

In addition to absorption, transmission losses in glasses are also caused by scattering, the energy deflected from the initial light path being lost. Shroeder (1977) in an extensive review has applied the theory of scattering to vitreous materials. Essentially scattering in any material is caused by small changes in density and hence permittivity ($\varepsilon \propto n^2$) of the material.

Lines (1984a and b) has shown that the scattering loss coefficient α is given by:

$$\alpha = 8\pi^3 <(\Delta\varepsilon)^2> v_\varepsilon/3\lambda^4$$

where λ is the wavelength, and $<(\Delta\varepsilon)^2> v_\varepsilon = \int \Delta\varepsilon(r)\, \Delta\varepsilon(0)\, d^3r$, v_ε being a correlation volume, and $\Delta\varepsilon(r)$ a spatial fluctuation in permittivity. We note immediately that the scattering loss is proportional to λ^{-4}.

Changes in ε can have several causes. In a crystalline material, thermal energy will induce changes in density and therefore induce scattering. More formally both optical and acoustic phonons will scatter incoming photons, as Raman and Brillouin scattering respectively. Both of these mechanisms are inelastic and involve changes in the frequency of the scattered light, the shift increasing for larger scattering angles. Since these are thermally induced an ideal crystal at absolute zero will not exhibit scattering. Pinnow *et al.* (1983) defined the Brillouin scattering to be:

$$\alpha_B = 8\pi^3 n^8 p^2 kT\beta_T/3\lambda^4 \tag{5.5}$$

where α_B is the Brillouin scattering coefficient, n is refractive index, p is the photoelastic coefficient, k is the Boltzmann constant, T is temperature in Kelvin and β_T is the static isothermal compressibility. Raman scattering has a similar form with a temperature dependence given by coth $(1/T)$. However, in general this is small compared to α_B and so will be ignored.

Vitreous materials also exhibit the above scattering mechanisms but in addition suffer from Rayleigh scattering, which results from static changes in density frozen into the glass on cooling from the melt. The significant parameter is fictive temperature T_f, that is the temperature corresponding to that at which the fluctuations were frozen in, i.e. near T_g. Because T_f is higher than the measurement temperature T, Rayleigh scattering is much larger than Brillouin although the two are related. Shroeder (1977) proposed the following relationship:

$$\alpha_S = \alpha_B(1 + R_{LP})$$

where R_{LP} is the Landau–Placzek ratio which is the intensity ratio of Rayleigh (I_R) to the total Brillouin ($2I_B$) scattered light. R_{LP} is proportional to T_f/T so the higher the fictive temperature the higher the Rayleigh scattering, an important point for halide glasses which have low values of T_f. In fact R_{LP} for vitreous silica is about 24 whereas for a perfect crystal it is of course zero. Hence silica has a much higher total scattering coefficient than quartz even at room temperature.

Since Rayleigh scattering in vitreous materials is dominant over Raman and Brillouin scattering, we can approximate total scattering:

$$\alpha_S \sim \alpha_R$$

In order to determine the Rayleigh coefficient an approximation can be made by using (5.5), and substituting T_f for T and β_{T_f} for β_T, so that:

$$\alpha_R = 8\pi^3 n^8 p^2 k T_f \beta_{T_f} / 3\lambda^4$$

Lines (1984a) has expressed this equation in terms of more accessible parameters and determined values of α_R. If we rewrite the equation to:

$$\alpha_S = B/\lambda^4 \text{ (dB km}^{-1}\text{)}$$

then Table 5.2 lists the results of Lines' calculations for several compositions. Crystalline SiO_2 has a much lower scattering loss than the glassy form for reasons already discussed. BeF_2 has a particularly low scattering loss due to its low T_f and low refractive index, and looks especially attractive for low-loss communication fibres. $ZnCl_2$ has a relatively high scatter loss despite its low T_f because n is high. For comparison GeS_2 is also shown with the highest value of B. A ZrF_4–BaF_2 based glass is shown in Table 5.2 as having a low value of scattering. However multicomponent glasses suffer from an additional scattering mechanism caused by small concentration fluctuations (Pinnow *et al.*, 1983). Lines determined the magnitude of the effect as:

$$\begin{aligned} \alpha_{conc}/\alpha_R &= 1.8 \text{ for } BaZr_2F_{10} \\ &= 13 \text{ for a (Ge,Bi,Tl)O glass} \end{aligned}$$

For a five or six component ZrF_4 based glass α_{conc} will probably be even larger than for $BaZr_2F_{10}$, but scattering is still likely to be comparable to that of vitreous SiO_2.

Other predictions of scattering in ZrF_4 glasses include $B = 0.40$ dB km^{-1} by Poignant (1981) and 0.112 by Shibata *et al.* (1981). However the latter estimate ignored concentration fluctuations.

Experimental measurements have only been made on ZrF_4 based glasses and data on BeF_2 and $ZnCl_2$ are not available. Unfortunately most results

Table 5.2 Calculated values of scattering loss in dB km^{-1} at 1 μm (B), taken from Lines (1984a and b)

Material	T_f (K)	n_D	B (μm^4 dB km^{-1})
SiO_2 (quartz)	–	1.45	0.019
SiO_2 (glass)	1473	1.45	0.47
GeO_2	840	1	0.79
BeF_2	600	1.27	0.064
$BaZr_2F_{10}$	600	1.51	0.12 (0.34*)
$ZnCl_2$	400	1.7	1.10
GeS_2	760	2	18.5

* Including concentration fluctuations.

also include the effects of macroscopic inclusions which frequently give a high background scatter and mask the true intrinsic scattering. Nevertheless Tran *et al.* (1982c and 1983b) have measured Rayleigh scattering in both bulk glasses and fibre at short wavelengths and recently Tran *et al.* (1986) obtained values in the 2.4–3.2 μm region. Their values for B were 0.41–0.69 dB km^{-1} in bulk glass and 3.16 in fibre and they confirmed the $1/\lambda^4$ dependence. Ohishi *et al.* (1985a) also found a somewhat higher value for B in a fluoride fibre, viz. 1.25. The lowest value of B found for a fibre to date was recently reported as 1.16 by Moore *et al.* (1986). The origin of the discrepancy between bulk and fibre is unclear but Tran suggested the formation by drawing induced scattering centres whose size is smaller than λ. Tran *et al.* (1986) also noted increased scattering from PbF_2 containing glasses, and commented on the increased melting times to achieve homogeneity. Schroeder *et al.* (1984) measured Rayleigh scattering in bulk specimens prepared by induction heating that were reportedly free of macroscopic inclusions. Although they did not measure wavelength dependence, values of R_{LP} were reported in the range 14–30 for a ZBLA glass and values of B were in the range 0.14 to 0.30 dB km^{-1} (cf. 0.66 in SiO_2), confirming that ZrF_4 glasses have low Rayleigh scattering.

Tran *et al.* (1984) pointed out that other optical applications than low-loss fibres can require low scattering, e.g. tactical weapons systems and wavefront sensing.

5.8 MINIMUM INTRINSIC LOSSES

Having considered the intrinsic losses that are present in halide glasses it is now possible to construct minimum loss V-curves. At the wavelengths of interest, absorption as a result of electronic transitions (Urbach edge) can be neglected. By combining the scattering data given by Lines with that of multiphonon edges shown in the references, Fig. 5.9 shows V-curves for SiO_2, GeO_2, BeF_2, $ZnCl_2$ and As_2S_3. Vitreous $ZnCl_2$ has the lowest ultimate predicted loss of about 0.0034 dB km^{-1} at 4.6 μm but BeF_2 is only just higher with a loss of 0.005 dB km^{-1} at a much shorter wavelength of 2.1 μm. Because of its low scattering loss BeF_2 still apppears to be a promising candidate and will probably be exploited further as techniques are developed to cope with its toxicity. Moreover its loss is only 0.014 dB km^{-1} at 1.55 μm, a factor of 10 better than SiO_2, where systems have already been extensively developed. GeO_2 shows a factor of two improvement over SiO_2, while the chalcogenide glasses are completely ruled out as candidates for low loss fibre. The ZrF_4 based glass has a minimum of 0.013 dB km^{-1} at 2.5 μm. This is not as low as originally anticipated, due mainly to the necessary presence of AlF_3, but is sufficiently low to make them attractive. The results are summarized in Table 5.3.

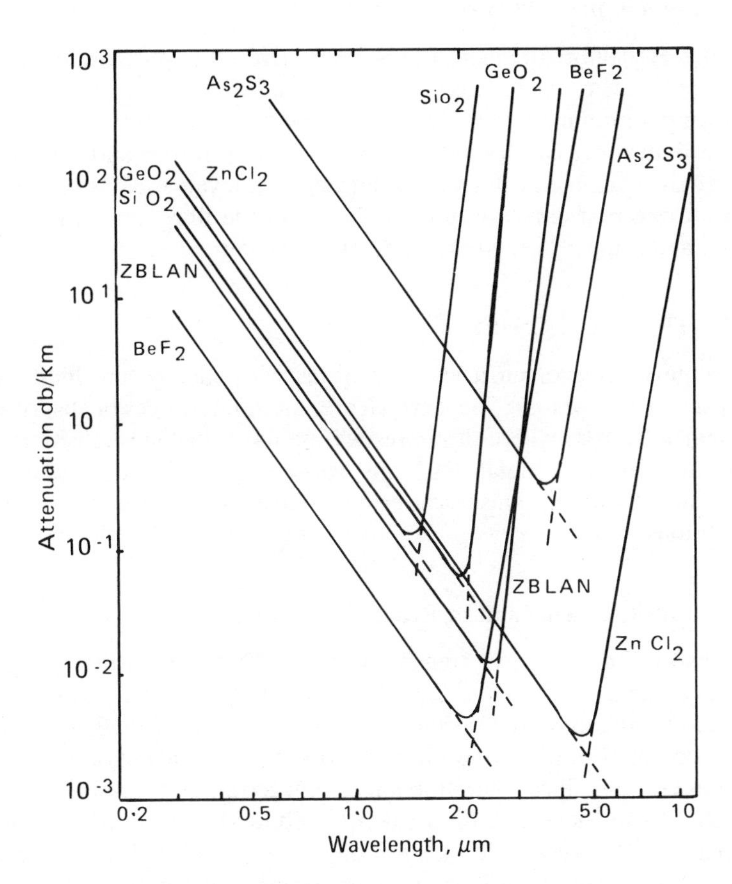

Fig. 5.9 Absorption and scattering losses for a range of optical materials plotted on a log–log scale which gives straight lines for Rayleigh scattering but a slight curvature to the multiphonon edge (after France *et al.*, 1986)

Table 5.3 Predicted minimum absorption losses for a range of glasses and the wavelengths for maximum transmission

Material	*Minimum loss* (dB km^{-1})	*Wavelength* (μm)
As_2S_3	0.33	3.6
SiO_2	0.14	1.6
GeO_2	0.06	2.0
ZBLAN fluoride	0.013	2.5
BeF_2	0.005	2.1
$ZnCl_2$	0.0034	4.6

In the best fluorozirconate fibres manufactured to date losses as low as 1 dB km^{-1} over short lengths have been observed with loss curves displaying a minimum near 2.5 μm but many extrinsic sources of loss still exist and these are considered next. Suitable transmitters and detectors are not widely available at this wavelength but Eknoyan *et al.* (1985) recently reported the first data transmission in a fluoride fibre at a wavelength of 2.6 μm and using a Ti:$LiNbO_3$ modulator as a source.

5.9 EXTRINSIC LOSSES

The requirements of ultra low loss fibres will place severe limits on all extrinsic sources of loss. Those which are potential sources of absorption in the near infra-red, where the fibres will be used, include transition metal (TM) and rare earth metal (RE) ions, various molecular species, dissolved gases and hydroxyl ions, scattering centres and fibre core-diameter fluctuations.

5.9.1 Transition metal and rare earth absorption

The absorption spectra of transition metal (TM) ions in fluoride glasses have been analysed by Ohishi *et al.* (1983) and France *et al.* (1986b), who showed that the ions are in octahedral co-ordination although with varying degrees of distortion. As a result their extinction coefficients are lower than those in vitreous silica where the ions are in tetrahedral co-ordination. On the other hand because of the weaker ligand field strength of fluoride ions compared with oxygen, the absorption peaks are shifted further into the infra-red, i.e. closer to ultimate fibre operating wavelengths.

The oxidation state of the ions affects the absorption observed and is one method by which some control can be achieved. As a result some authors have experimented with the use of an oxidizing atmosphere for melting as opposed to the neutral conditions of N_2 (O_2 by France *et al.* (1985) and NF_3 by Nakai *et al.* (1985a)). The effect of its use of the Fe and Cu absorption is particularly dramatic since Fe^{3+} has a half completed shell of d electrons and Cu^+ has a completed shell. For these reasons absorptions for Cu^+ and Fe^{3+} in the infra-red are weak, being spin forbidden. Under oxidizing conditions therefore iron is converted to Fe^{3+} and its optical absorption in the infra-red is eliminated. While Cu is then present in its oxidized state, which does absorb, the tail of its absorption is not as large as that for Fe^{2+} and therefore oxidizing conditions are preferred.

Apart from control by oxidation state it is also important to minimize the levels of impurity present, not only of Fe and Cu but also of other significant contaminants such as Co, Ni and Cr. Table 5.4 lists the absorptions of important TM ions at 2.55 μm obtained by extrapolation of absorption spectra of deliberately doped bulk specimens. Attention must be paid to

Table 5.4 The absorptions at 2.55 μm of a range of TM ions in a HMF fluoride glass (France *et al.*, 1986b; Ohishi *et al.*, 1983) and the levels of these contaminants observed in a fibre showing total absorption losses at at 2.55 μm of 1 dB km^{-1} (France *et al.*, 1986b)

Species	*Loss at 2.55 μm* (dB km^{-1} ppb^{-1})	*Observed conc.* (ppb)
Fe^{2+}	2.6×10^{-2}	<0.1
Fe^{3+}	$<10^{-3}$	~100
Cu^{+}	$<10^{-6}$	–
Cu^{2+}	10^{-4}	40
Co^{2+}	2.8×10^{-2}	15
Ni^{2+}	0.4×10^{-2}	20
Nd^{3+}	2.0×10^{-2}	20

contamination of raw materials, via melt atmosphere and via the crucible material using conventional melting.

Purification methods being adopted include sublimation, re-crystallization, ion exchange, electrolytic processing, liquid distillation and chemical vapour purification; purity levels down to ppb are now being claimed as in Table 5.4. One difficulty is how to measure impurity levels as low as this in order to assess the success of the purification method. While spectroscopic measurements on the fibres themselves provide one tool and the ultimate measure of success, it is difficult to separate out the many overlapping component peaks and to obtain reliable estimates of their separate contents. Other tools that have been used include mass spectrometry, atomic absorption, emission spectrography and ion sensitive electrode techniques. More extensive information on purification and analysis is contained in a range of reviews and papers within the conference proceedings edited by Lucas and Moynihan (1985) and Almeida (1986).

Apart from the TMs, rare earth ions (RE) also have absorptions in the infra-red. Because these involve f–f transitions they are much less affected by environment than are the d–d transitions of the TMs. Their peaks are therefore much sharper. As a result only a few ions have absorptions in significant parts of the infra-red spectrum (Ohishi *et al.*, 1983). Again typical absorption levels of various ions are given in Table 5.4. The most serious is Nd which has a strong absorption at 2.5 μm, just where the main transmission window is anticipated to be. This element is a common contaminant of LaF_3, a RE commonly used because of its significant stabilizing action in most HMF formulations. RE purification requires special techniques such as chromatography or ion exchange techniques because of their chemical similarity.

The optical transparency, low linear and non-linear refractive indices and high damage thresholds of many halide glasses make them suitable for consideration as host materials for TM and RE ions in optically pumped laser applications. Indeed BeF_2 glasses have been widely studied for this purpose and the initial discovery of BaF_2–ZrF_4 glasses arose out of an investigation to find crystalline fluoride hosts for Nd ions. For this reason many authors have analysed in detail not only the absorption spectra but also transition probabilities, fluorescence characteristics and stimulated emission cross-sections of a range of RE ions in halide glasses. Weber (1983) has reviewed the optical characteristics of Nd in fluoroberyllate, fluoroaluminate, HMF and chloride glasses and compared them with oxides and Lucas (1985) has discussed the behaviour of a wider range of RE ions.

The possibility of producing deliberately doped halide glasses for sensor applications has also been widely considered. For example, one application examined in detail has been the temperature dependence of the Eu^{3+} absorption pattern which gives rise to the possibility of a fibre optic thermometer with a temperature sensitivity quoted as 0.5 K in the temperature range 77 to 150 K, the sensitivity and range depending to some extent on doping level (Ohishi and Takahashi, 1986). El-Bayoumi *et al.* (1983) doped BaF_2–MnF_2–ThF_4 glasses with a wide range of RE fluorides and found that the resulting glasses had a low Verdet constant but exhibited strong paramagnetic behaviour.

5.9.2 Molecular absorptions

A variety of absorptions have been interpreted as arising from dissolved molecular species.

Hydroxyl ions are a major contaminant of fluoride glasses and a considerable amount of preparative effort is aimed at their removal (see section 5.3). Their fundamental absorption peak is at 2.9 μm and weaker combination bands are observed at 2.24 and 2.42 μm. France *et al.* (1984) reported measurements on a particularly dry ZBLAN glass fibre where the size of the OH^- peak was reduced to 20 dB km^{-1}, corresponding to only 2 ppb OH. These authors also suggested the fundamental consisted of a series of overlapping peaks and by fitting the peak shape to a sum of gaussians the extrapolated absorption spectrum had a transmission window at 2.55 μm fortuitously coinciding with the predicted low loss minimum and contributing an added absorption of less than 0.001 dB km^{-1}. They also commented that the OH peak had a significant tail extending on the long wavelength side of the peak which they attributed to H bonding. Shelby *et al.* (1986) commented that the presence of NaF in the glass apparently made the glasses more hygroscopic, increasing the size of the OH peak. Mitachi *et al.* (1986) have determined extinction coefficients for OH^- as 16.4 $l\,mol^{-1}\,cm^{-1}$ in a ZrF_4–BaF_2–GdF_3–AlF_3 glass and as 36.6 $l\,mol^{-1}\,cm^{-1}$ in an AlF_3–YF_3–BaF_2–CaF_2 glass using a combination of thermogravimetry and by following the IR absorption through hydration and dehydration. Fontenau *et al.* (1985) found values of 19.5 $l\,mol^{-1}\,cm^{-1}$ for a ZBLA glass and 31 $l\,mol^{-1}\,cm^{-1}$ for a BTYbZn glass.

Carter *et al.* (1986b) have interpreted absorptions at 4.252 and 4.755 μm in bulk glasses as arising from CO_2 and CO peaks respectively. Much weaker and quite sharp peaks were also observed in fibres in the fluoride glass transmission window at 2.675 and 2.775 μm and these were also attributed to CO_2 absorptions by reference to the absorption spectrum of gaseous CO_2. These gases were thought to arise principally from contaminated batch materials. The CO_2 absorptions could be removed by working under reducing conditions but materials purification was considered the most effective means of elimination.

NH_4^+ absorptions at 2.96 μm and 3.04 μm have also been identified in fibres, arising from the use of NH_4HF_2 as a fluorinating agent but are not expected to cause additional loss at the transmission window (France *et al.*, 1984). Other molecular species studied include SO_4^{2-} and PO_3^- but these have no significant absorptions below 4.5 μm (Poulain and Saad, 1984b).

The effect of oxide impurities in a ZBL glass has been reported by Hu Hefang and Mackenzie (1986) who deliberately doped glasses with BaO. These authors found a shift in the IR edge to shorter wavelengths and an additional absorption shoulder at 1350 cm^{-1}. They interpreted their observations as arising from multiphonon processes involving the vibration of F–Zr–O bonds at 680 cm^{-1}. They also reported that oxide impurities increase T_g, T_c and viscosity. Trégoat *et al.* (1985) have shown that RAP with CS_2 decreased the oxygen content of a BaF_2–ThF_4 glass so improving its transmission in the 8–10 μm range. Other authors have considered the effect of oxide on scattering (see section 5.9.4).

5.9.3 Radiation resistance

Since the defects produced by radiation damage absorb most strongly in the UV/visible part of the spectrum, halide glass fibres might be expected to be less affected by radiation than silica, their optimum transmission being further into the infra-red. Nevertheless at the very low losses postulated for such fibres any extrinsic absorptions potentially could reduce their transparency. Preliminary studies showed that fluoride glasses were discoloured by irradiation (e.g. Levin *et al.*, 1983). More recently Cases *et al.* (1985) and Friebele and Tran (1985a), using a combination of ESR and optical absorption spectroscopy, identified Zr^{3+}, Hf^{3+}, F_2^-, F^0 species present in HMF glasses after X-ray and γ-ray irradiation. These gave rise to absorptions at 463 nm (Zr^{3+}) and 290 nm (F_2^- and F^0), similar to equivalent defects produced in halide crystals. No defects were found associated with constituents such as La^{3+}, Al^{3+} and Li^+. Tanimura *et al.* (1985a), who irradiated with electrons, used similar techniques together with polarized photobleaching experiments to identify the nature of the defect configurations and commented that they were produced by radiolysis.

By extrapolating the Zr^{3+} peak observed in bulk samples assuming a gaussian shape, Tanimura *et al.* (1985b) suggested that in their irradiated samples absorptions were as high as 10^3 dB km^{-1} at 2 μm. Cases *et al.* (1985) found the Zr^{3+} absorption to be strongly inhibited in the presence of Pb and suggested that this was because Pb^+ was formed preferentially. However, a peak was also produced at 811 nm whose origin was interpreted as a Pb^{3+} centre. Ohishi *et al.* (1985b) found that the latter centre increased absorption out to 3300 nm but beyond this wavelength Pb doped glasses were more radiation-hard than undoped materials. These authors found that coloration of irradiated fluoride fibres was suppressed by iron or titanium doping, but cerium doping was ineffective. Friebele and Tran (1985a) pointed out that other absorptions may be present in the IR, e.g. associated with defect clusters, and therefore to reduce the zirconium absorption might not be all that is required. Zr^{3+} could also be removed by thermal bleaching requiring temperatures up to 400 K. Bansal and Doremus (1986) have studied the kinetics of this process. Colour centres associated with fluorine were only produced in low temperature experiments and were annealed out below 200 K. The only direct measurements on fibres were made by Ohishi *et al.* (1985b), who observed induced losses of between 1050 and 2300 dB km^{-1}10^{-6} roentgens in the 2.55 to 3.65 μm wavelength range when ZrF_4 fibres were exposed to γ-radiation, comparable with those induced in silica at 1.5 nm.

Glasses melted using a CCl_4 reactive atmosphere process contain significant amounts of Cl^- and measurements on such materials (Tanimura *et al.*, 1985a) or using glasses deliberately doped with chlorides (Friebele and Tran, 1985b; Griscom and Tran, 1985) have shown that FCl^- or Cl_2^- defects are also easily formed. They give rise to absorptions at 259 nm and 318 nm respectively and are considerably more thermally stable than the equivalent fluoride defects. At the same time the concentration of Zr^{3+} defects is also enhanced. For radiation resistance therefore chloride ions are undesirable.

5.9.4 Extrinsic scattering losses

Signal attenuation as a result of extrinsic scattering has been a significant problem in fluoride fibres. Generally the scattering losses can be expressed by an equation of the form:

$$\alpha_s = B\lambda^{-4} + C\lambda^{-2} + D$$

By comparing fibres of similar compositions but different scattering characteristics, Ohishi *et al.* (1985a) concluded that the extrinsic scattering could be accounted for in terms of a wavelength independent and a λ^{-2} term, although Tran *et al.* (1983a) did find an increased value of B in fibre form compared with bulk glass (see section 5.7). The origin of extrinsic scattering is partly non-uniformities in fibre geometry, which contribute to the wavelength independent loss, and partly discrete scattering centres.

Kanamori *et al.* (1986) studied the nature of the scattering centres in fibres and their loss characteristics. They found two kinds of defects, those above 1 μm consisting of microcrystallites, impurity particle inclusions and bubbles and whose number could be reduced by improved preform fabrication conditions, and those scattering centres below 1 μm which gave rise to losses following a $1/\lambda^2$ law (Mie scattering). Fibres manufactured under optimum conditions gave losses of 3 dB km^{-1} at 2.55 μm of which 2 dB km^{-1} could be attributed to scattering, the remainder being absorption.

Lu *et al.* (1984) and Lu and Bradley (1985) have identified many of the microcrystals which appear in ZBLAN glasses; LaF_3, AlF_3 and ZrF_4 are commonly present and barium fluorozirconates have been observed nucleated on small platinum particles. The significance of the thermal history of the glass melt was stressed by Ohishi *et al.* (1982) who have found that if a ZBGA melt was cooled through the zone of maximum crystal growth (400°C) at faster than a certain critical rate, then only Rayleigh scattering was observed. Similarly crystal growth can occur while reheating a glass, even at temperatures below the T_c observed using a DSC (Tokiwa *et al.*, 1985a). To reduce these effects requires careful attention to the preparation process and optimization of the glass compositions (see section 5.2). Tokiwa *et al.* (1985b) have proposed glass compositions for both core and cladding glass to allow crystal-free fluoride fibre preparation. Grodkiewicz and Van Uitert (1985) have made a similar study of a BeF_2 based glass.

Apart from devitrification, contamination of the melt can give rise to defects. Carter *et al.* (1985) found ZrO_2 particles in fibres whose origin was thought to be attack of the silica liner by volatilized ZrF_4. Mitachi *et al.* (1985) measured typical oxygen levels in preforms using a cyclotron and found typical values of 300 ppm. By deliberately doping melts with oxides they showed that observed losses were related to oxide content. At 600 ppm of oxide the excess scattering loss was 100 dB km^{-1} and was wavelength independent. Nakai *et al.* (1985c) measured light scattered from a fluoride melt as a function of temperature and deduced that oxide impurities were present even at that stage. Poulain and Saad (1984a) also reported that when refining melts at high temperature and for extended times, scattering losses increased, a result consistent with increasing oxygen levels. To reduce such effects improved levels of material purity and better fabrication conditions are necessary.

5.10 REFRACTIVE INDEX AND DISPERSION

Besides their extended IR transmission, halide glasses have unique refractive index and dispersion characteristics which make them of interest

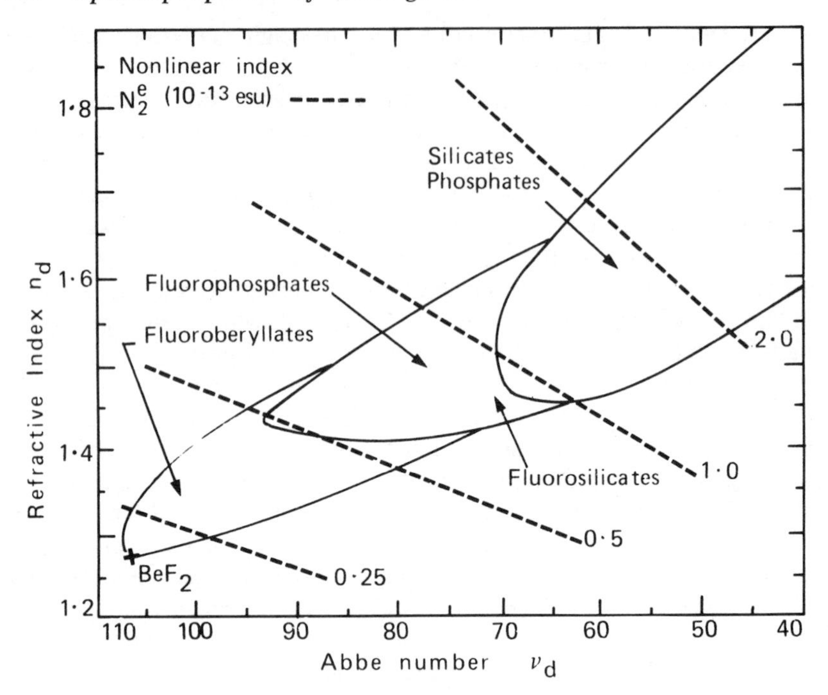

Fig. 5.10 Plot of refractive index against dispersion for a range of optical glasses (after Baldwin *et al.*, 1981).

for optical components. In particular fluoroberyllate glasses are distinguished by a low index of refraction and low dispersion (high Abbé number). These properties are summarized in Fig. 5.10. BeF_2 has the lowest n_D (1.2747) and highest Abbé number (106.8) of any inorganic material. Additions of modifying fluorides increase n_D and reduce ν.

The instantaneous index of refraction of a material is defined as:

$$n = n_0 + \Delta n$$

where n_0 is the linear index (e.g. n_D) and Δn is a change of index induced by a high power optical beam. This index is important when choosing laser optics or host glasses for lasing materials since changes in index at high laser powers can lead to self-focusing of the beam and eventual catastrophic failure. Δn is usually expressed as $n_2\langle E^2\rangle$ where $\langle E^2\rangle$ is the time averaged field intensity and n_2 is the non-linear index. Calculated values of n_2 are also plotted in Fig. 5.10. It can be seen that in particular BeF_2 has the lowest values of n_2 (0.23×10^{-13} esu).

Fluorozirconate glasses lie in an intermediate position between BeF_2 and SiO_2 in the Abbé diagram. Their indices lie in a range 1.49 to 1.53 depending on exact composition, with Abbé numbers in the range 65 to 85. Figure 5.11

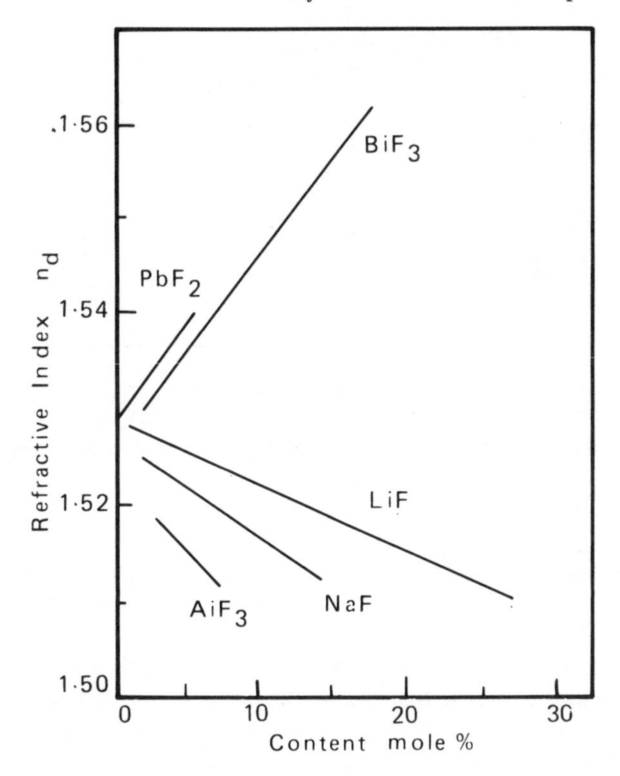

Fig. 5.11 Compositional dependence of HMF refractive index (after Takahashi *et al.*, 1981).

shows how the index can be modified by adjusting the composition. The addition of heavy polarizable fluorides increases the index and lighter less polarizable materials reduces it. The refractive index difference between core and cladding in optical waveguides can therefore be obtained by doping the core with PbF_2 or by ensuring an excess of NaF or AlF_3 in the cladding. If ZrF_4 is replaced by HfF_4 the refractive index also decreases slightly (Bendow *et al.*, 1981) and this accounts for the frequent use of HfF_4 in cladding glasses. Chloride glasses have higher indices than oxides with $ZnCl_2$ having a value of 1.7. These materials lie at the top right hand side of Fig. 5.10 between oxide and chalcogenide glasses.

In general refractive index will vary with temperature. Compositional tailoring however presents the possibility of designing glasses showing very small thermal distortion behaviour. Bendow (1984), using experimental data and theoretical estimates, suggests that HMF glasses are promising candidates for ultralow thermal distortion from the near UV through to the mid-infra-red. Measurements of the temperature dependence of refractive

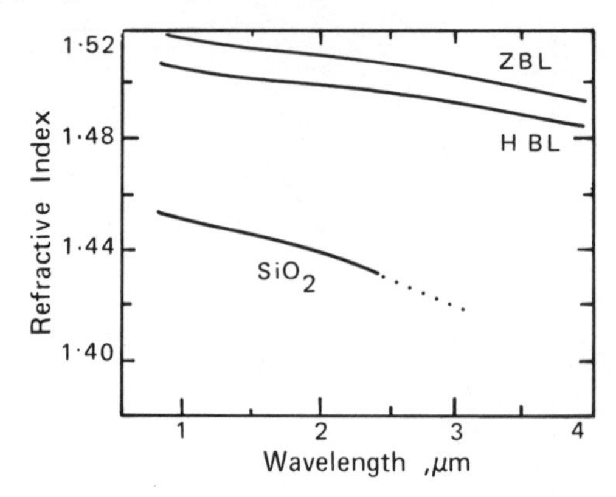

Fig. 5.12 Refractive index against wavelength for two HMF glasses (after Bendow *et al.*, 1981).

index have been made by Greason *et al.* (1985) who found values for $\partial n/\partial T$ of (14.2–14.8) $\times 10^{-6}$ °C^{-1} with very little dispersion for a fluorozirconate glass.

The variation of refractive index versus wavelength, or dispersion, has been determined for some halide glasses. In particular Bendow *et al.* (1981) measured refractive index in fluorozirconate and fluorohafnate glasses as a function of wavelength and the results are shown in Fig. 5.12. Fleming *et al.* (1985) have studied a BeF_2 based glass. Few data are available for $ZnCl_2$ glasses. Extrapolation of dispersion to longer wavelengths is frequently made by fitting a Sellmeier curve to n of the form:

$$n(\lambda) = \Sigma\, x_j\, \lambda^{(2j-6)} \tag{5.6}$$

For the case of the transmission of high rate bit streams along optical waveguides, the pulse broadening is related to the material dispersion M, where:

$$M = -(\lambda/c)\, \mathrm{d}^2 n/\mathrm{d}\lambda^2 \tag{5.7}$$

Therefore M can be determined from a knowledge of the coefficients x_j in (5.6). Of particular interest is the zero dispersion point, λ_0 where $M = 0$. At this point the bandwidth of a fibre made from the particular material will have its highest value, and if possible this should coincide with the minimum loss wavelength. From (5.7) this is equivalent to $\mathrm{d}^2 n/\mathrm{d}\lambda^2 = 0$, or the point of inflection in curves such as Fig. 5.12.

Material dispersion for SiO_2 and a ZrF_4 glass is shown in Fig. 5.13 and Fleming *et al.* (1985) have measured a BeF_2 based glass. Estimates of λ_0 can also be made by re-expressing (5.6) in terms of the Sellmeier energies introduced by Wemple (1979). Lines (1984a and b) has determined these

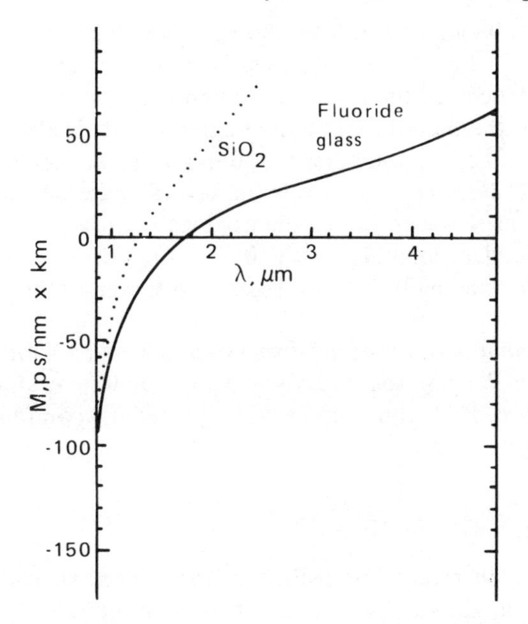

Fig. 5.13 Material dispersion for SiO_2 and a HMF glass (after Jeunhomme *et al.*, 1981).

values and they are listed in Table 5.5 for several materials. λ_0 for a fluorozirconate glass is at 1.63 μm, roughly similar to GeO_2 and significantly short of its minimum loss wavelength. However, the total dispersion in a fibre waveguide results from a combination of material dispersion and waveguide dispersion. Since the waveguide dispersion is of opposite sign to material dispersion and can be modified by changing the waveguide parameters, a careful choice of these parameters can shift the zero total

Table 5.5 Estimates of the zero dispersion wavelength for a range of glasses taken from Lines (1984a and b)

Material	λ_0 (μm)
SiO_2	1.27
GeO_2	1.69
BeF_2	1.03
Zr_2BaF_{10}	1.63
$ZnCl_2$	3.26
GeS_2	4.10

dispersion wavelength out to the 2–3 μm region. The values measured by Fleming *et al.* for doped BeF_2 glasses were higher than those calculated for pure BeF_2 and included the range of minimum loss.

The slope of the material dispersion curve $dM/d\lambda$ is also of importance since this will affect the total dispersion in a fibre waveguide. Nassau and Wemple (1982) have calculated this for several materials and shown that halide glasses have lower values than oxides (21 psec $nm^{-1}km^{-1}\mu m^{-1}$ in ZBLA fluoride glass compared with 105 in SiO_2). Brown and Hutta (1985) have shown that adding NaF to a ZBLA glass increases the gradient of the dispersion curve.

A measurement of the material dispersion curve for a fluoride glass in the range 1.1 μm to 2.1 μm was recently made by Monerie *et al.* (1985) using a monomode fibre. Eknoyan *et al.* (1985) reported data on fibre bandwidths measured on actual fibre.

5.11 OTHER PROPERTIES

Many authors have reported densities, T_g values and expansion coefficients etc. for particular glasses. In general T_g values for fluoride glasses tend to be in the range 200°C to 500°C while for the other halides the values fall below 200°C because they are less strongly bonded. Also, because the bonding is weak compared with silicates, expansion coefficients tend to be high – typically 150–200 $\times$ 10^{-7} °C^{-1} in fluoride glasses and much higher in other halides. Similarly characteristics such as microhardness decrease rapidly as the bonding becomes weaker through the sequence oxides, fluorides and other halides. Since many of the components are elements with high atomic weights, the glass densities are generally high, e.g. 4.5 g cm^{-3} for ZBLAN glasses. Systematic studies on fluorides have been carried out by Frischat *et al.* (1984) and Shelby *et al.* (1984 and 1986).

5.11.1 Viscosity

One way to categorize the effect of different fluorides on physical properties is according to whether they have a network forming, a stabilizing or a network-modifying rôle by analogy with silicate systems (Baldwin and Mackenzie, 1979a and b). On this basis fluorides such as LaF_3, AlF_3 and YF_3 tend to have a stabilizing action and therefore increase T_g and decrease α (Frischat *et al.*, 1984). Alkali fluorides on the other hand, as network modifiers, tend to decrease T_g.

Systematic studies have been carried out by Shelby *et al.* (1984 and 1986) who examined the effect of replacing BaF_2 with NaF in glasses of two compositions, one a ZBLAN glass and another a ZnBYbT composition. In both cases replacing BaF_2 by NaF on a molar basis decreased the viscosity but also lowered the expansion coefficient, in contravention to the effect in silicates, although in the latter case the Si : O ratio is maintained constant by a molar substitution whereas in the former case the Zr : F ratio decreases. The effect of LiF was greater in lowering viscosity than NaF. These authors found transformation range viscosities to be well fitted by an Arrhenius relationship with activation energies in the range 600–850 kJ mol^{-1}. However no particular trend with composition was reported.

The effect of different compositions on the slope of the temperature–viscosity curve has also been studied systematically by Tran *et al.* (1982b) who examined the low temperature behaviour of a range of compositions to discover which had the best working properties, the very steep curves found in the simple ZBLA system making low temperature fabrication very difficult. These authors found that additions such as LiF, BiF_3 and PbF_2 significantly reduced the slope of the curve giving a glass with more favourable working properties, i.e. a longer working range.

Fluorozirconate melts at high temperatures are very fluid. Some high temperature viscosity data are available and Moynihan *et al.* (1986) have tried to fit the form of the temperature–viscosity curve to the available high and low temperature data in order to allow extrapolation into the intermediate temperature region below the liquidus where crystallization rates are very fast. The Fulcher equation gives only a poor fit and Moynihan *et al.* have shown that a better although still not perfect fit is obtained with a Cohen–Grest type equation similar to that applicable for metal alloys, salts and organic melts.

Because the T_g values for the fluoride glasses are low the possibility that relaxational effects might be significant even at operating temperature must be considered. Moynihan *et al.* (1984) found that substantial relaxation occurred in a few hours at temperatures as low as 60°C below T_g; the process was extremely non-linear and self-retarding. Using the Narayanaswamy–Tool approach they predicted the rates of physical aging for a rapidly cooled ZBLA glass at temperatures close to ambient.

5.11.2 Durability

Fluorozirconate glasses are relatively stable to attack from humid atmospheres and can be left indefinitely in laboratory conditions. They do however slowly dissolve in water. ThF_4 based glasses are reported to be significantly more stable. BeF_2 develops tacky surfaces after 60 h exposure to moist atmospheres, and $ZnCl_2$ will take up water within a matter of hours.

Because of the importance of durability of fluorides over long periods of time, particularly if they are to be used in fibre form, the mechanisms of attack have been extensively examined. Fjeldly *et al.* (1984) have followed the penetration of atmospheric water into fluorohafnate and fluorozirconate glasses at different humidities by measuring surface oxygen content using

Auger spectroscopy. Penetration was shown to be slow even after a 6 month exposure to 80% RH. Trégoat *et al.* (1985) have shown that penetration of atmospheric water into a BaF_2–ZnF_2–YbF_3–ThF_4 glass at temperatures up to T_g is diffusion controlled. Leach rates are very high compared with conventional silicates; Simmons *et al.* (1983) have determined values some 10^5 times higher than for borosilicates. Doremus *et al.* (1985) have proposed a complex model to describe the observed behaviour in an aqueous environment involving water penetration, followed by solution, particularly of the more soluble components such as NaF and AlF_3, and reprecipitation of phases such as ZrF_4, and barium fluorozirconates and hydroxyfluorides. Maximum durability is therefore likely to be achieved by avoiding the more water soluble components although these are normally important in controlling other properties such as temperature–viscosity relationships and stability. Barkatt and Boehm (1984) have suggested that where HMF glasses are prepared in the presence of a large excess of NH_4HF_2 their durability is impaired and dramatic improvements can be achieved by remelting. Burman *et al.* (1984) have also observed that surfaces drawn direct from the melt without subsequent preparation are more durable. Successful coatings of MgF_2, SiO_2 and Al_2O_3/SiO_2 have been applied using ion assisted deposition techniques to fluoride glass substrates at near ambient temperatures and have been shown to have good adhesion and to improve durability (McNally *et al.*, 1986).

Jiang Zhong *et al.* (1986) determined the durability of some $ZnCl_2$ based glasses.

5.11.3 Strength

Strength measurements on HMF fibres have given, typically, values of 500 MN m^{-2} (1% strain) (France *et al.*, 1983). Such a value is clearly lower than the expected ultimate value and arises because of a number of damage mechanisms causing surface degradation. Lau *et al.* (1985a) have shown that fibre drawing temperature significantly influences strength, and have observed the presence of small crystals acting as stress raisers at the glass surface. Ohishi *et al.* (1984) have found that water in the atmosphere during fibre pulling can also encourage crystal growth. Lau *et al.* (1985b) have also found that a humid atmosphere will cause the strength to approximately halve in 4 days and pointed that the Teflon coatings often used for jacketing fibres are not an adequate protection as a moisture barrier. Tran *et al.* (1984b) report the successful application of diamond-like coatings to provide a hermetic seal. Nakata *et al.* (1985) used a protective jacket of chalcogenide glass to reduce attack in a humid atmosphere. Microhardness values for fluorides are also lower than for silica so that mechanical damage will be more severe. This will be even more true for the other halides.

Interestingly Vaughn and Risbud (1984) doped fluoride glasses with a small quantity of nitride and reported an improvement in microhardness.

Several authors have made more fundamental measurements of the fracture toughness of fluorides and of stress corrosion characteristics of the glasses. Tran *et al.* (1984b) and Pantano (1986) have provided excellent summaries of work in these two areas. On this basis Tran *et al.* concluded that ultimate strengths of up to 3500 MN m^{-2} might be achieved. They also reviewed possible strengthening mechanisms which included surface crystallization, ion-exchange and compressive coatings.

5.12 FABRICATION

Fabrication of bulk optical components can follow traditional fabrication routes subject to the condition that glass stability will often limit the thickness of pieces which can be obtained. Turk (1981) has investigated the possibility of using a hot pressing technique to produce prepressed optically finished components. An optically finished tungsten carbide die was used and pressing was carried out in an argon atmosphere at 315°C using a BaF_2–ZrF_4–ThF_4 glass with successful results. Attempts to hot forge smaller pieces of a ZBLA glass into a larger component were unsuccessful, however, the original boundaries remaining visible.

Most interest has centred around the desire to manufacture halide fibres. The simplest approach has been to cast rods, to coat with a vacuum shrunk PTFE coating and to use the latter as the cladding. Fluorine produced by breakdown of the polymer during fibre drawing is claimed to reduce crystallization. The choice of polymer cladding is severely limited by the high expansion coefficient of the glass and the need to obtain a match. An alternative approach again to produce simple unclad fibres was proposed by Tran *et al.* (1982b). They pulled fibres using two crucibles one placed above the other. Melting took place in the top crucible while the lower one was kept at lower temperature (317°C) suitable for fibre pulling. At the appropriate time the melt flowed from the top crucible into the lower one which also had a hole in it from which fibres could be pulled. Using this technique fibres up to 1km long and with a diameter of 125 μm were pulled.

More sophisticated techniques have had to be developed for making core-clad fibres. The most common technique utilizes the built in casting approach. After pouring the cladding composition carefully into a mould to avoid trapping bubbles the melt is allowed to cool for a few seconds, the central hot liquid still remaining is then tipped out and replaced by the core glass composition (Mitachi *et al.*, 1981). An improvement of this approach utilizes rotational casting to improve the cylindrical symmetry of the preform produced (Tran *et al.*, 1982a). Typical core to cladding diameter ratios produced by this technique are 1 : 5. Once a suitable preform has been made it is then pulled using conventional techniques although the low T_g and steep temperature–viscosity curve mean that a furnace with a very narrow hot zone and excellent temperature control are required.

Recently Tokiwa *et al.* (1985a) have described the application of the double crucible method to produce monomode fibre in lengths up to 500 m as well as multimode fibre. Poignant (1985) has also paid attention to the manufacture of single mode material using a similar technique and Monerie *et al.* (1985) have described the manufacture and characterization of monomode fibre. An alternative approach has been the use of a reactive vapour phase transport process where halides have been diffused into the centre of a flouride glass tube so building up a concentration profile. Single mode and multimode graded index preforms have been made in this way (Tran *et al.*, 1984b).

5.13 CONCLUSIONS

A massive investment of effort into the study of halide glasses for a wide range of applications has taken place over the last few years. This chapter has concentrated on glasses for fibre optics and has shown how improvements in fabrication techniques and glass properties have been made which have allowed fibres to be manufactured which, in short lengths at least, are approaching the transparency of the best SiO_2 fibres even though to date only conventional melting techniques have been used. Already fibres can be made which are commercially useful for specialist applications. The search to develop ultralow loss materials will continue although the achievement of ultimate transparencies will probably require the development of suitable vapour phase routes.

Apart from fibre applications these new categories of glasses have a range of possible applications as bulk materials particularly where mid-infra-red transparency is required and there is no doubt that interest in them will continue for many years to come.

REFERENCES

Almeida, R. M. (ed.) (1986) *NATO ARW on Halide Glasses for Infrared Fiberoptics, Vilamoura, Portugal*, Martinus Nijhoff, Dordrecht, in press.
Almeida, R. M. and Mackenzie, J. D. (1982) *J. Mater. Sci.* **17**, 2533–8.
Angell, C. A., Changle Liu and Sundar, H. G. K. (1985) *Mater. Sci. Forum*, **5**, 189–92.
Angell, C. A. and Wong, J. (1970) *J. Chem. Phys.*, **53**, 2053–66.
Baldwin, C. M., Almeida, R. M. and Mackenzie, J. D. (1981) *J. Non-Cryst. Solids*, **43**, 309–44.
Baldwin, C. M. and Mackenzie, J. D. (1979a) *J. Non-Cryst. Solids*, **31**, 441–5.
Baldwin, C. M. and Mackenzie, J. D. (1979b) *J. Am. Ceram. Soc.*, **62**, 537–8.
Banerjee, P. K., Bendow, B., Drexhage, M. G. *et al.* (1981) *J. Phys. (Paris), Colloq.*, **42**, C6, 75–7.

Bansal, N. P. and Doremus, R. H. (1983) *J. Am. Ceram. Soc.*, **66**, C132–C133.
Bansal, N. P. and Doremus, R. H. (1985) *J. Mater. Sci.*, **28**, 2794–800.
Bansal, N. P. and Doremus, R. H. (1986) *Mater. Res. Bull.*, **21**, 281–8.
Bansal, N. P., Doremus, R. H., Bruce, A. J. and Moynihan, C. T. (1983) *J. Am. Ceram. Soc.*, **66**, 233–8.
Bansal, N. P., Doremus, R. H. and Moynihan, C. T. (1985) *Mater. Sci. Forum*, **5**, 211–8.
Barkatt, A. and Boehm, L. (1984) *Mater. Lett.*, **3**, 43–5.
Bendow, B. (1973) *Appl. Phys. Lett.*, **23**, 133–4.
Bendow, B. (1977) *Am. Rev. Mater. Sci.*, **7**, 23–53.
Bendow, B. (1984) *Proc. SPIE Int. Soc. Opt. Eng.*, **505**, 81–9.
Bendow, B., Banerjee, P. K., Drexhage, M. G. *et al.* (1982a) *J. Am. Ceram. Soc.*, **65**, C8–C9.
Bendow, B., Banerjee, P. K., Drexhage, M. G. *et al.* (1983) *J. Am. Ceram. Soc.*, **66**, C64–C66.
Bendow, B., Banerjee, P. K. Mitra, S. S. *et al.* (1984) *J. Am. Ceram. Soc.*, **67**, C136–C138.
Bendow, B., Brown, R. N., Drexhage, M. G. *et al.* (1981) *Appl. Opt.*, **20**, 3688–90.
Bendow, B., Galeener, F. and Mitra, S. S. (1982b) *1st Int. Symp. on Halide and Other Non-Oxide Glasses, Cambridge.*
Bendow, B., Galeener, F. and Mitra, S. S. (1985) *J. Am. Ceram. Soc.*, **68**, C92–C95.
Berkey, G. E. and Sarkar, A. (1982) *Topical Meeting on Optical Fibre Communications, OFC2, Pheonix, Arizona*, p. 54.
Brown, R. W., Bendow, B., Drexhage, M. G. and Moynihan, C. T. (1982) *Appl. Opt.*, **21**, 361–3.
Brown, R. N. and Hutta, J. J. (1985) *Appl. Opt.*, **24**, 4500–2.
Burman, C., Lanford, W. A., Doremus, R. H. and Murphy, D. (1984) *Appl. Phys. Lett.*, **44**, 845–6.
Busse, L. E., Lu, G., Tran, D. C. and Sigel, G. H. (1985) *Mater. Sci. Forum*, **5**, 219–28.
Carter, S. F., France, P. W., Moore, M. W. and Williams, J. R. (1985) *Mater. Sci. Forum*, **5**, 397–403.
Carter, S. F., France, P. W., Moore, M. W. *et al.* (1986a) *Phys. Chem. Glasses*, in press.
Carter, S. F., France, P. W. and Williams, J. R. (1986b) *Phys. Chem. Glasses*, **27**, 42–7.
Cases, R., Griscom, D. L. and Tran, D. C. (1985) *J. Non-Cryst. Solids*, **72**, 51–63.
Cheng Ji-jian and Bao Shan-zhi (1985) *J. Phys. (Paris), Colloq.*, **46**, 449–53.
Deutsch, T.F. (1973) *J. Phys. Chem. Solids*, **34**, 2091–4.
DiStefano, T. H. and Eastman, D. E. (1971) *Phys. Rev. Lett.*, **27**, 1560–2.
Doremus, R. H., Murphy, D., Bansal, N. P. *et al.* (1985) *J. Mater. Sci.*, **20**, 4445–53.
Drexhage, M. G. (1985) *Treatise on Materials Science and Technology. Glass IV* (eds M. Tomozawa and R. Doremus), Academic Press, New York.
Drexhage, M. G., Bendow, B., Brown, R. N. *et al.* (1982) *Appl. Opt.*, **21**, 971–2.
Durteste, Y., Monerie, M. and Lamouler, P. (1985) *Electron. Lett.*, **21**, 723–4.
Eknoyan, O., Moeller, R. P., Bulmer, C. H. *et al.* (1985) *IOOC–ECOC '85, Venice, Italy*, Technical Digest Vol. 1, 743–6.
El-Bayoumi, O. H., Drexhage, M. G., Bruce, A. J. *et al.* (1983) *J. Non-Cryst. Solids*, **56**, 429–34.
El-Bayoumi, O. H., Suscavage, M. J. and Bendow, B. (1985) *J. Non-Cryst. Solids*, **73**, 613–24.
Esnault-Grosdemouge, M. A., Matecki, M. and Poulain, M. (1985) *Mater. Sci. Forum*, **5**, 241–55.

Fjeldly, T. A., Hordvik, A. and Drexhage, M. G. (1984) *Mater. Res. Bull.*, **19**, 685–91.
Fleming, J. W., Grodkiewicz, W. H., Modugno, S. A. and Van Uitert, L.G. (1985) *Mater. Sci. Forum*, **5**, 361–9.
Fontenau, G., Trégoat, D. and Lucas, J. (1985) *Mater. Sci. Bull.* **20**, 1047–51.
France, P. W., Carter, S. F. and Harris, E. A. (1986a) *Phys. Chem. Glasses*, in press.
France, P. W., Carter, S. F. and Parker, J. M. (1986b) *Phys. Chem. Glasses*, **27**, 32–41.
France, P. W., Carter, S. F., Moore, M. W. and Williams, J. R. (1985) *Electron. Lett.*, **21**, 602–3.
France, P. W., Carter, S. F., Moore, M. W. and Williams, J. R. (1986c) *NATO ARW on Halide Glasses for Infrared Fiberoptics, Vilamoura, Portugal* (ed. R.M. Almeida), Martinus Nijhoff, Dordrecht, in press.
France, P. W., Carter, S. F. and Williams, J. R. (1984a) *J. Am. Ceram. Soc.*, **67**, C234–C244.
France, P. W., Carter, S. F., Williams, J. R. and Beales, K. J. (1983) *2nd Int. Symp. on Halide Glasses*, Troy, New York, Paper 11.
France, P. W., Carter, S. F., Williams, J. R. *et al.* (1984b) *Electron. Lett.*, **20**, 607–8.
Friebele, E. J. and Tran, D. C. (1985a) *J. Non-Cryst. Solids*, **72**, 221–32.
Friebele, E. J. and Tran, D. C. (1985b) *J. Am. Ceram. Soc.*, **68**, 279–81.
Frischat, G. H., Knaust, J. and Chandreshekar, G. V. (1984) *Glastech. Ber.*, **57**, 173–6.
Galeener, F. (1978) *Solid State Commun.*, **25**, 405–8.
Galeener, F. and Lucovsky, G. (1970) *Structure and Excitations of Amorphous Solids*, American Institute of Physics, New York, p. 223.
Gannon, J. R. (1981) *Proc. SPIE Int. Soc. Opt. Eng.* (Infrared Fibres (0.8–12 μm), **266**, 62–8.
Goldschmidt, V. M. (1926) *Skr. Nor. Vidensk. Akad. Oslo, I: Math-naturwiss. Kl.*, No. 8, 7–156.
Greason, P., Detrio, J., Bendow, B. and Martin, D. J. (1985) *Mater. Sci. Forum*, **6**, 607–10.
Griscom, D. L. and Tran, D. C. (1985) *J. Non-Cryst. Solids*, **72**, 159–63.
Grodkiewicz, W. H. and Van Uitert, L. G. (1985) *J. Non-Cryst. Solids*, **74**, 223–8.
Hu Hefang and Mackenzie, J. D. (1986) *J. Non-Cryst. Solids*, **80**, 495–502.
Hutton, J. J., Suscavage, M. J., Drexhage, M. G., and El-Bayoumi, O. H. (1984) *SPIE. Infrared Optical Materials and Fibres III*, **484**, 83–8.
Izawa, T., Shibata, N. and Takeda, A. (1977) *Appl. Phys. Lett.*, **31**, 33–5.
Izumitani, T. and Hirota, S. (1985) *Mater. Sci. Forum*, **6**, 645–7.
Jeunhomme, L., Poignant, H. and Monerie, M. (1981) *Electron. Lett.*, **17**, 808–9.
Jiang Zhonghong, Hu Xinyuan and Hou Lisong (1986) *J. Non-Cryst. Solids*, **80**, 543–49.
Kanamori, T., Hattori, H., Sakaguchi, S. and Ohishi, Y. (1986) *Jpn. J. Appl. Phys. Part 2*, **25**, L203–L205.
Kanamori, T., Oikawa, K., Shibata, S. and Manatabe, T. (1981) *Jpn. J. Appl. Phys.*, **20**, L326–L328.
Kanamori, T. and Takahashi, S. (1986) *Jpn. J. Appl. Phys. Part 2*, **24**, L758–L60.
Kaufman, L., Agren, J., Nell, J. and Hayes, F. (1982) *1st Int. Symp. on Halide and other Non-Oxide Glasses, Cambridge.*
Kaufman, L. and Birnie, D. (1983) *2nd Int. Symp. on Halide Glasses*, Troy, New York, Paper 17.
Kawamoto, Y. (1984) *Phys. Chem. Glasses*, **25**, 88–91.
Kondrat'ev, Yu. N., Petrovskii, G.T. and Raaben, E.L. (1969) *Zh. Prikl. Spectrosk.*, **10**, 69–72.

Lau, J., Nakata, A. M. and Mackenzie, J. D. (1985a) *J. Non-Cryst. Solids*, **70**, 233–44.
Lau, J., Nakata, A. M. and Mackenzie, J. D. (1985b) *J. Non-Cryst. Solids*, **74**, 229–36.
Lecoq, A. and Poulain, M. (1980a) *Verres Réfract.*, **34**, 333–42.
Lecoq, A. and Poulain, M. (1980b) *J. Non-Cryst. Solids*, **41**, 209–17.
Levin, K. H., Tran, D. C., Ginther, R. J. and Sigel, G. H. (1983) *Glass Technol.*, **24**, 143–5.
Lines, M. E. (1984a) *J. Appl. Phys.*, **55**, 4052–7.
Lines, M. E. (1984b) *J. Appl. Phys.*, **55**, 4058–63.
Lu, G. and Bradley, J. (1985) *Mater. Sci. Forum*, **6**, 551–9.
Lu, G., Fisher, C. F., Burk, M. J. and Tran, D. C. (1984) *Ceram. Bull.*, **63**, 1416–8.
Lucas, J. (1985) *J. Less-common Metals*, **112**, 27–40.
Lucas, J. and Moynihan, C. T. (1985) *Mater. Sci. Forum*, **5/6**, Proc. 3rd Int. Conf. on Halide Glasses, Rennes, June 1985.
Ma Fu Ding, Lau, J. and Mackenzie, J. D. C. (1984) *J. Non-Cryst. Solids*, **80**, 538–42.
MacFarlane, D. R., Matecki, M. and Poulain, M. (1984) *J. Non-Cryst. Solids*, **64**, 351–62.
Matecki, M. (1983) *Mater. Res. Bull.*, **18**, 293–300.
Matecki, M., Poulain, M. and Poulain, M. (1983) *Proc. 2nd. Int. Symp. on Halide Glasses*, Troy, New York, Paper 27.
McNally, J. J., Al-Jumaily, G. A., McNeill, J. R. and Bendow, B. (1986) *Appl. Opt.*, **25**, 1973–6.
Mitachi, S., Miyashita, T. and Kanamori, T. (1981) *Electron. Lett.*, **17**, 591–2.
Mitachi, S., Sakaguchi, S., Shikano, H. *et al.* (1985) *Jpn. J. Appl. Phys.*, **24**, L827–L828.
Mitachi, S., Sakaguchi, S. and Takahashi, S. (1986) *Phys. Chem. Glasses*, **27**, 144–6.
Miya, T., Terunuma, Y., Hosaka, T. and Miyashita, T. (1979) *Electron. Lett.* **15**, 106–8.
Monerie, M. (1986) *Electron. Lett.*, **22**, 999–1000.
Monerie, M., Alard, F. and Maze, G. (1985) *Electron. Lett.*, **21**, 1179–81.
Moore, M. W., Carter, S. F., France, P. W. and Williams, J. R. (1986) *ECOC '86, Barcelona, Spain, 22/25 Sept.*, pp. 299–302.
Moynihan, C. T. (1986) *NATO ARW on Halide Glasses for Infrared Fiberoptics, Vilamoura, Portugal* (ed. R. M. Almeida), Martinus Nijhoff, Dordrecht, in press.
Moynihan, C. T., Bruce, A. J., Gavin, D. L. *et al.* (1984) *Polym. Eng. Sci.*, **24**, 1117–22.
Moynihan, C. T., Drexhage, M. G., Bendow, B. *et al.* (1981) *Mater. Res. Bull.*, **16**, 25–30.
Moynihan, C. T., Mossadegh, R., Gupta, P. K. and Drexhage, M. G. (1986) *Mater. Sci. Forum*, **6**, 655–64.
Nakai, T., Mimura, Y., Shinbori, O. and Tokiwa, H. (1985a) *Jpn. J. Appl. Phys.*, **24**, L714–L716.
Nakai, T., Mimura, Y., Tokiwa, H. and Shinbori, O. (1985b) *Jpn. J. Appl. Phys. Part 1*, **24**, 1658–60.
Nakai, T., Mimura, Y., Tokiwa, H. and Shinbori, O. (1985c) *J. Lightwave Technol.*, **LT-3**, 565–8.
Nakai, T., Mimura, Y., Tokiwa, H. and Shinbori, O. (1986) *J. Lightwave Technol.*, **LT-4**, 87–9.
Nakata, A., Lau, J. and Mackenzie, J. D. (1985) *Mater. Sci. Forum*, **6**, 717–20.

Nassau, K. and Wemple, S. H. (1982) *Electron. Lett.*, **18**, 450–1.
Neilson, G. F., Smith, G. L. and Weinberg, M. C. (1985) *J. Am. Ceram. Soc.*, **68**, 629–32.
Nishii, J., Kaite, Y. and Yamagishi, T. (1985) *J. Non-Cryst. Solids*, **74**, 411–5.
Ohishi, Y., Kanamori, T. and Mitachi, S. (1982) *Mater. Res. Bull.* **17**, 1563–72.
Ohishi, Y., Kanamori, T., Mitachi, S. and Takahashi, S. (1985a) *Appl. Opt.* **24**, 3227–30.
Ohishi, Y., Mitachi, S., Kanamori, T. and Manabe, T. (1983) *Phys. Chem. Glasses*, **24**, 135–40.
Ohishi, Y., Mitachi, S. and Takahashi, S. (1984) *Mater. Res. Bull.*, **19**, 673–9.
Ohishi, Y., Mitachi, S., Takahashi, S. and Miyashita, T. (1985b) *IEE Proc. J Optoelectronics*, **132**, 114–18.
Ohishi, Y. and Takahashi, S. (1986) *Appl. Optics*, **25**, 720–3.
Pantano, C. G. (1986) *NATO ARW on Halide Glasses for Infrared Fiberoptics, Vilamoura, Portugal* (ed. R. M. Almeida), Martinus Nijhoff, Dordrecht, in press.
Parker, J. M. (1986) *NATO ARW on Halide Glasses for Infrared Fiberoptics, Vilamoura, Portugal*, (ed. R. M. Almeida), Martinus Nijhoff, Dordrecht, in press.
Parker, J. M., Ainsworth, G. N., Seddon, A. B. and Clare, A. (1986) *Phys. Chem. Glasses*, in press.
Parker, J. M., Seddon, A. B. and Clare, A. (1986) *Phys. Chem. Glasses*, in press.
Phillip, H. R. (1966) *Solid State Commun.*, **4**, 73–5.
Pinnow, D. A., Rich, T. C., Ostermayer, F. W. and DiDomencio, M. (1983) *Appl. Phys. Lett.*, **22**, 527–9.
Poignant, H. (1981) *Electron. Lett.*, **17**, 973–4.
Poignant, H. (1982) *Electron. Lett.*, **18**, 199–200.
Poignant, H. (1985) *Electron. Lett.*, **21**, 1179–81.
Poulain, M. (1981) *Nature, London,* **293**, 279–80.
Poulain, M. (1983) *J. Non-Cryst. Solids*, **56**, 1–14.
Poulain, M. and Lucas, J. (1978) *Verres Réfract.*, **32**, 505–13.
Poulain, M., Poulain, M., Lucas, J. and Brun, P. (1975) *Mater. Res. Bull.*, **10**, 243–6.
Poulain, M., Poulain, M and Matecki, M. (1981) *Mater. Res. Bull.*, **16**, 555–64.
Poulain, M. and Saad, M. (1984a) *Proc. SPIE Int. Soc. Opt. Eng.*, **505**, 165–70.
Poulain, M. and Saad, M. (1984b) *J. Lightwave Technol.*, **LT-2**, 599–602.
Rawson, H. (1967) *Inorganic Glass-forming Systems*, Academic Press, London.
Robinson, M. and Pastor, R. C. (1982) *1st. Int. Symposium on Halide and Other Non-oxide Glasses, Cambridge.*
Rupprecht, G. (1964) *Phys. Rev. Lett.*, **12**, 580–3.
Saissy, A., Botineau, J., Macon, L. and Maze, G. (1985) *J. Phys. Paris, Lett.*, **46**, L289–L294.
Sarhangi, A. (1986) *NATO ARW on Halide Glasses for Infrared Fiberoptics, Vilamoura, Portugal* (ed. R. M. Almeida), Martinus Nijhoff, Dordrecht, in press.
Sarhangi, A. and Thompson, D. A. (1986) *12th ECOC, Barcelona, Spain, 22/25 Sept.*, Vol. 3, pp. 62–8.
Savage, J. A. (1982) *1st. Int. Symp. on Halide and Other Non-oxide Glasses, Cambridge.*
Savage, J. A. (1985) *Infrared Optical Materials and Their Antireflection Coatings*, Adam Hilger, Bristol, UK.
Schroeder, J., Fox-Bilmont, M., Pazol, B. *et al.* (1984) *Proc. SPIE. (Infrared Optical Materials and Fibres III)*, **484**, 61–71.

Shelby, J. E., Lapp, J. C. and Suscavage, M. J. (1986) *J. Appl. Phys.*, **59**, 3412–6.
Shelby, J. E., Pantano, C. G. and Tesar, A. A. (1984) *J. Am. Ceram. Soc.*, **67**, C164–C165.
Shibata, S., Horiguchi, M., Jinguji, K. *et al.* (1981) *Electron. Lett.*, **17**, 775–7.
Shroeder, S. (1977) *Glass I: Interaction with electromagnetic radiation.* Treatise on Materials Science and Technology, Vol. 12 (eds. M. Tomozawa and R. H. Doremus), Academic Press, New York, pp. 158–222.
Sigel, G. (1971) *J. Phys. Chem. Solids*, **32**, 2373–83.
Sigel, G. (1977) *Treatise on Materials Science and Technology, Vol 12* (eds. M. Tomozawa and R. H. Doremus), Academic Press, New York.
Simmons, C. J., Azali, S. A. and Simmons, J. H. (1983) *2nd. Int. Symp. on Halide Glasses*, Troy, New York, Paper 47.
Simmons, C. J. and Simmons, J. H. (1985) *J. Am. Ceram. Soc.*, **68**, C258–C259.
Sinclair, R. N., Desa, J. A., Etherington, G. *et al.* (1980) *J. Non-Cryst Solids*, **42**, 107–15.
Sparks, M. G. and DeShazer, L.G. (1981) *Proc. SPIE Int. Soc. Opt. Eng. (Infrared Fibres (0.8 μm–12 μm)*, **266**, 3–9.
Strom, U., Freitas, J. A., Devaty, R. P. and Tran, D. C. (1984) *Proc. SPIE Int. Soc. Opt. Eng.*, **484**, 74–7.
Sun, K. N. (1949) US Patent 2 466 509.
Suscavage, M. J. and El-Bayoumi, O. H. (1985) *J. Am. Ceram. Soc.*, **68**, C256–C257.
Szigeti, B. (1950) *Proc. Roy. Soc. London, Ser. A*, **204**, 51–62.
Takahashi, S., Shibata, S., Kanamori, T. *et al.* (1981) *Advances in Ceramics Vol. 2, Physics of Fiber Optics,* American Ceramic Society, Columbus, Ohio, p. 74.
Tanimura, K., Ali, M., Feuerhelm, L. F. *et al.* (1985a) *J. Non-Cryst Solids*, **70**, 397–407.
Tanimura, K., Sibley, W. A., Suscavage, M. and Drexhage, M. (1985b) *J. Appl. Phys.*, **58**, 4544–52.
Tokiwa, H., Mimura, Y., Nakai, T. and Shinbori, O. (1985a) *Electron. Lett.*, **21**, 1131–2.
Tokiwa, H., Mimura, Y., Shinbori, O. and Nakai, T. (1985b) *J. Lightwave Technol.*, **LT-3**, 574–8.
Tokiwa, H., Mimura, Y., Shinbori, O. and Nakai, T. (1985c) *J. Lightwave Technol.*, **LT-3**, 569–73.
Tran, D. C., Burk, M. J., Sigel, G. H. Jr. and Levin, K. H. (1984a) *Tech. Dig. Conf. on Optical Fibre Communication, New Orleans, LA*, Paper TUG2. Opt. Soc. America, Washington DC.
Tran, D. C., Fisher, C. F. and Sigel, G. H. (1982a) *Electron. Lett.*, **18**, 657–8.
Tran, D. C., Fisher, C. F. and Sigel, G. H. (1983a) *2nd Int. Symp. on Halide Glasses*, Troy, New York, Paper 50.
Tran, D. C., Ginther, R. J. and Sigel, G. H. (1982b) *Mater. Res. Bull.*, **17**, 1177–84.
Tran, D. C., Levin, K. H. Fisher, C. F. *et al.* (1983b) *Electron. Lett.*, **19**, 165–6.
Tran, D. C., Levin, K. H., Ginther, R. J. and Sigel, G. H. (1982c) *Electron. Lett.*, **18**, 1046–8.
Tran, D. C., Levin, K. H., Ginther, R. J. and Sigel, G. H. (1986) *Electron. Lett.* **22**, 117–9.
Tran, D. C., Sigel, G. H. and Bendow, B. (1984b) *J. Lightwave Technol*, **2**, 566–86.
Trégoat, D., Fontenau, G. and Lucas, J. (1985) *Mater. Res. Bull.*, **20**, 179–85.
Trégoat, D., Fontenau, G., Moynihan, C. T. and Lucas, J. (1985) *J. Am. Ceram. Soc.*, **68**, C171–C173.
Trégoat, D., Liepman, M. J., Fonteneau, G. *et al.* (1986) *J. Non-Cryst. Solids*, **83**, 282–96.

Turk, R. R. (1982) *Proc. SPIE Int. Soc. Opt. Eng. (Emerging Optical Materials)*, **297**, 204–11.
Urbach, F. (1953) *Phys. Rev.*, **92**, 1324.
Vaughn, W. L. and Risbud, S. H. (1984) *J. Mater. Sci. Lett.*, **3**, 162–4.
Walrafen, G. E., Hokmabadi, M. S., Guha, S. and Krishnan, P. N. (1985) *J. Chem. Phys.*, **83**, 4427–43.
Weber, M. J. (1983) *Wiss. Z. Friedrich Schiller Univ. Jena*, **32**, 239–51.
Wemple, S. H. (1979) *Appl. Opt.*, **18**, 31–5.
Williams, R. T., Nagel, D. J., Klein, P. H. and Webber, M. J. (1981) *J. Appl. Phys.*, **52**, 6279–84.
Yamane, M., Kawazoe, H., Inoue, S. and Maeda, K. (1985) *Mater. Res. Bull.* **20**, 905–11.

6

Applications of microporous glasses

N. Ford and R. Todhunter

6.1 INTRODUCTION

In this chapter we review some of the applications of glasses in which microporosity of the glass plays an important part. This is not a comprehensive review of all applications since many of them are hidden in the depths of patent specifications. However, we illustrate the variety of applications that are possible and some of the problems that are encountered.

Firstly, we discuss briefly the process of phase separation and the mechanisms that give rise to it. We discuss the Vycor process, perhaps the earliest industrial exploitation of phase separation and then, in some detail, the use of porous glass membranes for reverse osmosis, which has long been associated with Warwick University. This is followed by a review of a number of applications to demonstrate their variety.

6.2 PHASE SEPARATION

Phase separation in glass has been known for many years; one of the earliest studies was by Grieg (1927). Since then it has been the subject of many investigations and there is a whole area of glass science devoted to this phenomenon in different glass-forming systems. In this section we discuss the phenomenon as it affects microporous applications. For a more detailed discussion the reader is referred to a number of reviews (Cahn and Charles, 1965; James, 1975; Vogel, 1977; Tomazawa, 1979 and others).

There are essentially two types of phase separation, stable and metastable immiscibility. Stable immiscibility occurs when the glass separates into two or more distinct phases at temperatures greater than the liquidus temperature. On cooling, the resulting glass shows distinct regions of differing composition. These manifest themselves as distinct layers or a milkiness in the glass. Metastable immiscibility occurs at temperatures below the liquidus temperature. An example of a system that exhibits stable immiscibility is the $CaO–SiO_2$ system between about 2 and 27 wt % CaO.

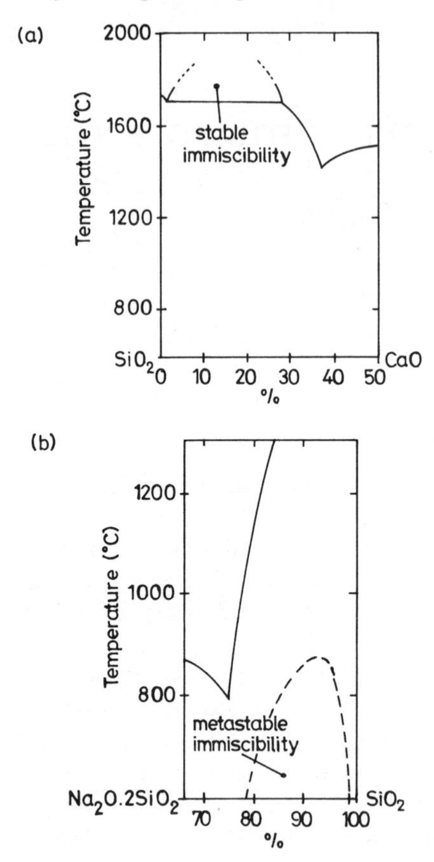

Fig. 6.1 Regions of phase separation in the $CaO–SiO_2$ and $NaO_2–SiO_2$ systems: (a) separation into two liquid phases in the $CaO–SiO_2$ system; (b) sub-liquidus phase separation in the $NaO_2–SiO_2$ system. Data from Phillips and Muan (1959), Kracek (1939) and Haller *et al.* (1974).

Above the liquidus temperature of 1705°C the melt separates into two liquid phases (Fig. 6.1(a)). Metastable immiscibility can be illustrated by the $Na_2O–SiO_2$ system (Fig. 6.1(b)), in which the region of immiscibility is well below the liquidus temperature. It is metastable immiscibility that is primarily of importance in this chapter.

The mechanism of phase separation in a system is determined by the free energy–composition curve for that system. Two distinct mechanisms operate which in turn give rise to two distinct microstructures.

Consider a binary system of components X and Y where the molar concentration of X is given by C and that of Y by $(1-C)$. At a temperature T greater than the critical temperature, T_c, the melt will be uniform and single phase; this results from a balancing of the decrease in energy of the system due to increasing order and the increase in energy due to disordering, the latter being temperature dependent. This implies that the melt free energy–composition curve must be of the form shown in Fig. 6.2(a).

If then an inhomogeneity of composition C_i develops in a melt of composition C_0, the change in free energy will be given by the gap between the tangent to the point on the free energy curve for the composition C_0, and the energy curve at the composition C_i shown as ΔF in Fig. 6.1(a). If the interval ΔF is located above the energy curve then $\Delta F < 0$ and if it is located below the energy curve $\Delta F > 0$. Thus it can be seen that in this case $\Delta F > 0$ and hence a compositional fluctuation of the type discussed gives an increase

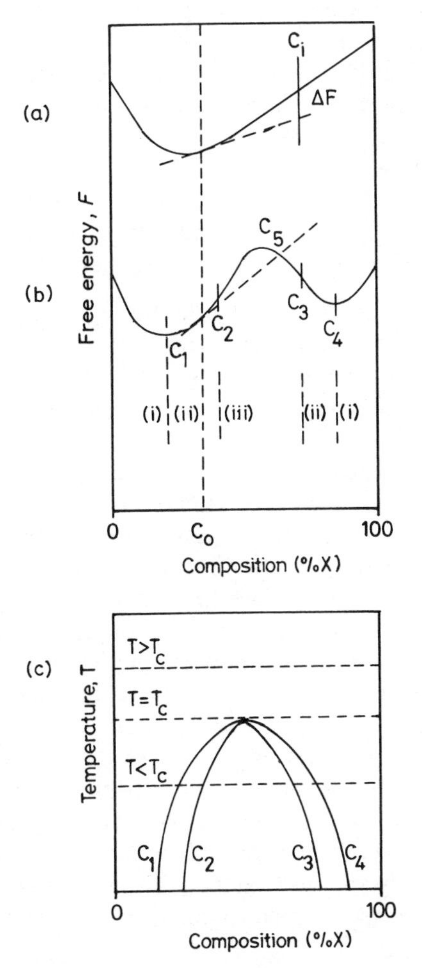

Fig. 6.2 Free energy–composition curve for the binary system $(C)X + (1-C)Y$.

in energy and is unstable. More generally, if $\partial^2 F/\partial C^2 > 0$ then $\Delta F > 0$ for all points on the curve, as is the case in Fig. 6.2(a).

If the temperature, T, is now reduced below T_c, the energy against composition curve of a melt that undergoes phase separation will be of the form given in Fig. 6.2(b). This curve can be divided into three regions.

1. For compositions $C_0 < C_1$ and $C_0 > C_4 : \Delta F > 0$, as above, and phase separation is impossible.
2. For compositions $C_1 < C_0 < C_2$ where C_2 is a point of inflection at which $\partial^2 F/\partial C^2 = 0$, if we consider a region of inhomogeneity developing of composition $C_i < C_5$, where C_5 is the point of intersection of the tangent to the curve at C_0 and the curve, then ΔF, given by the gap between the tangent and the curve, is greater than zero. Hence the region of inhomogeneity will be unstable and tend to redissolve. If the composition of the inhomogeneity is greater than C_5 then $\Delta F < 0$ and the change in its composition towards that of C_4 gives a decrease in free energy, thus separation is energetically favourable for inhomogeneities in this domain.

   For compositions $C_3 < C_0 < C_4$ the situation is analogous to that for $C_0 < C_1$ and $C_0 > C_4$.
3. For compositions $C_2 < C_0 < C_3$ the development of an inhomogeneity with any composition gives a decrease in energy since $\partial^2 F/\partial C^2 < 0$.

To summarize, the effect that these different regions of stability have on the development of phase separation is as follows.

If the original composition, C_0, is given by $C_0 < C_1$ and $C_0 > C_4$; the melt is stable and at no point does the energy decrease on the formation of a region of inhomogeneity.

For $C_1 < C_0 < C_2$ and $C_3 < C_0 < C_4$ the change in energy resulting from the formation of a second phase depends on the composition of that second phase. As illustrated, for compositions greater than C_5, in a melt of composition C_0, a net decrease in energy can occur. Thus it can be considered that there is a potential barrier to phase separation. In this region phase separation progresses by a nucleation and growth process where growth is only possible when the inhomogeneities attain a critical size. As a result a droplet-type microstructure dispersed in a matrix phase is formed. This is the region of binodal phase separation.

For $C_2 < C_0 < C_3$ there is no barrier to the formation of an inhomogeneity and then even the smallest compositional fluctuation will grow. Therefore the melt will tend to phase-separate at many points throughout its volume and it is energetically favourable to form a two phase system, where the phases are of compositions C_1 and C_4. This is the region of spinodal phase separation and the microstructure tends to be a finely interconnected continuous structure.

If one now considers the effect on the points C_1, C_2, C_3 and C_4 in the free energy–temperature curve they will trace out loci which are coincident at the critical temperature T_c. Thus a plot of the loci as a function of temperature and composition will be of the form given in Fig. 6.2(c) and the loci will trace the boundaries between the regions of binodal and spinodal phase separation and stable phase formation.

6.2.1 The effect of phase separation on glass properties

As many properties of materials are dependent on microstructure one would expect phase separation to alter the properties of the glass. This is, in fact, observed to be true and often the effect depends upon whether the phase separation is binodal or spinodal in nature. To predict the properties of a phase separated glass one needs to know the compositions of the phases and their distribution. It is then possible to predict, to a limited extent, the properties of the phase separated glass from data for the corresponding single phase glasses.

The viscosity of glasses is observed to increase with phase separation in systems where the glass separates to give a high viscosity matrix phase, such as the high silica phase that develops in the Na_2O–B_2O_3–SiO_2 system, but not in those systems where phase separation gives a high viscosity phase dispersed in a phase of lower viscosity. In some systems phase separation plays a role in subsequent crystallization processes and a number of studies have been made of this effect (e.g. Natagawa and Izumitani, 1969).

Chemical durability is strongly dependent on the microstructure of the phase separation. If an interconnected microstructure of a chemically soluble phase is formed within a chemically durable phase, then the more soluble phase will be preferentially attacked. This is the basic process behind the production of microporous glasses and the Vycor-type processes. If a chemically durable phase is formed dispersed in a less durable phase then the resultant glass may have a lower durability than the parent glass, whilst a low durability phase dispersed in a more durable phase would be expected to give increased durability over the parent glass. Pyrex glass relies upon this for its high chemical durability. Pyrex is a phase separated borosilicate glass in which the phase separation is a fine droplet phase dispersed in a high silica matrix (Doremus and Turkalo, 1979). Conversely, alumina is added to commercial soda–lime–silica glass to reduce phase separation as the matrix phase is high in soda and would give rise to poor weathering characteristics.

As the electrical conductivity of glasses is controlled mainly by the diffusion of alkali ions under the influence of an electric field, the bulk conductivity of a phase separated glass will be primarily determined by the alkali content of the continuous phase. Thus it will increase for glass that separates to give an alkali-rich matrix or an interconnected microstructure.

6.3 THE VYCOR PROCESS

Perhaps the earliest practical application of phase separation was that by Hood and Nordberg (1938, 1940, 1943) who developed the Vycor process for the production of high silica ware by a porous glass route. They identified a range of compositions in the R_2O–B_2O_3–SiO_2 system which, when suitably heat-treated, exhibited phase separation with each phase continuous (Fig. 6.3). Subsequent removal of the soluble alkali–borate phase by acid leaching resulted in a highly siliceous (>95%) skeleton with a mutually interconnected pore structure.

The immiscibility regions for the systems R_2O–B_2O_3–SiO_2 where R = Na, K or Li have been determined by several workers and are well documented (Volf, 1961). The phase separation is generally thought to proceed by a spinodal mechanism but it has also been suggested that a nucleation and growth process could be responsible, with coalescence occurring in the growth stages to give an interconnected structure (Haller *et al.*, 1970).

The tendency for borosilicate glasses to phase separate can be controlled by the addition in small quantities of other elements into the glass. For example, small additions of Al_2O_3 are known to retard the separation process considerably (Simmons and Macedo, 1971) and are useful in exercising control over otherwise unstable systems. Furthermore, as Al_2O_3 can partially enter the SiO_4 network, the physical properties of the interconnected porous phase can be enhanced. Additions of strongly

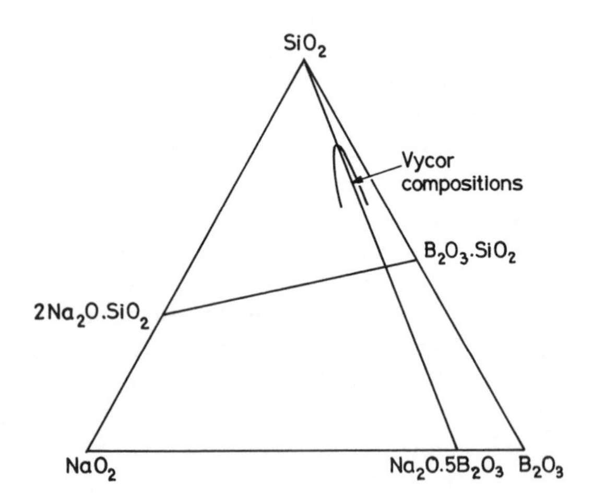

Fig. 6.3 Region of phase separation in Vycor glasses after Volf (1961) and Hood and Nordberg (1938, 1940, 1943).

polarizing elements such as the transition series have been shown to strongly enhance phase separation.

Small quantities of P_2O_5, V_2O_5, MoO_3 and WO_3 shift the immiscibility region with the result that the more soluble high-borate phase contains less silica and this leads to enhanced leaching (Tomozawa and Takamore, 1980).

The accumulation of these additional species in the leachable phase eradicates any detrimental effect they may have on the final leached glass. Any resulting deviations from the optimum line as a result of compositional modifications can generally be accommodated by variation of temperature or time during the heat-treatment process.

Thermal heat-treatment of Vycor-type glasses has a very strong influence on the phase separation process and hence the pore size distribution and morphology of the leached glass network. The liquidus temperature of the insoluble silica phase is around 1000°C with the onset of phase separation at 750°C. Heat-treatment between 750°C and 1000°C results in both separation and devitrification although the resulting structure is one of droplets of the soluble phase in a continuous silica-rich matrix. To develop a fully interconnected structure heat-treatments of a few hours to several days at temperatures between 500 and 600°C are required and there is evidence that prolonged heat-treatment above 600°C will result in the coarsening of a finely interconnected system to that of droplets by diffusion-controlled agglomeration of the soluble phase. In many cases the phase-separated glasses exhibit opalescence and techniques have been developed (Kuline, 1955) to monitor microstructural development by studying the degree of opalescence as a function of heat-treatment. Other techniques commonly used to this end are electron microscopy and differential thermal analysis (DTA). For an in-depth treatment of the Vycor process the reader is referred to Volf (1961).

Leaching is typically carried out with 3N HCl or 5N H_2SO_4 at about 100°C and proceeds at a rate of around 1 mm per day. Following leaching the glass is thoroughly washed and carefully dried. The leaching properties can be enhanced by additions of S, Sb_2O_3, ZrO_2 and TiO_2. In practice, the leaching of a phase-separated glass results in swelling or shrinkage which induce intolerable stresses within the porous structure leading to breakage. Early work in the area by Hood and Nordberg (1940) resulted in the determination of the so-called optimum line in the ternary phase fields along which the stresses are reduced to a minimum.

This swelling takes place due to H_3O^+ ions replacing Na^+ ions in the initial stages of leaching, then as the B_2O_3 is leached the silica skeleton shrinks. The leached glass is then sintered at 900–1200°C with a shrinkage of about 35% to give a fully compacted glass. The unsintered, leached glass is also marketed as Vycor 7930, or thirsty Vycor, which is hygroscopic and has a pore size of 2–4 nm.

6.4 REVERSE OSMOSIS

6.4.1 Introduction

Where two solutions with differing solute concentrations are separated by a semipermeable membrane, the gradual equilibration of the system by solvent flow across the membrane will establish a finite pressure known as the osmotic pressure. In a reverse osmosis process the osmotic pressure of a system is opposed and exceeded, usually by a mechanical pump, to reverse the natural flow of solvent and increase the solute concentration differential. This has obvious applications, most commonly in the desalination of sea water to produce a potable supply.

Solute rejection may be defined as a coefficient R, given by:

$$R = \frac{C_f - C_p}{C_p} \times 100$$

where C_f and C_p are feed and product concentrations respectively. The total throughput of the system or flux is defined as the product volume per square metre of external surface area per hour and has the units m h^{-1}.

The application of microporous glass membranes to reverse osmosis was first suggested by Hood and Nordberg (1940) using phase separated and leached alkali borosilicate glasses. In addition Nordberg (1944) reported alkaline-earth-alkaliborosilicates containing zirconia which could be similarly leached to produce a porous integral structure, and further development of this system (Hood and Nordberg, 1950) resulted in the alkali-free, high resistivity glass known commercially as E-glass.

Salt rejections from porous Vycor membranes were first reported by Kraus *et al.* (1966), and Phillips (1974) found that the rejection could be improved by partially collapsing the pores through a thermal treatment near the sintering temperature. They achieved an 88% rejection with a 0.2 m h^{-1} flux for a membrane with a 2 nm average pore radius. McMillan and Matthews (1976) attempted to minimize pore size by the inclusion of Al_2O_3 in the parent glass which is known to retard phase separation and enhance control of pore size. They reported a 60% rejection for a 2.6 nm radius. Several workers (Belfort, 1972; Ballou and Wydevan, 1972; Littman and Gutter, 1968) investigated salt rejection from Corning 7930 Vycor and recorded rates of 60–80% for a 2 nm pore radius and a dramatic increase to 90–95% for average pore distributions below 2 nm although corresponding flux rates were as low as 10^{-3} m h^{-1}. Maddison (1980) made a thorough investigation of porous E-glass membranes and reported internal surface areas of 500 m^2g^{-1} and a binodal pore distribution with peaks at 1 nm and 2 nm radius. Rejections of up to 60% were achieved at fluxes of 0.5 m h^{-1}. Some of the above data are summarized in Fig. 6.4.

Clearly, optimization of glass composition, heat-treatment and leaching

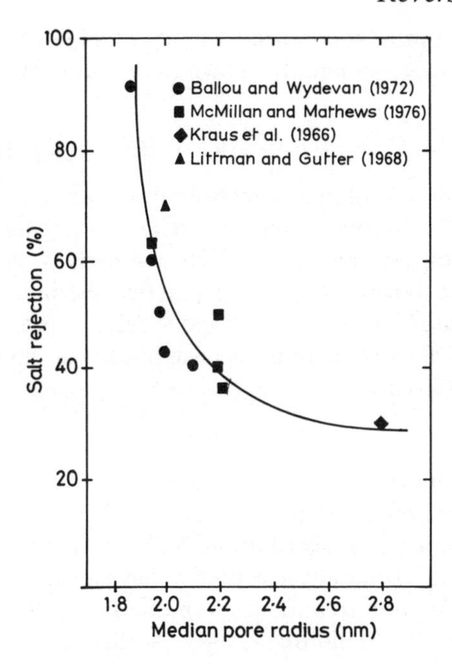

Fig. 6.4 The salt rejection of porous glass membranes as a function of the median pore size. A compilation of data from Ballou and Wydevan (1972), McMillan and Matthews (1976), Kraus *et al.* (1986) and Littman and Gutter (1968).

conditions are important in establishing a pore structure which will yield a high rejection (>80%) and a high flux, although in practice one of these is generally improved at the expense of the other (Fig. 6.4). The application of various coatings to pore surfaces has gone some way to improving both reverse osmosis performance and the mechanical strength of an inherently weak membrane and are discussed separately.

6.4.2 Membrane pore characterization

Several techniques are available for the quantitative and semi-quantitative analysis of porous glasses. Of these, inert gas adsorption–desorption at low temperatures provides the best method for surface area determinations and, after certain assumptions are made regarding pore morphology, computational techniques can be used to estimate pore size distribution. Typical gas adsorption apparatus is described in several texts (McMillan and Matthews, 1976; Tran, 1976) and monitors the change in mass of an initially out-gassed sample due to the surface adsorption of a suitable gas (e.g. N_2) over a range of pressures – typically 10^{-5}–10 torr. From these data a plot of adsorbed gas volume versus relative pressure can be made and total surface area calculated from a generalized theory of physical adsorption after Brunauer *et al.* (1935), commonly known as the BET theory and expressed by the equation:

$$V = V_m Cp/[(p_0 - p)(1 + (C - 1) p/p_0)]$$

where V is the adsorbed volume at pressure p, V_m the volume of an adsorbed monolayer and C a derived constant. This expression brings together discreet adsorption mechanisms which themselves generate the five common isotherms shown in Fig. 6.5. In practice, the value of C for nitrogen is small and if measurements are confined to relative pressures in the range 0.05–0.35 then the above equation can be applied to all cases and the surface area, S, estimated from:

$$S = V_m aN/M$$

where a is the volume of one gas molecule, N Avogadro's number and M the molar volume of adsorbate.

Estimation of pore size distribution relies on many assumptions and approximations but is generally accepted to have an accuracy of about 10% (Everett, 1958). A full treatment of models adopted and calculation techniques is given by Maddison (1980). Additional techniques for pore analysis include electron microscopy, water adsorption (Huang *et al.*, 1972)

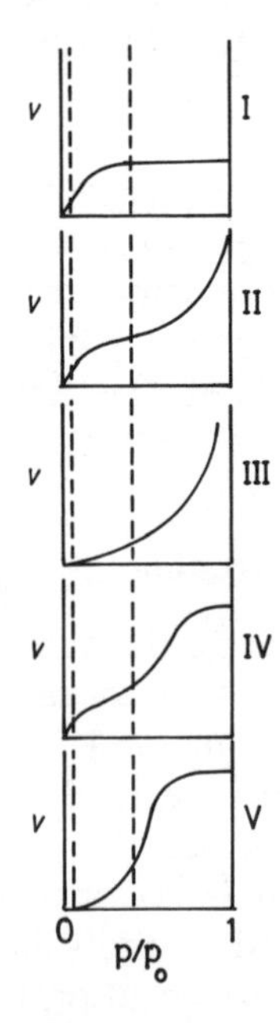

Fig. 6.5 The five common adsorption isotherms classified following Brunauer *et al.* (1940). Relative pressures of 0.05 and 0.35 are indicated.

and zinc ion adsorption (Kozawa, 1974). However, these are at best semi-quantitative.

6.4.3 Experimental work at Warwick University

Recent work at Warwick University under the supervision of the late P.W. McMillan concentrated on three main areas:

1. assembly of multifibre reverse osmosis cells
2. the application of coatings to pore walls
3. diversification away from simple desalination of sea-water

(a) Hollow glass-fibre production and treatment

Porous membranes are normally preferred in the form of hollow fibres for two reasons. Firstly, they are capable of withstanding the high hydrostatic pressures required for reverse osmosis and secondly, they give higher product fluxes due to their high external surface area to volume ratio. They are generally produced either by drawing down larger diameter tube or direct pulling from the melt and attention should be paid to the respective glass thermal histories resulting from these techniques. Both of these methods were used at Warwick although much of the later work preferred the latter. In this case a heated platinum bushing and bubble-tube were used as shown schematically in Fig. 6.6. The parent glass was a commercially

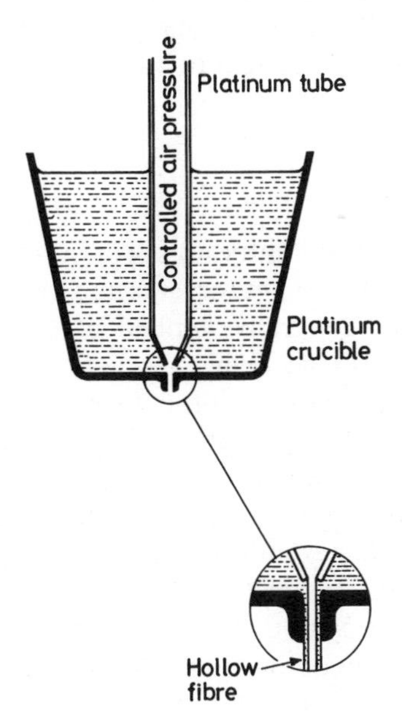

Fig. 6.6 Hollow fibre production technique for reverse osmosis fibres.

available borosilicate of a Vycor-type composition remelted from feeder cullet. Overall fibre diameter and wall thickness were controlled by adjusting both bubble pressure and drawing speed and a range of fibres were produced with diameters varying from 100 to 600 μm and 10–100 μm wall. Dimensions were accurately maintained by this method over continuous fibre lengths of up to several kilometres.

Fibres were heat-treated in the range 500–600°C and leached in 3M constant boiling HCl (70–80°C) under solvent extraction to enhance flushing and prevent build-up of leached products which are known to generate high internal stresses. This effect is markedly reduced by the addition of NH_4Cl to the leaching solution although the reason for this is unclear (Egndi *et al.*, 1969). After washing was carried out, some fibres were exposed to an additional leach in 1M NaOH to remove potential build-up in pores. All fibres were washed in solutions of decreasing acidity and temperature to prevent chemical and thermal shock and dried slowly under controlled relative humidity to prevent cracking.

Multifibre cells of up to 1000 fibres have been assembled as shown schematically in Fig. 6.7 and a system developed for sealing off fibres broken during this process. Once rehydrated, porous fibres have an increased mechanical strength although breakages occurred after several hours of operation. Reverse osmosis results were obtained on a test rig with a recirculating pump operating at about 100 bars. A feed solution of 3.5% NaCl was used and rejection of approximately 50% recorded at an estimated flux of 0.3 m h^{-1}. Several hours of operation were possible on a number of cells from which up to 100 ml of permeate were collected. However, an increase in fibre strength was sought to improve cell reliability during both assembly and operation.

(b) Surface treatment of porous membranes

To improve fibre strength the internal pore surfaces may be coated. The coatings are generally of two types: organic and inorganic.

Organic coatings are generally applied as an aqueous or alcoholic solution, trimethylchlorosilane being a typical example. However, these coatings are generally of a hydrophobic nature whereas the reverse osmosis process requires that they be hydrophillic. This conversion can be achieved via an oxidation or sulphonation process.

Inorganic coatings are commonly applied from metal chlorides as hot-end coatings in the glass container industry and when exposed to water vapour convert to the oxide so improving the mechanical strength (Williams, 1975). $TiCl_4$ and $SnCl_4$ coatings can be applied to pore surfaces in various ways. The sorption characteristics of the liquid phase are generally poor and result in pore filling and drastic reduction in reverse osmosis flux.

Vapour phase techniques are almost exclusively preferred with the fibre

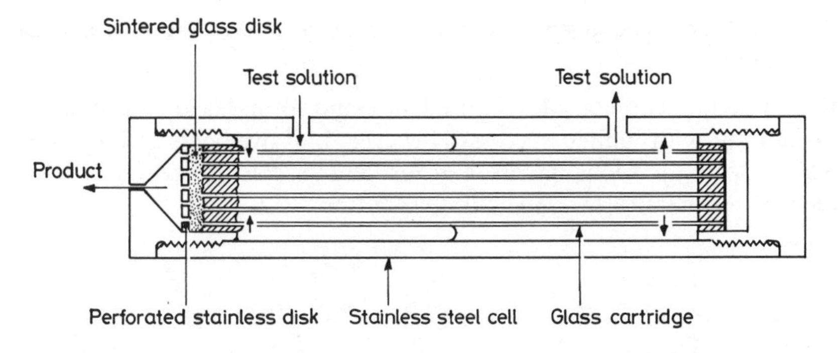

Fig. 6.7 Multifibre reverse osmosis test cell.

pretreatment and type of carrier gas having a profound effect on success rate. Ideally a monomolecular coating is desired to control pore radius reduction and prevent the build-up of unreacted material. In addition it is important to remove all free and surface bound water to prevent oxide precipitation in the pores. A low concentration of $TiCl_4$ in high purity, dry nitrogen has proved successful in increasing fibre strengths by up to 200% and reducing pore radii by 0.4–0.8 nm. The addition of thermal treatments has enhanced the efficiency of this process. The resulting reverse osmosis performance showed an increase of 20–30% on rejection rates although some reduction in flux was noted which may be due to partial pore filling or contamination by hydrophobic species.

Additional strengthening mechanisms tried were micro-crack blunting by HF etching and crack healing via thermal treatments both of which were relatively unsuccessful.

(c) Diversification of feed solutions

Although much of the early reverse osmosis work concentrated on saline feed solutions there has been much commercial interest shown in the application of porous glass fibres in the separation of organic systems from aqueous solution. Large scale desalination plants using organic porous fibres currently operate worldwide and have great advantages over glass in terms of ease of production, cost and overall robustness. However, such cells would be incapable of handling many of the organic solvents found in effluent from commercial chemical processes. In this respect, the chemical durability of glass fibres offers potential application in this field.

Simulated effluent in aqueous solution was provided by both ICI and BP for testing and included the following for which the rejection rates are given:

1. 3V/V% acetic acid (37%)
2. 3V/V% methanol (50%)

3. 3V/V% ethylene glycol (−42%) i.e. concentration of feed rather than dilution.

In addition, preliminary tests on ethanol and formic acid solutions were carried out with limited success. However, improvements in these areas would open up wide applications for porous glass membranes in the recovery of chemicals and in pollution control.

6.5 ANTIREFLECTION COATINGS AND OPTICAL WAVEGUIDES

Engineers and scientists involved in the construction of complex optical systems have long been interested in antireflective surfaces. Much of this

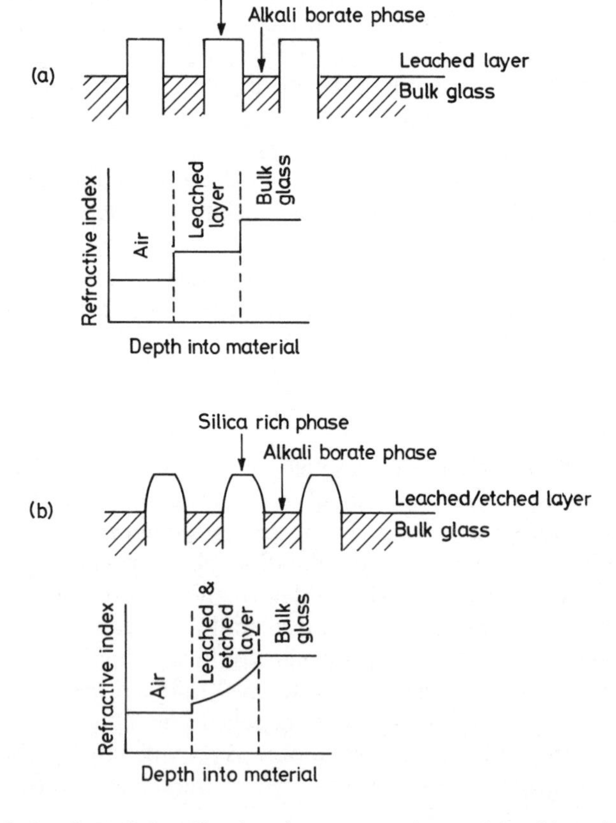

Fig. 6.8 (a) Gradient index film by phase separation and leaching; (b) by phase separation, leaching and etching.

effort has concentrated on deposited coatings in a variety of forms, but it has been appreciated for many years that a gradient index surface which is integral with the glass has significant advantages. Indeed the first patent for a gradient index antireflection surface process was granted to Taylor (1904). These early techniques were based upon an acid etch process.

In more recent times a process has been developed which is based upon a phase separation route. Using a borosilicate glass (the system upon which a large number of common optical glasses are based) that gives spinodal phase separation when heat-treated, it is possible to form a porous surface by acid leaching. This is illustrated schematically in Fig. 6.8(a).

Minot (1976) found that it was possible to reduce the reflectivity to less than 0.5% over the wavelength range 350–2500 mm using this technique. Samples of Corning 7740, a commercial borosilicate glass, were heat-treated for 3 h at temperatures between 600 and 660°C. Following heat treatment their surfaces were leached in mineral acid for 30 s to 35 min at 45 or 80°C. On removal from the solution, interference fringes were observed. Optimum film formation, resulting in a reflectivity of 0.5%, was obtained in 5 min at 45°C and 90 s at 80°C.

The heat treatment temperature was found to have a significant effect on the near-infra-red reflectivity. In a sample heat treated at 600°C the reflectivity rose from 0.5% at 1 μm to approximately 5–6% at 2.5 μm, whereas samples heat treated at 630°C to 660°C show a constant reflectivity of 0.5% up to 2.5 μm.

Minot (1977) also considered the angular reflectivity of the films produced. The single surface reflectivity was found to rise from about 0.25% for near normal incidence to about 7% for an angle of incidence of 70°. This compares with 4% and 17% for untreated glass at near normal and 70° incidence respectively. The low reflectivity at high angles of incidence represents a considerable improvement over discrete thin film coatings which are effective at near normal incidence only.

Elmer and Martin (1979), also working on Corning 7740 glass, investigated the use of $NH_4F \cdot HF$ in combination with leaching mineral acids to enlarge the pores in the graded index surface layer. This work demonstrated that the two-surface reflectivity could be reduced to 0.2% under optimum conditions. They attributed this to the fact that the porosity of the film was not constant but decreased into the bulk (Fig. 6.8(b)) since the conditions for silica dissolution were more favourable at the surface. With an optimum film-forming solution it was found that the film thickness was relatively insensitive to time because the leach front travelled at a rate slightly greater than that of the silica dissolution at the surface. This results in the leach front gradually moving ahead of the etch surface and thus the film thickness only increases slowly, the rate of increase being a function of HF concentration. Using this process Elmer and Martin found optimum treatment times of 3 h for a solution at 90°C and 20 h for a solution at 20°C.

The Hoya Corporation have developed a borosilicate glass, ARG-2, specifically as an optical glass for graded index antireflection film purposes. Samples of this glass, of unspecified composition, were heat-treated for 3 h at 570°C to induce interconnected phase separation and then etched and leached in a solution of $NH_4F \cdot HF$ and HNO_3. A single surface reflectivity of less than 0.5% was reported by Asahara and Izumitani (1980) after leaching for 10 min at 70°C.

A method of producing a step index fibre for optical waveguide applications, using a phase separated and leached glass, was reported by Simmons *et al.* (1979) and also in a patent by Macedo and Litovitz (1976). A porous glass preform was made by melting an approximately $B_2O_3 \cdot 2SiO_2$ glass, with a little alkali, and phase separated by heat treating at 550°C for 2 h to give two interconnected phases. The more soluble borate phase, containing most of the alkali, boron and impurities, was leached out in a nitric acid solution at 95°C to leave a purified silica skeleton (~94 mol % SiO_2).

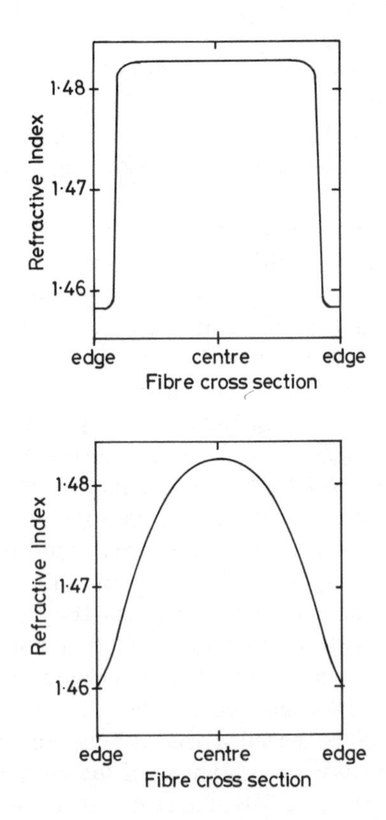

Fig. 6.9 Two possible refractive index profiles for leached and doped optical fibres by Simmons *et al.* (1979a).

These porous silica glass preforms can now be modified by the addition of refractive index modifying dopants. In the example given by Simmons *et al.* the modifying salt was $CsNO_3$ which decomposes to Cs_2O below the sintering temperature of the porous silica. The process consisted of impregnating the preform uniformly by immersing in a saturated solution of $CsNO_3$ dissolved in water at 105°C. Cooling the solution causes the salt to precipitate in the pores. The dopant is removed from the outer layer by soaking the preform in clean water at ~0°C to slowly remove the precipitated salt from the outer layer.

The preform is dried, heated to decompose the $CsNO_3$ to Cs_2O and sintered to consolidate the silica network, resulting in a glass rod with approximately 15 wt % Cs_2O in the core and less than 0.5 wt % Cs_2O in the outer layer. This preform can then be pulled into fibre. Typical refractive index profiles are shown in Fig. 6.9. Different doping profiles can be obtained by varying the conditions and solvent used in leaching the $CsNO_3$.

An advantageous side effect of this doping is that the high-Cs_2O containing core has a higher expansion coefficient than the Cs_2O deficient outer layer. Thus on cooling the core tends to shrink more than the outer layer resulting in the development of a residual compressive stress in the surface. This gives an increased strength to the fibres.

6.6 RESISTANCE THERMOMETERS AND SUPERCONDUCTING MATERIALS

Low temperature thermometry from room temperature down to a few kelvins is often achieved using carbon resistance thermometers. These thermometers have been commercially available for many years. They are based upon the resistance against temperature characteristics of carbon films often fabricated by forming a colloidal suspension of carbon particles in a suitable medium, usually an organic. The characteristics of such a device are dependent on the carbon particle size and it is found that decreasing particle size gives increasing sensitivity and resistivity (Terry, 1968).

Lawless (1972) found that by impregnating carbon into porous glass the particle size was decreased and the sensitivity increased, whilst the resistivity decreased. Lawless measured the thermometric properties of a porous alkali–borosilicate glass impregnated with carbon by a process patented by Ellis (1951). This process results in a pore size of about 3.5–4.5 nm, giving a particle size smaller, by a factor of 5, than previous work.

Rectanglular plates of the impregnated glass had Nichrome–gold electrodes vacuum deposited on to the ends and copper leads attached. The resulting device was then sealed into Pt cans after vacuum baking. This is necessary due to the residual unfilled porosity of the material. The temperature–resistance characteristics were measured between 2 and 400 K and compared with those for a commercial Ge thermometer and two types of commercial carbon resistors (Fig. 6.10). As can be seen the impregnated glass thermometer gave a smooth variation over the entire temperature range with a higher sensitivity than the other types of thermometer. Although there was a large variation in the device-to-device measurements, the repeatability of a particular device was found to be good.

By 1982 these carbon impregnated porous glass thermometers were readily available commercially and were found to have another advantage over previous types of thermometer in that they have a small and correctable error due to magnetoresistance, and thus are useful in fields in excess of 19 T (Sample *et al.*, 1982).

A variation on this application is the development of high critical magnetic field superconducting materials by impregnating a porous glass with a suitable alloy (Watson, 1971). In the porous glass the metal is thought to take the form of granular particles connected by electron tunnelling, (Hindley and Watson, 1969). Alloys and metals such as indium, lead–bismuth and lead–bismuth–antimony have been investigated. These composite materials have superconducting properties similar to niobium–titanium alloys, with the advantage of reduced cost, but improvements in the critical current density are required to make them technologically attractive.

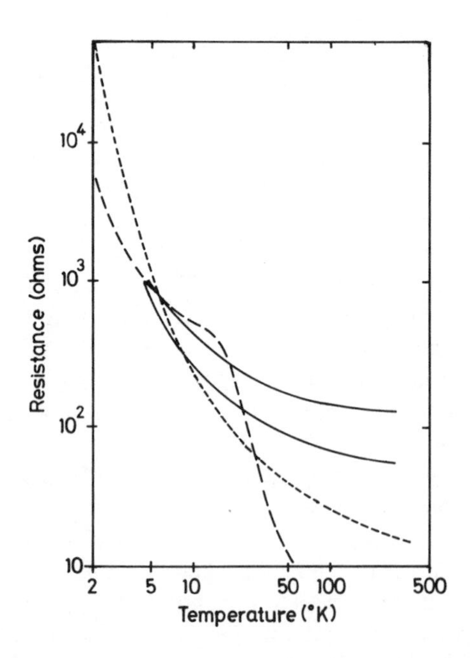

Fig. 6.10 Resistance–temperature characteristics of a carbon-impregnated porous glass thermometer (---) compared against a commercial GE thermometer (–––) and two carbon resistance thermometers (———). After Lawless (1972).

6.7 NUCLEAR WASTE DISPOSAL

Safe storage of radioactive waste is becoming an immense problem. Currently waste is stored in solution in underground tanks, which is not satisfactory as the long-term possibility of leakage is of real concern. A number of methods of converting this waste into a solid, chemically stable, inert form for burial have been devised, such as calcining the waste, mixing it with glass powder and then firing at 1100–1200°C to form a solid mass. This high firing temperature causes the volatilization of some products and is problematic; thus a technique that operates at or below 1000°C would be useful.

A number of workers have suggested encapsulation in porous glass as a solution. Simmons *et al.* (1979b) suggested a variation of their waveguide process whereby the waste in the form of a solution of nitrates, or other salts that decompose to the oxide at temperatures around 700°C, is impregnated into a phase separated and leached borosilicate glass. On removal from the liquor and cooling, the salts precipitate out of the solution, in the pores of the glass, and then by immersing in clean solvent the surface layer can be purged. Thus after heat treating to decompose the salts to their oxides and sintering, the waste is encapsulated in a predominantly silica network with a fully compacted, pure silica layer at the surface.

Simmons *et al.* demonstrated the feasibility of this using mixtures of Fe, Sr, Y, Zr, Cs, Ba, La, Ce and Nd in the nitrate form to simulate nuclear waste. In particular they studied the release of Cs_2O from the compacted glass, as this offers the potential worst case of the above elements in terms of mobility within the glass and solubility, thus giving an indication of the containment possibilities. The results they obtained were certainly competitive with other forms of containment.

Tingley *et al.* (1982) used a porous silica rod to investigate the encapsulation of radioactive krypton. They loaded and sintered the porous silica rods by heating them to 900°C under 40 MPa of krypton. By this technique they managed to achieve krypton concentrations in the glass of the order of 20–30 cm^3 of gas (calculated at STP) per cm^3 of glass. This is thought to be sufficient for storage of short lived gaseous products.

Pratten and McMillan (1984) investigated the incorporation of magnesium, aluminium and iron oxides into a special phase-separated and leached borosilicate glass. This gave porosities of between 25 and 54%, resulting in up to 35 wt % of oxide being taken up in the fully sintered material. Sintering was done at 1000°C and was accompanied by partial crystallization resulting in a dense glass-ceramic material that was mechanically and chemically stable.

6.8 REFRACTORY FOAMS

A number of processes have been described for producing glass foams (Marceau, 1967; Slayter, 1964; Schott, 1971; Mackenzie, 1973). Most of these processes require high forming temperatures and have high thermal expansion coefficients making them expensive to produce and unsuitable for insulation purposes. Refractory glasses can withstand high temperatures and have low thermal expansion coefficients, but they require high melting and processing temperatures.

Techniques for producing foamed refractory glasses at relatively low temperatures have been developed by Elmer (1971) and Johnson (1976). That of Johnson is typical of the use of a phase separated glass in this application. The process involves melting of a phase separable glass, usually a borosilicate composition, at normal glass melting temperatures. The glass is then quenched to form a frit and powdered to a particle size smaller than 5 mesh. This powder is heated in a mould with a foaming agent at a temperature above the glass softening point, between 650°C and 850°C, for a period between a few minutes to hours depending on the mould size. The foamed glass is then heat-treated at about 500°C for a period of 1/2 h to 4 h to induce phase separation and then cooled. In practice the foaming and heat treatment to induce phase separation are carried out consecutively as a continuous process.

The phase separated, foamed glass is then leached in warm dilute acid to remove the borate phase, which can take between 1 h and 2 days depending upon particle size, proportion of leachable phase and pore size. The leached glass is then dried to remove the leaching solution.

Refractory foams produced by this method have a skeletal structure of substantially pure silica enclosing the voids which were previously occupied by the borate phase. The addition of boron to the silica glass in the initial stage reduces the melting point, thus saving on energy costs, and the boron recovered during the leaching can be recycled.

The density of these foams can be as low as $0.075\ g\ cm^{-3}$ and although the foams are weak they are very useful for lightweight, high temperature insulation where structural strength is not required. The thermal expansion coefficient is very low, around $4\text{–}10 \times 10^{-7}\ ^{\circ}C^{-1}$. The foams can withstand temperatures up to just below the melting point of silica without deformation or loss of strength. Chemical durability is substantially higher than that of typical foamed glasses.

6.9 ENZYME IMMOBILIZATION AND CATALYST SUPPORTS

Enzymes, which are essentially biological catalysts, are used extensively in food and biological based industries. They can be added to the process in the form of an aqueous solution but this is very wasteful and, as enzymes are expensive to produce, some form of immobilization is essential. Immobilization is not a new idea and materials such as charcoal (Nelson and Griffin, 1916) and organics (Katchalski and Bor-Eli, 1960; Mitz and Summiaria, 1961) have previously been used. In the early 1960s work was begun on investigating the suitability of porous glass membranes for dialysis purposes (Messing, 1969). Porous glass tubes were used but the work was abandoned as the porous glass was found to retain proteins from the solution.

It was found that the protein–glass bond was so strong that it could not be broken by strong acids, ammonium hydroxide or a variety of ionic strength buffer solutions. Some of the proteins used in these experiments were enzymes and thus the possibility of using porous glass to immobilize enzymes was discovered.

Porous glass has significant advantages over other forms of immobilization as it is thermally and chemically stable, mechanically strong, resistant to biological attack and is dimensionally stable. Also the porous glass carrier can be regenerated by heating to sufficiently high temperature to burn off the biological material.

For this application a small pore diameter with a narrow pore distribution is required, the pore diameter being about twice the size of the major axis of the enzyme unit cell (Messing, 1974). The enzyme can be attached to the glass by immersing in a enzyme solution, by a polymerization process or by a silane coupling, the latter being the preferred technique. There are numerous patents and papers relating to enzyme immobilization in glass (e.g. Weetall, 1969; Messing, 1970a, 1970b).

The use of porous glass membranes for dialysis is still being pursued by some workers and experiments have now reached animal trials (von Baeyer *et al.*, 1982a, 1982b). Porous glass membranes were manufactured by the technique developed by Schnabel and Vaulont (1978) for reverse osmosis membranes using a sodium borosilicate glass of a Vycor type. A mean pore radius of 5 to 25 μm was used in the form of capillaries. These experiments have shown that porous glass membranes are competitive with cellulose acetate membranes but the retention of proteins is still a problem.

Porous glass as a support for chemical catalysts has been suggested for many years and there are many patents in this field. Their use has been limited because of the difficulties in producing glass with a high pore volume and good thermal stability. In general these were based on Vycor-type glasses which would produce pore volumes of 0.1–0.3 cm^3 g^{-1} and give a skeleton of approximately 96% silica. Hammel (1976) developed a low silica glass which on careful heat treatment and leaching would give a large pore volume and a skeleton of ~99% silica. This has the advantage of being thermally stable up to ~1000°C. Pore volumes of the order of 0.6 cm^3 g^{-1} were obtained with a surface area of 70–200 m^2 g^{-1}.

REFERENCES

Asahara, U. Y. and Izumitani, T. (1980) *J. Non-Cryst. Solids.*, **42**, 269.
Ballou, E. V. and Wydevan, T. (1972) *J. Colloid. Interface Sci.*, **41**, 198.

Belfort, G. (1972) PhD thesis, University of California.

Brunauer, S., Emmett, P. H. and Teller, E. (1935) *J. Am. Ceram. Soc.*, **60**, 309.

Brunauer, S. *et al.* (1940) *The Adsorption of Gases and Vapours*, Vol. 1, Princeton University Press, New Jersey.

Cahn, J. W. and Charles, R. J. (1965) *Phys. Chem. Glasses*, **6**, 181.

Doremus, R. H. and Turkalo, A. M. (1979) *Science*, **164**, 418.

Egndi, K., Tasaha, K. and Tarumi, S. (1969) *J. Ceram. Soc. Jpn*, **22**, 301.

Ellis, R. B. (1951) US Patent 2 556 616.

Elmer, T. H. (1971) US Patent 3 592 619.

Elmer, T. H. and Martin, F. W. (1979) *Am. Ceram. Soc. Bull.*, **58**, 1092.

Everett, D. H. (1958) *The Structure and Properties of Porous Materials*, Butterworths, London.

Greig, J. W. (1927) *Am. J. Sci.*, **13**, 133.

Haller, W., Blackburn, D. and Simmons, J. H. (1974) *J. Am. Ceram. Soc.*, **57**, 120.

Haller, W., Blackburn, D., Wagstaff, E. and Charles, R. (1970) *J. Am. Ceram. Soc.*, **53**, 34.

Hammel, J. J. (1976) US Patent 3 972 720.

Hindley, N. K. and Watson, J. H. P. (1969) *Phys. Rev.*, **183**, 30.

Hood, H. P. and Nordberg, M. E. (1938) US Patent 2 106 744.

Hood, H. P. and Nordberg, M. E. (1940) US Patent 2 221 709.

Hood, H. P. and Nordberg, M. E. (1943) US Patent 2 315 329.

Hood, H. P. and Nordberg, M. E. (1950) US Patent 2 494 259.

Huang, R. T., Demirel, T. and McGee, T. D. (1972) *J. Am. Ceram. Soc.*, **55**, 399.

James, P. F. (1975) *J. Mater. Sci.*, **10**, 1802.

Johnson, J. D. (1976) US Patent 3 945 816.

Katchalski, E. and Bor-Eli, A. (1960) *Nature, Lond.*, **188**, 856.

Kozawa, A. (1974) *Electrochemistry of Manganese Dioxide*, Vol 1, Marcel Dekker, New York.

Kraceck, F. C. (1939) *J. Am. Ceram. Soc.*, **61**, 2869.

Kraus, K. A. *et al.* (1966) *Science*, **151**, 194.

Kuline, K. (1955) *Silikat. Tech.*, **6**, 690.

Lawless, W. N. (1972) *Rev. Sci. Instrum.*, **43**, 1743.

Littman, F. E. and Gutter, G. A. (1968) Office of Saline Water, Report 379.

Macedo, P. B. and Litovitz, T. A. (1976) US Patent 3 938 974.

Mackenzie, J. D. (1973) *Proc. Symp. on Utilisation of Glass Waste, University of Mexico, 24/25 June 1973*, p. 293.

Maddison, R. (1980) PhD thesis, University of Warwick.

Marceau, W. E. (1967) US Patent 3 325 264.

McMillan, P. W. and Matthews, C. E. (1976) *J. Mater. Sci.*, **11**, 1184.

Messing, R. A. (1969) *J. Am. Ceram. Soc.*, **91**, 2370.

Messing, R. A. (1970a) *Enzymologia*, **38**, 39.

Messing, R. A. (1970b) *Enzymologia*, **38**, 370.

Messing, R. A. (1974) *Biotechnol. Bioeng.*, **16**, 897.

Minot, M. J. (1976) *J. Opt. Soc. Am.*, **66**, 515.

Minot, M. J. (1977) *J. Opt. Soc. Am.*, **67**, 1046.

Mitz, M. A. and Summiaria, L. (1961) *Nature (London)*, **189**, 576.

Nakagawa, K. and Izumitani, T. (1969) *Phys. Chem. Glasses*, **10**, 179.

Nelson, J. M. and Griffin, E. G. (1916) *J. Am. Ceram. Soc.*, **38**, 1109.

Nordberg, M. E. (1944) *J. Am. Ceram. Soc.*, **27**, 299.

Phillips, S. V. (1974) *Desalination*, **14**, 209.

Phillips, B. and Muan, A. (1959) *J. Am. Ceram. Soc.*, **42**, 414.

Pratten, N. A. and McMillan, P. W. (1984) Warwick University Internal Report GR/B/23007.

Sample, H. H., Brandt, B. L. and Rubin, L. G. (1982) *Rev. Sci. Instrum.*, **53**, 1129.
Schnabel, R. and Vaulont, W. (1978) *Desalination*, **24**, 249.
Schott, L. A. (1971) US Patent 3 628 937.
Slayter, J. (1964) US Patent 3 151 966.
Simmons, J. H. and Macedo, P. B. (1971) *Disc. Faraday Soc.,* **50**, 155.
Simmons, J. H. *et al.* (1979a) *Appl. Opt.*, **18**, 2732.
Simmons, J. H. *et al.* (1979b) *Nature (London)*, **278**, 729.
Taylor, H. D. (1904) UK Patent 29 561.
Terry, C. (1968) *Rev. Sci. Instrum.*, **39**, 925.
Tingley, G. L., Lytle, J. M., Gray, W. J. and Wheeler, K. R. (1982) *J. Am. Ceram. Soc.*, **65**, 5.
Tomosawa, M. (1979) *Treatise on Material Science and Technology*, Vol. 17, Academic Press, New York.
Tomozawa, M. and Takamore, T. (1980) *J. Am. Ceram. Soc.*, **63**, 276.
Tran, S. M. (1976) University of Warwick Internal Report.
Volf, M. B. (1961) *Technical Glasses*, Chapter 10, Pitman, London.
Vogel, W. (1977) *J. Non-Cryst. Solids.*, **25**, 170.
von Baeyer *et al.* (1982a) *Trans. Am. Soc. Artif. Intern. Organs*, **28**, 488.
von Baeyer *et al.* (1982b) *J. Memb. Sci.*, **11**, 275.
Watson, J. H. P. (1971) *J. Appl. Phys.*, **42**, 46.
Weetall, H. H. (1969) *Nature (London)*, **233**, 959.
Williams, H. P. (1975) *Glass Technol.*, **16**, 34.

7

Glass-ceramics in substrate applications

G. Partridge, C. A. Elyard and M. I. Budd

7.1 INTRODUCTION

Glass-ceramics are now becoming well established as engineering materials, enabling the ranges covered by 'conventional' glasses and sintered ceramics to be extended and diversified considerably. Investigation into their use in a wide range of technical and engineering applications is proceeding throughout the world and indeed, in certain areas, for example microwave radomes, vacuum envelopes, laser envelopes, telescope mirrors and domestic cooker tops and cooking ware, their use is now well established.

Glass-ceramics are inorganic materials, generally but not necessarily silicate-based materials, which are initially prepared as glasses and which, in bulk form, are shaped by glass-forming techniques. They are then processed further by suitable heat-treatment to develop, firstly, nuclei in the glass and susequently crystal phases (McMillan, 1979).

Figure 7.1 illustrates a typical heat-treatment cycle for such a glass-ceramic with nucleation and crystallization temperature holds (more holding stages may be included as necessary to develop the required structure and properties). The heat-treatment process is so designed that the microstructure of the resultant body is one in which one or more crystal phases exist (together with a residual glassy phase) in a closely interlocking structure with mean crystal sizes generally in the region of 1 μm, although, in some cases, the mean crystal size can be considerably less (Partridge, 1982).

Alternatively the glass-ceramic can be processed via a powder route (e.g. die or isostatically pressed, slip cast, tape cast, or as a powder coating) and 'sintered' to achieve full density. Further heat-treatment, as above, may be necessary to convert the sintered body to glass-ceramic, or the required crystallization may take place in the one firing schedule.

The extensive range of properties which can be realized with glass-ceramics make them suitable for consideration as substrate materials in applications covering a wide range of frequencies and where either thick or

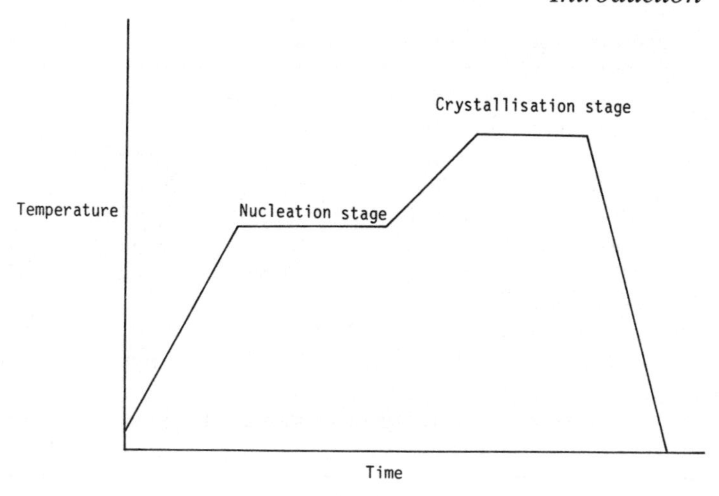

Fig. 7.1 Typical heat-treatment schedule for the production of glass-ceramics showing nucleation and crystallization holding stages.

thin film circuitry is required. In many cases, they provide attractive alternatives to other substrate materials, e.g. glasses and ceramics.

Such substrate materials which have been used previously include fused silica, borosilicate glasses and alumina and beryllia ceramics. The glasses, whilst of relatively low dielectric constant and thus suitable for use at high frequencies, find only limited use because of their mechanical weakness. Alumina ceramics, in particular, have found extensive use in substrate applications, and with modern fabrication methods such as green state tape casting and extrusion, a high quality substrate with good flatness and surface finish can be produced relatively cheaply. Alumina has a dielectric constant in the region of 10 and this limits its use at frequencies in excess of 10–20 GHz. The attainable strength and brittleness of alumina ceramics also limit the size of substrates which can be handled safely. In many applications the thickness of the substrate is required to be about 0.5 mm, which puts a realistic limit on the size of an alumina substrate of about 100 × 100 mm. If larger sizes are required, then greater thicknesses have to be used to ensure adequate strength. Beryllia has also found use in substrate applications, and has the advantage over alumina in that it has a much lower dielectric constant of ~6.5 and a higher thermal conductivity which is important where high thermal dissipation is required. However, it is significantly weaker than alumina, it is not easily worked to a good surface finish, and it is extremely toxic, especially in powder form. Manufacturers are, therefore, often reluctant to specify it in their products.

New ceramic materials which are becoming available in substrate form include aluminium nitride. This material also offers high thermal

Table 7.1 Comparison of general properties of bulk substrate materials

	Glass-ceramic	*Fused silica*	*Borosilicate glass*	Al_2O_3	*BeO*
ε	5–8.5	3.5	4.5	~10	~6.5
$\tan\delta \times 10^4$	2 to >100	<10	<10	1 to 10	<10
Strength	Medium–high	Low	Low	High	Good
$\alpha \times 10^7\,°C^{-1}$ 20–500°C	0–180 (depends on comp.)	5	30–50	75–80	75–80
Thermal conductivity	Low	Very low	Very low	High	Very high
Surface finish possible	Very good	Very good	Very good	Good–very good	Moderate

conductivity but without the toxicity of beryllia. However, materials such as aluminium nitride are expensive at present and pose problems of incompatibility with many materials used in microelectronics, for example, thick film inks.

A comparison of the relevant properties of these materials, together with the ranges which can be achieved with glass-ceramics, is given in Table 7.1. Further details of the ranges of linear thermal expansion coefficients which can be obtained with the principal groups of glass-ceramics are shown in Fig. 7.2 and compared with those of alumina, beryllia and aluminium nitride ceramics. Further reference to this important feature of controllability of expansions will be made during the course of this chapter.

A further important feature of glass-ceramics is that they can generally be prepared with a fine crystal structure so that a high quality surface finish can readily be achieved by grinding and polishing, the surface finish approaching that which can be obtained on glasses. This enables both thick and thin film circuitry to be employed. At the same time glass-ceramics are considerably stronger than glasses and, in some cases, stronger than the sintered ceramics, thus easing handling problems.

The first part of this chapter describes the properties of several types of glass-ceramic in bulk form, and compares and contrasts these with properties of other substrate materials. The major disadvantage of glass-ceramics lies in their relatively poor thermal conductivities, which although slightly higher than those of glasses are still much lower than those of Al_2O_3, BeO and AlN ceramics, particularly the latter two. However, other possible applications of glass-ceramics lie in their use as thin coatings on metal plates, which are relatively good thermal conductors, offsetting to a large extent the low thermal conductivities of the glass-ceramics and providing strong, rigid

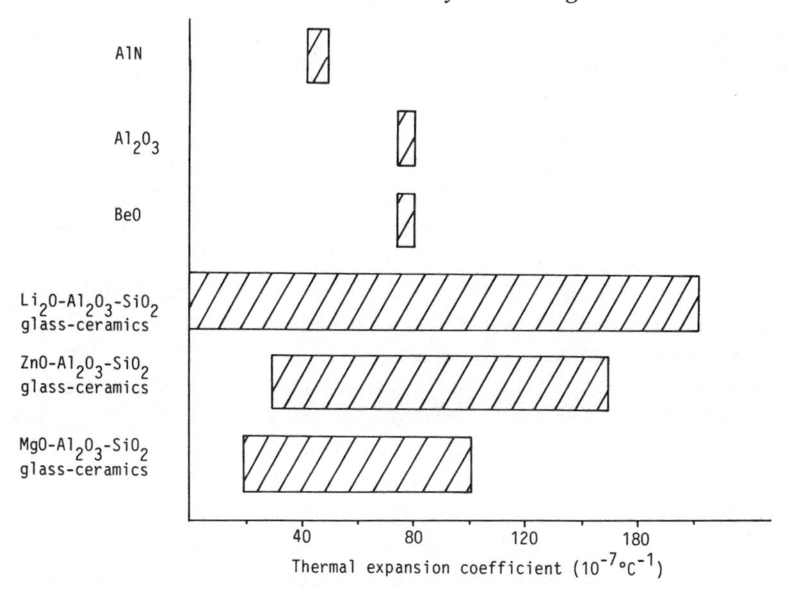

Fig. 7.2 Comparison of the ranges of linear thermal expansion coefficients possible with Al_2O_3, BeO, AlN and various glass-ceramics.

substrates with high thermal dissipation suitable for use in hybrid circuitry. These are discussed in the second part of this chapter.

The work on glass-ceramic substrates and glass-ceramic coated metal described in this chapter has been carried out principally at the GEC Engineering Research Centre, Stafford. Where work has been carried out by other organizations reference is made as appropriate.

7.2 BULK CRYSTALLIZED GLASS-CERAMICS

In this section, glass-ceramics which are prepared by the controlled crystallization of bulk glass compositions are discussed (as opposed to materials prepared from powder techniques which will be considered later). The bulk glass-ceramics which are described in this section are those from the $Li_2O–Al_2O_3–SiO_2$, $ZnO–Al_2O_3–SiO_2$ and $MgO–Al_2O_3–SiO_2$ systems which have been investigated by the authors. These cover a range of properties and thus illustrate what can be achieved in the field of bulk glass-ceramic substrates.

7.2.1 Preparation

Preparation of all these glasses of various types involves mixing the relevant raw materials and melting, generally in Pt/Rh crucibles on laboratory scale, (larger scale melting can be performed in suitable refractories such as zircon) at temperatures in the range 1200–1650°C. Following refining, the glasses can be cast or pressed using preheated metal moulds to form flat plates, blocks or cylinders. Annealing to relieve stress is accomplished at temperatures in the range 450°C to 700°C, the higher annealing temperatures generally being associated with the lower alkali and alkali-free materials.

Where necessary, test components are machined from the cast blanks; the test pieces and remaining blanks are then heat-treated to convert the glass shapes to glass-ceramics. The heat-treatment processes employed generally involve two holding temperatures; (essentially as shown in Fig. 7.1) the first for 'nucleation' purposes, at temperatures in the range 500°C to 900°C, and the second, for crystallization purposes, at temperatures in the range 800°C to 1250°C. The lower hold temperatures are associated with the alakli-containing materials and the higher ones with the alkali-free materials.

The nucleation/crystallization processes in the various glass-ceramics discussed have been studied by a variety of techniques which have been described previously (Partridge *et al.*, 1973). These studies enable heat-treatment procedures to be developed for the different glass-ceramics which optimize properties for specific applications, usually with particular reference to electrical and mechanical properties.

Typical properties achieved in the glass-ceramics are discussed in the following sections, in comparison with data on other ceramic substrate materials, for example alumina and beryllia, key properties of which are given in Table 7.2.

Table 7.2 Properties of selected alumina ceramics and beryllia ceramics

Compositional data	*Ceramic*			
	A1	*A2*	*A3*	*B1*
Al_2O_3	99	99	95	–
BeO	–	–	–	99
Properties, X-band				
ε				
20°C	9.31	9.66	9.02	6.6
400°C	9.88	10.30	9.47	–
$\tan\delta \times 10^4$				
20°C	10	1	7	5
400°C	10	2	9	
Density ($g\,cm^{-3}$)	3.73	3.84	3.64	2.95
Strength (MPa)	280	310	300	240
Thermal expansion coefficient, $\alpha \times 10^7$				
20–500°C	78	78	75	78

7.2.2 Li_2O–Al_2O_3–SiO_2 materials

Glass-ceramic materials of this type can be prepared within the glass-forming system shown in Fig. 7.3 and cover a wide range of thermal expansion coefficients ranging from in the region of zero to about 120×10^{-7} °C^{-1}. This enables low expansion materials which have very good resistance to thermal shock to be prepared, and enables other materials to be matched in thermal expansion characteristics to a number of metals and alloys which are employed in the microelectronics field. The materials which are thought most likely to find use as substrates are of medium to high thermal expansion, and typical properties of a number of these are given in Table 7.3.

The linear thermal expansion characteristics of these materials are particularly influenced by their alumina content. Those with low alumina content in which the principal crystal phases are lithium disilicate and silica (either as quartz or cristobalite), are of higher expansion. Increase in the Al_2O_3 content results in the presence of low expansion crystal phases such as β-spodumene ($Li_2O \cdot Al_2O_3 \cdot 4SiO_2$) or β-eucryptite ($Li_2O \cdot Al_2O_3 \cdot 2SiO_2$), producing glass-ceramics of low or even zero-expansion.

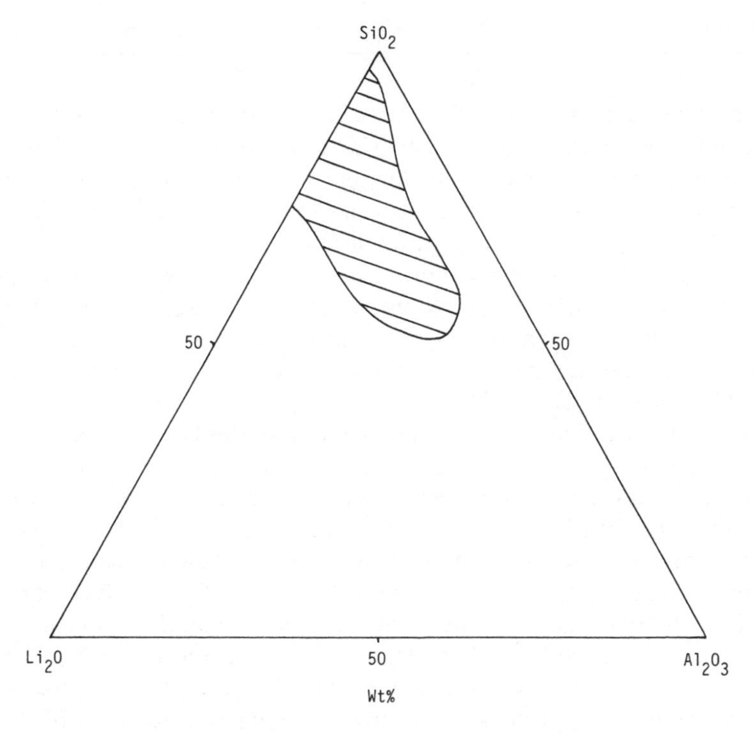

Fig. 7.3 Glass-ceramic forming region in the Li_2O–Al_2O_3–SiO_2 system.

Table 7.3 Properties of selected Li_2O–Al_2O_3–SiO_2 glass-ceramics

Compositional data	*Glass-ceramic*					
	L1	*L2*	*L3*	*L4*	*L5*	*L6*
Total alkali (%)	14	12	10	15	10	4
Al_2O_3 (%)	10	10	10	5	10	20
Alkaline earths	–	2% BaO	5%	–	9%	–
Nucleant	P_2O_5	P_2O_5	P_2O_5	P_2O_5	P_2O_5	TiO_2 + ZrO_2
Properties, X-band						
ε						
20°C	5.63	5.68	5.66	5.21	5.97	5.91
400°C	6.21	6.15	5.98	–	6.31	–
$\tan \delta \times 10^4$						
20°C	106	93	68	138	107	1300
400°C	208	168	–	–	110	–
Density ($g\,cm^{-3}$)	2.44	2.47	2.51	2.40	2.59	–
Strength (MPa)	255	230	160	380	175	–
Thermal expansion coefficient, $\alpha \times 10^7$						
20–500°C	52	45	45	93	50	~0

The changes in the overall composition also influence the types of nucleating agent which are effective in enabling fine grained materials to be obtained – an important factor, not only in providing the structure necessary for high strength, but also in permitting good surface finishes to be obtained – good enough for application of thin films if so required. In the lower Al_2O_3 content materials the most effective nucleating agent has been shown to be P_2O_5, whilst at higher Al_2O_3 content the most effective nucleating agent is TiO_2, in some cases alone, but in the very low expansion glass-ceramics in combination with other nucleants such as ZrO_2 (Partridge, 1982).

The dielectric properties of typical materials of this type are illustrated in Fig. 7.4. The dielectric constants are generally in the range 5–6 over a wide range of frequencies at room temperature. At microwave frequencies the dielectric constant increases with rise in temperature, as would be expected at lower frequencies, although the changes there would be expected to be greater than at microwave frequencies (McMillan, 1979). Reduction in the alkali content of the glass-ceramics, where the alkali is replaced by a divalent metal oxide (in particular BaO) leads to a reduction in the increase of dielectric constant with temperature ($\delta\varepsilon/\delta T$). In this type of glass-ceramic, the change in dielectric constant, ε, over the temperature range 20–400°C can be as low as 0.3.

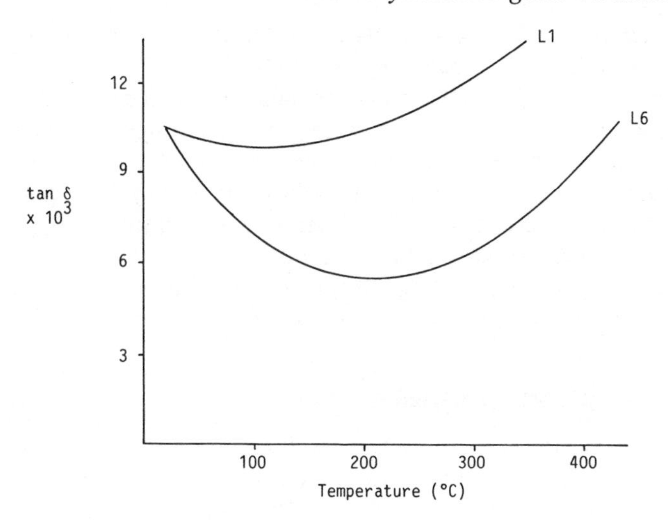

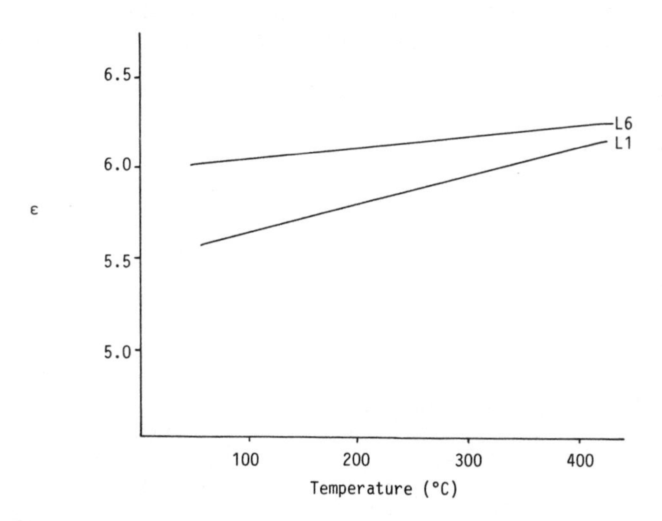

Fig. 7.4 Dielectric properties of selected Li_2O–Al_2O_3–SiO_2 glass-ceramics (at 9.36 GHz).

The dielectric losses in this type of glass-ceramic are not as low as with the other glass-ceramics discussed in this chapter. At microwave frequencies (X-band), loss angles are generally in the range 0.007 to 0.014 at room temperature. Interestingly the dielectric losses initially decrease with increase in temperature, passing through a minimum at about 200°C and then rising so that the figures at about 400°C are similar to those obtained at room temperature (Figure 7.4). As might be expected the lower loss materials are those which have the lowest alkali contents.

In comparison, the dielectric constants of the high purity alumina ceramics which have found considerable use as substrates lie in the range 9.3 to 9.7 at X-band, with $\delta\varepsilon$, 20–400°C in the range ~0.4 to 0.6. The change in ε of alumina with rise in temperature can be decreased by the use of dopants, for example a titanate, but whilst $\delta\varepsilon$ can be reduced to ~0, the dielectric constant itself is then increased significantly to ~11.7.

The loss tangents in the high purity and doped aluminas are low, less than 0.001, over the temperature range considered.

7.2.3 $ZnO–Al_2O_3$-SiO_2 materials

Glass-ceramic materials of this type with compositions falling within the range shown in Fig. 7.5 can be produced with thermal expansion coefficients ranging from about 30 to 150 × 10^{-7} °C^{-1}. High expansion glass-ceramics of this type result from the presence of large proportions of cristobalite. The materials for which cristobalite is present only in small proportions and for which the major crystal phase is willemite (Zn_2SiO_4) have expansion coefficients in the range 30 to 60 × 10^{-7} °C^{-1} and it is these materials which

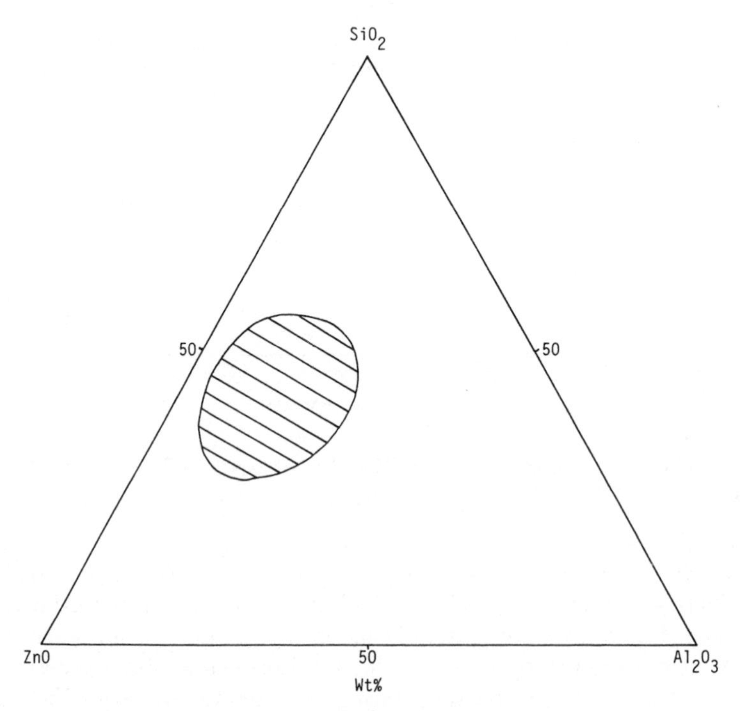

Fig. 7.5 Glass-ceramic forming region in the $ZnO–Al_2O_3–SiO_2$ system.

are being examined as potential substrates. Typical properties of a number of these are given in Table 7.4 (McMillan and Partridge, 1972).

As pointed out above, the thermal expansions of these materials are particularly influenced by their silica content and when sufficient silica is present to permit the development of a silica phase, usually cristobalite, the expansion rises rapidly, the expansion characteristic reflecting the marked volume increase which occurs on the α–β cristobalite phase change at about 270°C.

Work on nucleation of these materials has shown that TiO_2 in proportions in the region of 5% by weight is effective, although lower proportions of TiO_2 can be used in conjuction with other nucleants, for example P_2O_5. Fine grained materials are then obtainable in these glass-ceramics and they are more refractory than the Li_2O–Al_2O_3–SiO_2 materials reviewed previously, being capable of withstanding temperatures in excess of 1000°C, whereas many of the Li_2O–Al_2O_3–SiO_2 materials are restricted to temperatures below about 850°C.

The dielectric properties of typical materials of this type are given in Table 7.4 and illustrated in Fig. 7.6. The dielectric constants of these materials are generally in the range 6–7.5 over a wide range of frequencies. At microwave frequencies the changes occurring with temperature in typical materials are illustrated in Fig. 7.6. The addition of alkaline earth oxides, for example

Table 7.4 Properties of selected ZnO–Al_2O_3–SiO_2 glass-ceramics

Compositional data	*Glass-ceramic*				
	Z1	*Z2*	*Z3*	*Z4*	*Z5*
ZnO (%)	30	57	42	37	47
BaO (%)	15	–	15	20	5
Nucleant	TiO_2	TiO_2 + P_2O_5	TiO_2 + P_2O_5	TiO_2 + P_2O_5	TiO_2 + P_2O_5
Properties, X-band					
ε					
20°C	7.35	6.76	6.79	6.97	6.45
400°C	7.49	7.14	7.00	7.18	6.73
$\tan \delta \times 10^4$					
20°C	6	13	5	8	22
400°C	30	31	9	9	56
Density ($g\,cm^{-3}$)	3.46	3.79	3.70	3.72	3.68
Strength (MPa)	125	–	95	–	–
Thermal expansion coefficient, $\alpha \times 10^7$					
20–500°C	50	60	30	31	38

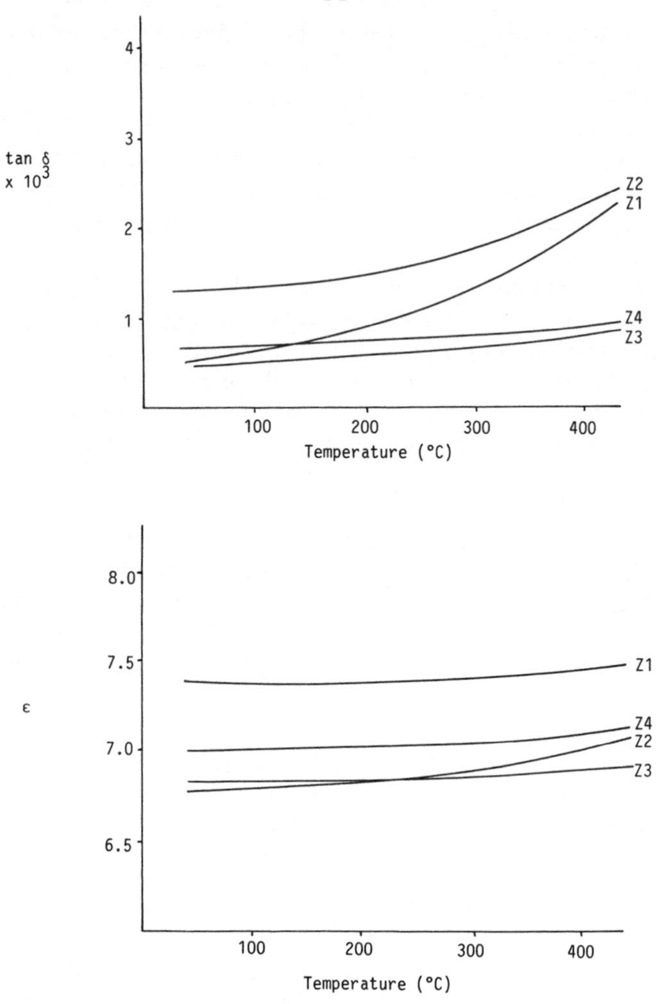

Fig. 7.6 Dielectric properties of selected $ZnO–Al_2O_3–SiO_2$ glass-ceramics (at 9.36 GHz).

BaO and/or SrO, tends to increase the room temperature dielectric constant but decreases the change of dielectric constant with temperature rise up to 400°C (McMillan and Partridge, 1972). Low changes of ε are possible with $ZnO–Al_2O_3–SiO_2$ materials and changes as low as 0.14, over the range 20–400°C, have been measured. The dielectric losses observed in these materials are very low at microwave frequencies, much lower than measured for the $Li_2O–Al_2O_3–SiO_2$ materials, as is to be expected in view of their very low alkali contents (generally limited to impurity levels only – typically less than about 0.1%). They are comparable with those obtained for the high alumina ceramics (Table 7.2). Unlike the Li_2O–Al_2O_3–SiO_2 materials which have densities of ~2.5 g cm^{-3}, these glass-ceramics have densities similar to those of the alumina ceramics (approaching 4 g cm^{-3}). Generally, they do not exhibit the high strengths which are possible with Al_2O_3 ceramics and which can also be achieved with the Li_2O–Al_2O_3–SiO_2 materials.

7.2.4 MgO–Al_2O_3–SiO_2 materials

Glass-ceramic materials of this type can be prepared from compositions falling within the range shown in Fig. 7.7 with thermal expansion coefficients ranging from about 20 × 10^{-7} °C^{-1} to about 100 × 10^{-7} °C^{-1}.

The two main types of glass-ceramic investigated in the MgO–Al_2O_3–SiO_2 system have been based on cordierite ($2MgO \cdot 2Al_2O_3 \cdot 5SiO_2$) and clino-enstatite ($MgO \cdot SiO_2$). Other crystal phases, including cristobalite (SiO_2), forsterite ($2MgO \cdot SiO_2$), mullite ($3Al_2O_3 \cdot 2SiO_2$), spinel ($MgO \cdot Al_2O_3$) and zirconia (ZrO_2) were produced in minor proportions in some materials.

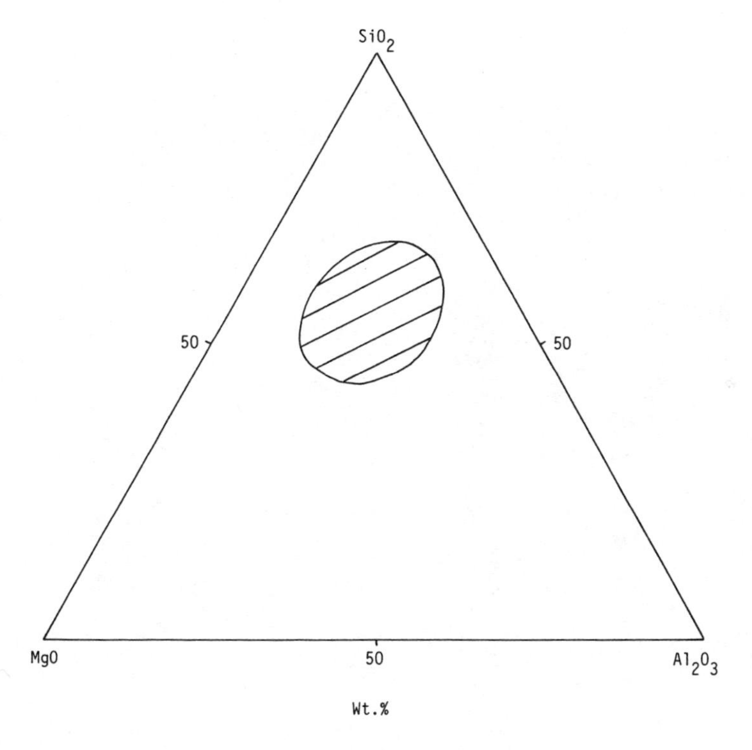

Fig. 7.7 Glass-ceramic forming region in the MgO–Al_2O_3–SiO_2 system.

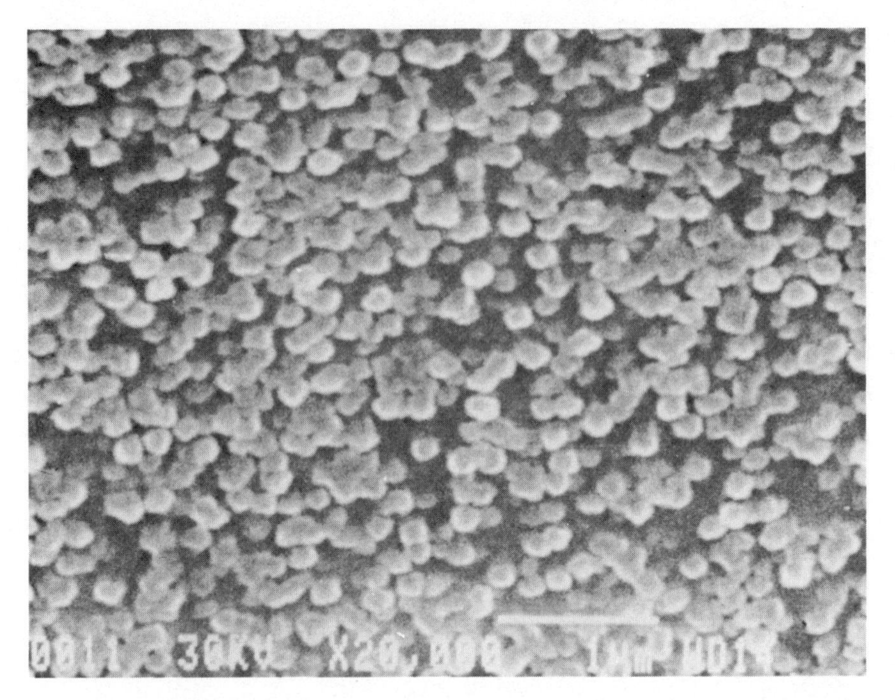

Fig. 7.8 SEM of etched fracture surface of partially heat-treated (850°C, 10 h) clino-enstatite based glass-ceramic showing high nucleus population.

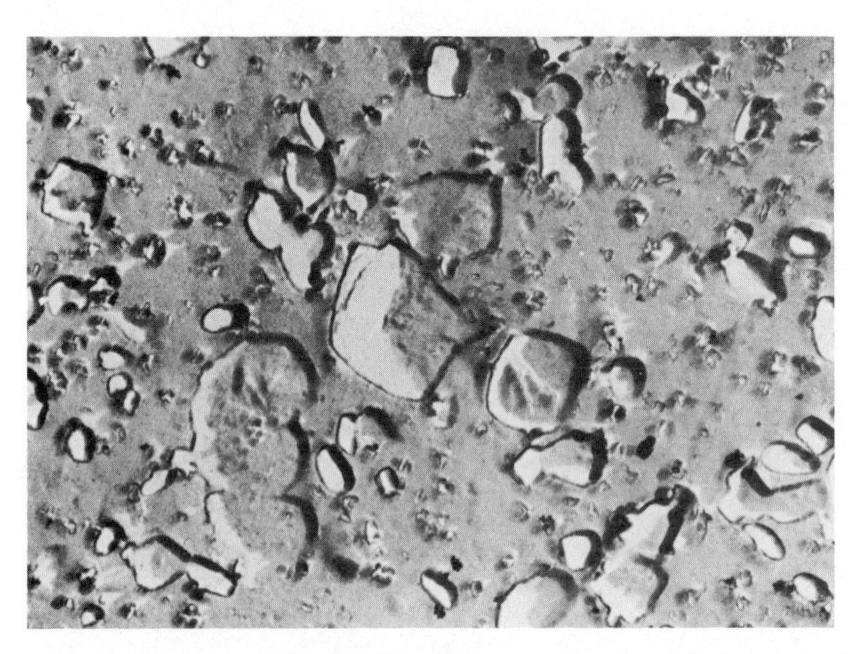

Fig. 7.9 TEM of Pt/C shadowed replica of fracture surface of partially heat-treated (850°C, 24 h) clino-enstatite based glass-ceramic. ——— 1 μm

The preferred nucleating agents for the cordierite based materials are TiO_2, or combinations of TiO_2 and ZrO_2. For the clino-enstatite glass-ceramics, ZrO_2 is an efficient nucleant. The cordierite based glass-ceramics have been nucleated at temperatures in the range 850–900°C and subsequently heated to temperatures of up to 1250°C to allow the development of the desired crystal phases. The crystallization process is complex and it has been suggested (Barry *et al.*, 1978) that, following the formation of the nucleating phase (a magnesium alumino-silicate structurally similar to petalite), a number of intermediate phases appear prior to the development of cordierite at between 1150°C and 1250°C.

The clino-enstatite based materials have been produced by subjecting the precursor glasses to a nucleation stage at temperatures between 550°C and 700°C and a crystallization stage at between 800°C and 1100°C. Tetragonal zirconia has been identified as the initial phase to appear in these glass-ceramics, and Fig. 7.8 shows that a high population of tetragonal zirconia nuclei is built up during the first stage of the heat-treatment. Figures 7.9 and 7.10 show how the microstructure develops as the heat-treatment progresses, and it can be seen that the final grain size is in the order of 1–2 μm. Thus, it appears that the tetragonal zirconia particles, which have a mean spacing of ~0.2 μm, are inefficient in providing growth sites for the clino-enstatite phase, yet the final microstructure is still relatively fine. The

Fig. 7.10 TEM of Pt/C shadowed replica of fracture surface of fully heat-treated clino-enstatite based glass-ceramic. ——— 1 μm

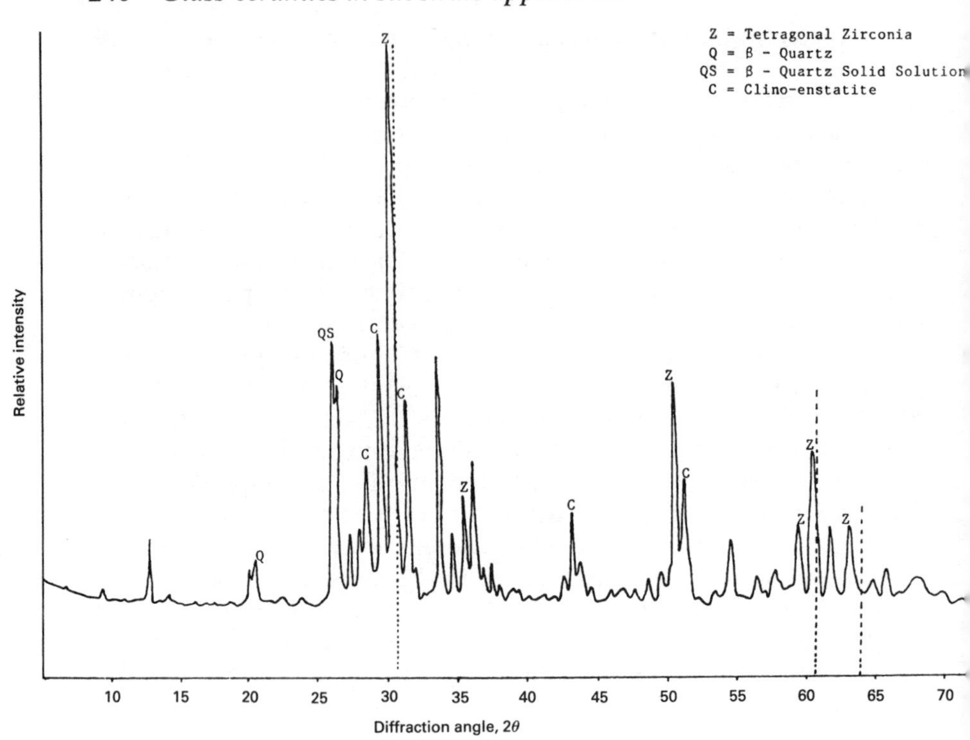

Fig. 7.11 X-ray diffraction data for clino-enstatite based glass-ceramic.

presence of the tetragonal form of zirconia is surprising as the large amount of MgO which is present would be expected to stabilize the zirconia in the cubic form. Verification of the presence of tetragonal zirconia has been obtained by X-ray diffractometry (Fig. 7.11), where the peaks marked Z correspond to those produced by tetragonal zirconia; the dashed lines indicate where the peaks would be expected were the zirconia present in its cubic form.

(a) Dielectric properties

The dielectric constants of both types of $MgO–Al_2O_3–SiO_2$ glass-ceramic are lower than that of alumina, particularly the cordierite based materials; consequently their use at high frequencies may be advantageous. The lower density and higher silica content of the cordierite based materials contribute to their low dielectric constant of about 6 at microwave frequencies. The dielectric properties of typical cordierite based glass-ceramics are given in Table 7.5 and illustrated in Fig. 7.12 (materials M1 and M3). By varying the

Table 7.5 Properties of selected $MgO–Al_2O_3–SiO_2$ glass-ceramics

Compositional data	*Glass-ceramic*				
	M1	*M2*	*M3*	*M4*	*M5*
MgO	15	12	14	12	22
Al_2O_3	20	26	30	26	22
Nucleant	TiO_2	TiO_2	TiO_2	TiO_2 + ZrO_2	ZrO_2
Properties, X-band					
ε					
20°C	5.65	5.91	6.39	5.60	8.5
400°C	5.78	5.90	6.29	5.65	8.8
$\tan\delta \times 10^4$					
20°C	3	5	5	4	60
400°C	20	43	43	25	172
Density ($g\,cm^{-3}$)	–	–	–	–	3.18
Strength (MPa)	350	162	127	270	750
Thermal expansion coefficient, $\alpha \times 10^7$					
20–500°C	53	32	24	31	82

composition and titania content it is possible to prepare glass-ceramics of this type for which $\delta\varepsilon/\delta T$ is zero (20–400°C) or indeed negative. The dielectric losses in these materials are low at room temperature, but at elevated temperatures tend to be higher than those which can be achieved with $ZnO–Al_2O_3–SiO_2$ materials or alumina ceramics.

The lower silica content and higher ZrO_2 content are considered to be responsible for the higher dielectric constants of the clino-enstatite based materials (ε 8–8.5). The losses in the clino-enstatite glass-ceramics are significantly greater than those of the cordierite materials at microwave frequencies, but reasonable low ($\sim 5 \times 10^{-4}$) at lower frequences, e.g. 1 MHz. Work carried out by the authors has shown that a requirement of a low loss material is the absence of alkali metal ions, so high purity raw materials are required to achieve this.

(b) Mechanical properties

The strength of the $MgO–Al_2O_3–SiO_2$ based glass-ceramics depends on whether cordierite or clino-enstatite develops as the major crystal phase. Cordierite glass-ceramics can possess strengths of up to ~250 MPa in the machined condition and the clino-enstatite materials have shown even higher strengths – up to 350 MPa. One possible explanation for the high

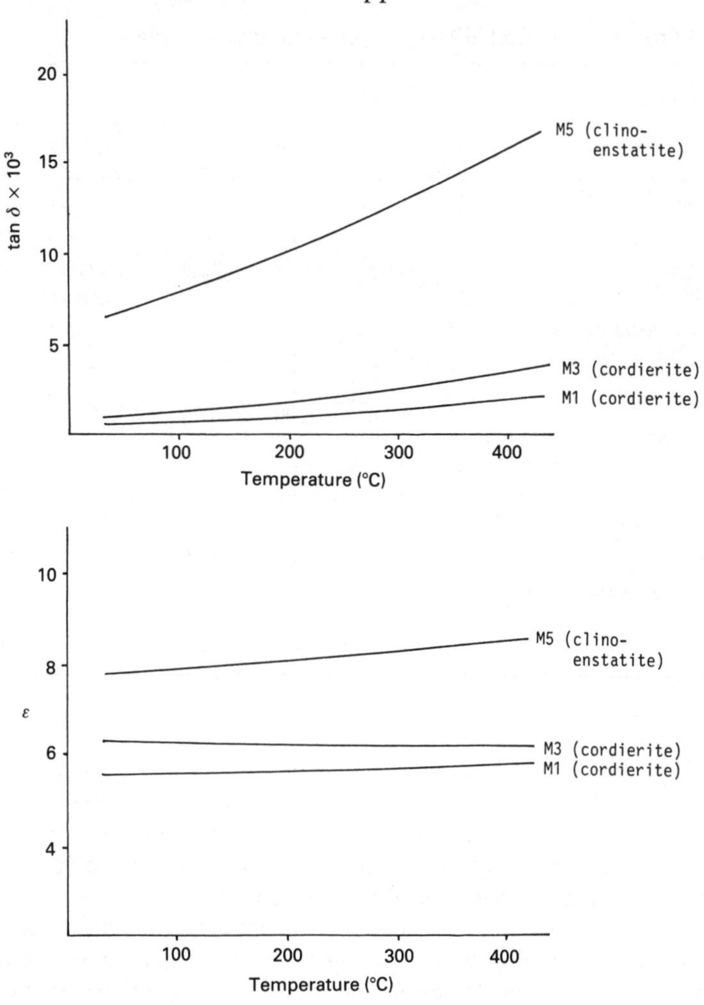

Fig. 7.12 Dielectric properties of selected $MgO–Al_2O_3–SiO_2$ glass-ceramics (at 9.36 GHz).

strength of the clino-enstatite materials is that the tetragonal zirconia particles cause transformation toughening (McMeeking and Evans, 1982). Another strengthening mechanism found in the clino-enstatite glass-ceramics arises from the development of different proportions of crystal phases at the surface of the material during heat-treatment. Figure 7.13 shows such a surface layer approximately 50 μm thick. The thermal expansion coefficient of this surface layer (measured on a sintered powder compact comprised essentially of surface phases), is in the order of 40×10^{-7} °C^{-1}. This is considerably lower than that of the bulk, and gives rise to

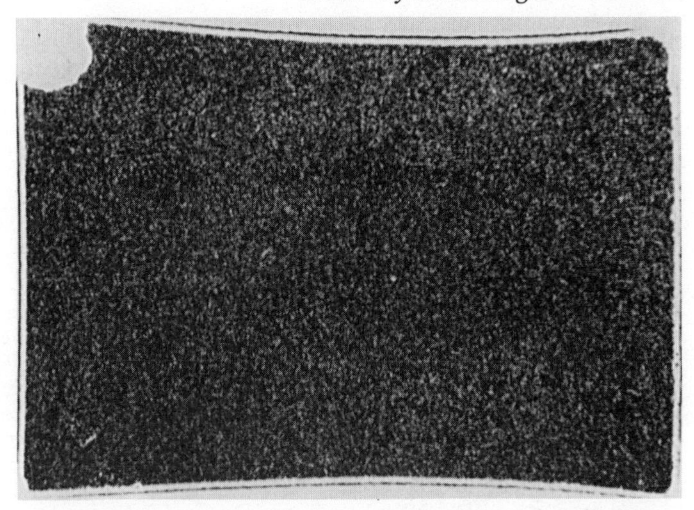

Fig. 7.13 Transverse section of clino-enstatite glass-ceramic bar (showing surface layer) heat-treated at 550°C for 20 h and at 910°C for 20 h after machining.

compressive stresses in the surface region. The strength of materials possessing this compressive surface layer can be 600–750 MPa and this should permit large area substrates to be produced and handled relatively safely.

(c) Thermal properties

The linear thermal expansion coefficients of a number of materials are given in Table 7.5 and the expansion curves are plotted in Fig. 7.14. It is clear that a bulk derived clino-enstatite based glass-ceramic is extremely well matched in expansion characteristics to both alumina and beryllia. This is advantageous since much of the coating and packaging technology associated with microelectronic circuitry has been developed to be matched in expansion to these materials. The cordierite material is closer to silicon in its expansion characteristics, as is the clino-enstatite based material obtained by heat-treating a powder compact (powder based materials are dealt with in more detail subsequently). Thus it is considered that these types of material may form a basis from which a silicon matching technology may be developed.

7.2.5 Compatibility with thick and thin film technology

The requirements for compatibility with thick and thin film methods of application of conductor tracks and resistive elements differ but both can be readily met by the bulk glass-ceramics described above.

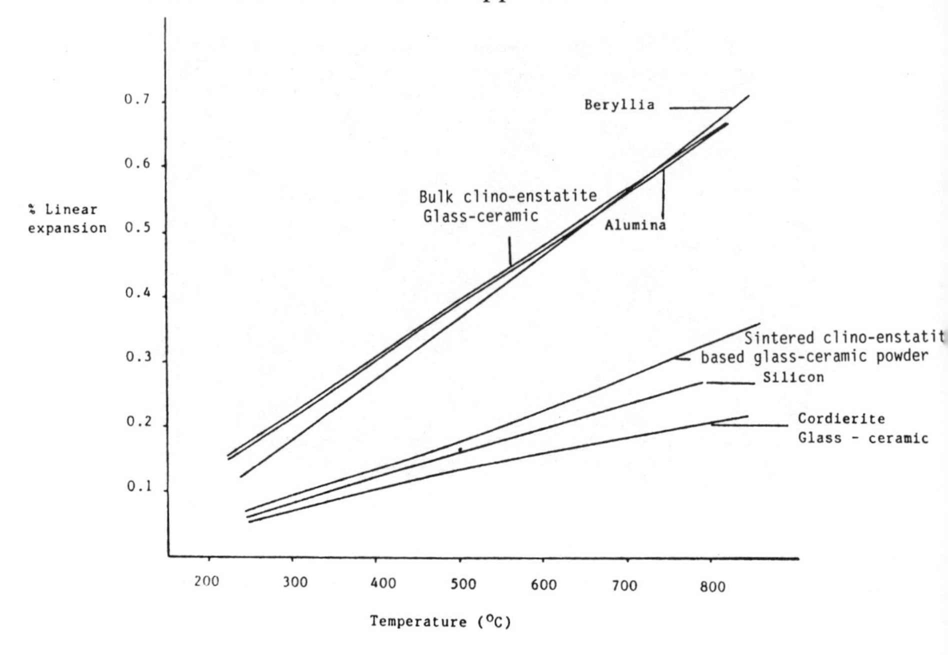

Fig. 7.14 Graph showing thermal expansion characteristics of alumina, beryllia and silicon and of matching glass-ceramics from the $MgO–Al_2O_3–SiO_2$ system.

(a) Thick film

Thick film circuitry is applied using suitable inks (containing organic binders and, in many instances, a proportion of a glass in finely divided powder form) by means of screen printing technology. This is a well established process and many proprietary inks are available which will provide conductor tracks in, for example, gold, palladium–silver and copper. The majority of commercially available inks have been developed to be compatible with alumina ceramics. It has now been established that a number of these inks are compatible with glass-ceramics of the types discussed.

In order to be suitable for thick film application it is generally necessary that the substrate is flat, although printing on to mono-curved surfaces is possible. The surface finish required on the glass-ceramic is similar to that which would be applicable to alumina ceramics (0–4 μm Ra). After screening on, the ink is dried and is then fired at temperatures in the region of 800–900°C. In general terms, the firing temperatures required for the inks on glass-ceramics are lower than those on alumina because the glassy phase in the glass-ceramics permits the reaction between glass-ceramic and ink (glass-phase) to occur more readily. Glass-ceramics of the $Li_2O–Al_2O_3–SiO_2$

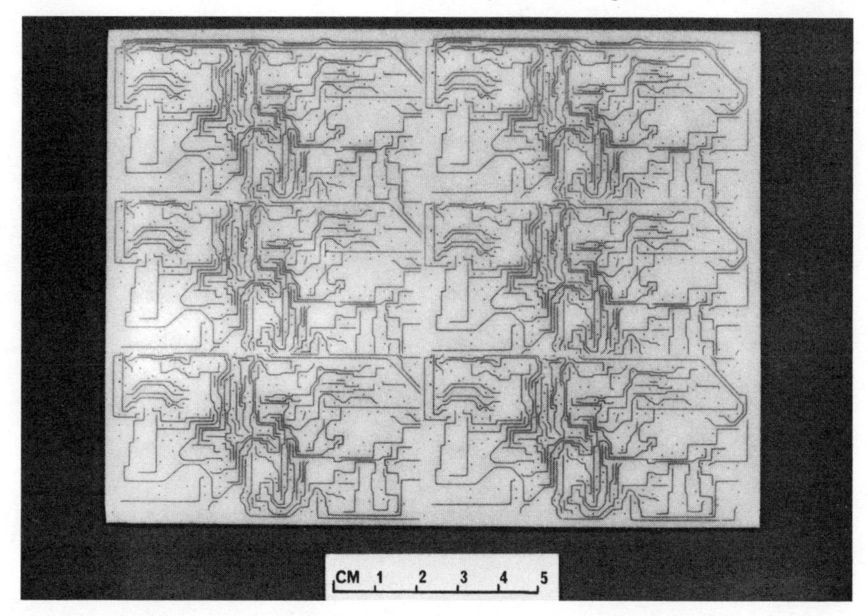

Fig. 7.15 Bulk glass-ceramic substrate with thick film circuitry. (Reproduced by kind permission of Marconi Electrical Devices Ltd, Microsystems Division.)

type may not be compatible with the higher firing inks because they cannot readily withstand temperatures greater than about 850°C (depending on composition) and 'swimming' of the ink may occur. Furthermore, inks containing silver are generally not desirable with the glass-ceramics of the Li_2O–Al_2O_3–SiO_2 type because of the possibility of silver diffusion occurring in the substrate. This results in a yellow discoloration, which may be cosmetically unacceptable on a basically white material, whilst in some cases there may be a significant deterioration in the electrical resistivity of the glass-ceramic, with uncertain long term effects.

The alkali-free bulk glass-ceramics, in which all phases including the glass are more refractory, are better than the alkali-containing materials in these respects, since they can accept the firing temperatures of all inks and are less susceptible to silver migration.

Figure 7.15 shows a bulk glass-ceramic substrate on which thick film circuitry has been successfully applied at the GEC Engineering Research Centre, Stafford.

(b) Thin film

The fine crystal structure of glass-ceramics enables good machined surface finishes to be obtained, much more readily than on, for example, high alumina content substrates. This arises, in part, from the very fine crystal

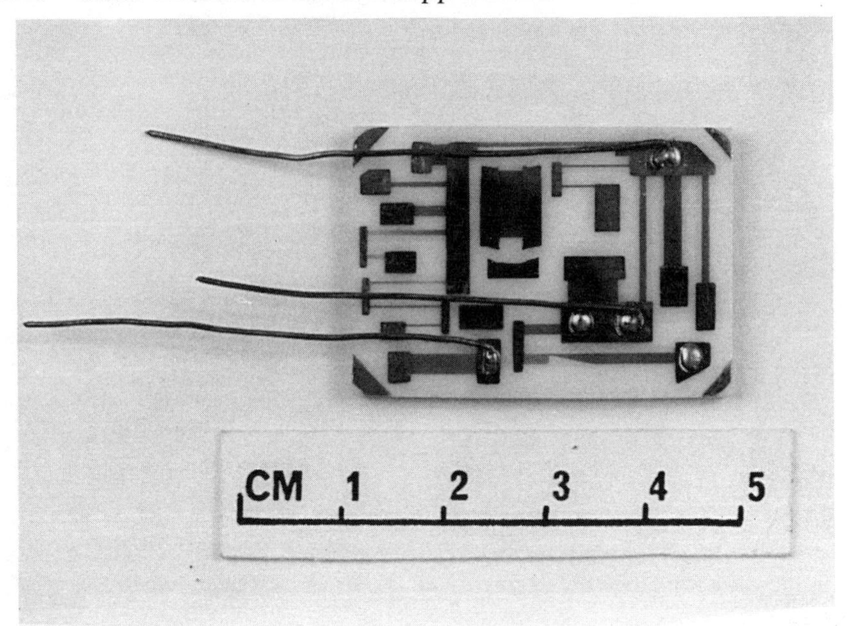

Fig. 7.16 Bulk Li_2O–Al_2O_3–SiO_2 substrate with thin film circuitry and soldered leads.

structure and lack of 'pull-out' of crystal during machining operations, and the glass-ceramics are not so hard as alumina so that stock removal is easier. Glass-ceramics can be prepared with surface finishes only a little inferior to those which can be obtained on glasses, and surface finishes better than 25 nm CLA are possible. Such a surface finish is satisfactory for the application of thin film circuitry. Thin film conductor tracks (e.g. gold) and resistive elements (e.g. Nichrome) are applied by vacuum evaporation techniques using suitable sources and masking. Figure 7.16 illustrates thin film circuitry which has been applied to a polished Li_2O–Al_2O_3–SiO_2 substrate. The quality of the adhesion of the circuitry to the substrate is illustrated by the ability to solder connecting wires to pads on the circuit.

7.3 BULK GLASS-CERAMICS VIA POWDER TECHNIQUES

In this section, glass-ceramics which are prepared by sintering powders of the glass composition are considered. This method of combining the powder particles to form a bulk material follows essentially the principles of conventional sintered ceramic technology. The difference lies in that the starting powder is a glass, and sintering and crystallization processes occur in the one firing cycle.

The advantages of this method are essentially that relatively thin films of the dielectric material can readily be obtained as described below in a process which is likely to be cheaper than the bulk approach described above where a significant proportion of substrate cost can result from machining operations. A problem is that, in common with other sintered materials, closed pores are present in the fired product, which reduce dielectric properties, dielectric breakdown strength, mechanical strength and capability of obtaining a high quality surface finish.

A number of types of glass-ceramic have been examined by the authors for the preparation of bulk materials via the powder route, including materials of the LiO_2–Al_2O_3–SiO_2, ZnO–Al_2O_3–SiO_2 and MgO–Al_2O_3–SiO_2 types discussed above. In a number of instances which are illustrated below the nucleation/crystallization phenomena in the materials differ from those experienced in the bulk glass-ceramics. This arises in part from the nature of the particle surface which results in a greater predominance of surface crystallization processes rather than bulk nucleated crystallization.

7.3.1 Preparation

Glasses of the various types are prepared in a similar manner to those for bulk materials. Following refining the glasses are cast into cold water or on to cold metal plates to provide frit. After drying (as necessary), the glass frit is reduced to powder by milling (for example in a porcelain jar mill with flint pebbles). The characteristics of the powder so obtained have been determined at Stafford by means of a Malvern 3600D laser diffractometer. The results obtained with this instrument are typically as shown in Fig. 7.17. This information is essential in determining the particle size distribution required for the preparation of the sintered glass-ceramic body, which is controlled in subsequent batches of the powder as part of quality assurance procedures.

Nucleation/crystallization processes in the various glass-ceramics have been studied using techniques described previously (Partridge *et al.*, 1973). These studies, together with sintering trials on powder compacts, enabled the conditions to be determined which would develop required properties in the materials. The required sintering conditions vary to some extent with the method of forming the 'green state' body, requiring that each case be treated individually in the final iterations of process development.

The powder processing routes for the preparation of glass-ceramic substrates which have been examined by the authors include:

1. Compaction by means of die pressing or isostatic pressing
2. Screen printing
3. Tape casting

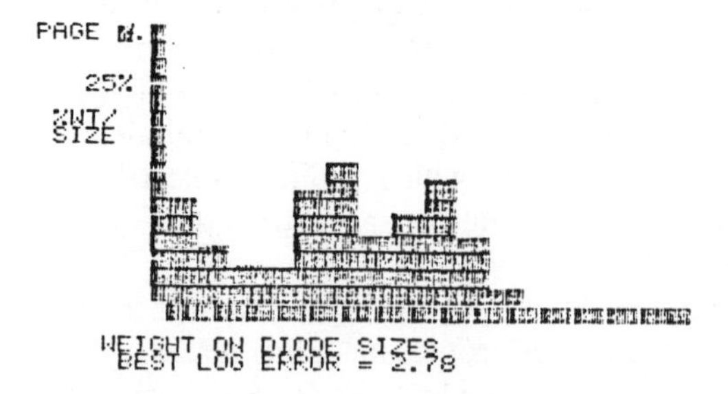

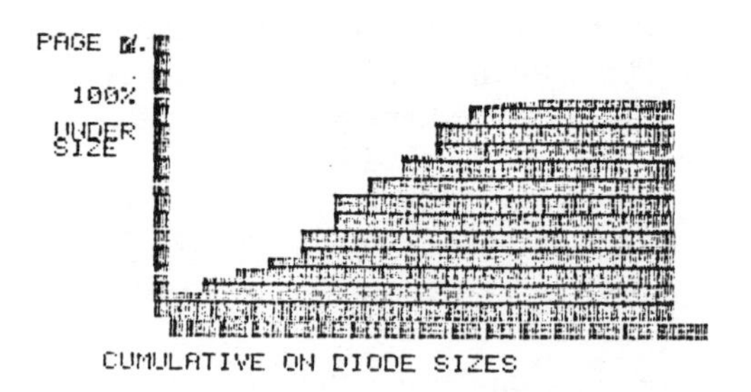

MALVERN 2600/3600 PARTICLE SIZER VA.

MALVERN INSTRUMENTS LTD, SPRING LANE, MALVERN, ENGLAND.

PRINTING RESULTS FROM DATA BLOCK 1

TIME 00-03-30 RUN NO. 2 LOG ERROR = 2.78

SAMPLE CONCENTRATION = 0.0020 % BY VOLUME
OBSCURATION = 0.30

SIZE BAND UPPER	LOWER	CUMULATIVE WT BELOW	WEIGHT IN BAND	CUMULATIVE WT ABOVE	LIGHT ENERGY COMPUTED	MEASURED
118.4	54.9	100.0	0.0	0.0	12	0
54.9	33.7	100.0	0.0	0.0	21	8
33.7	23.7	100.0	0.0	0.0	36	27
23.7	17.7	100.0	0.0	0.0	57	51
17.7	13.6	100.0	0.0	0.0	94	88
13.6	10.5	98.3	1.7	0.0	151	150
10.5	8.2	90.4	7.9	1.8	243	243
8.2	6.4	75.9	14.5	9.6	369	370
6.4	5.0	65.3	10.6	24.1	530	530
5.0	3.9	57.7	7.6	34.7	719	715
3.9	3.0	41.2	16.5	42.3	923	922
3.0	2.4	28.2	13.0	58.8	1164	1172
2.4	1.9	23.7	4.5	71.8	1431	1425
1.9	1.5	19.2	4.6	76.3	1720	1718
1.5	1.2	12.3	6.9	80.8	2047	2047

Fig. 7.17 Typical particle size data produced on Malvern Instruments 3600D particle size analyser.

(a) Compaction by pressing

This involves incorporation into the powder of a binder such as diethylene glycol monostearate (DGMS) or polyvinyl alcohol (PVA) which will hold the powder particles together in the desired shape prior to further processing.

(b) Screen printing

This process is utilized for the build up of multilayers on a previously prepared substrate, which can be a suitable organic film from which the screened material can be separated. The powder and organic binder (e.g. ethyl cellulose in terpineol) are mixed to form an ink which can be screened through a suitable mesh. This method offers the advantage of preparing thin layers in selected patterns and will be referred to later in the section on glass-ceramic coatings on metals.

(c) Tape casting

This process again involves mixing powder and suitable organic liquids to form a viscous substance which can be applied via a doctor blade process on to a suitable organic carrier film to form a continuous tape. The ceramic/binder mix 'sets' sufficiently to form a flexible tape which can be used to form multilayer structures after pre-processing such as punching holes through the tape. This method has been used with ceramics such as alumina (substrates) and $BaTiO_3$ (capacitors) but it is only recently that glass-ceramics have been investigated and utilized in this context.

(d) Firing

Following the preparation of the 'green state' glass-ceramic body the material has to be sintered to maximum density. As a first stage the binder has to be burned out at temperatures in the region of 500°C and during this process sufficient oxygen must be maintained in the furnace atmosphere to enable this to occur without leaving any deleterious deposits (e.g. carbon). Subsequently, as the temperature is raised to promote sintering, the furnace atmosphere is not necessarily oxidizing and in some cases is advantageously neutral as noted below. The firing continues to temperatures in the range 800°C to 1200°C, depending on composition. In this stage of the process sintering occurs by solid phase reactions between the particles and by flow as the glass particles soften (Fig. 7.18). At the same time crystallization in the particles is progressing by a combination of bulk and surface processes, particularly the latter in fine powders. By this means, the loosely bonded initial glass powder is converted to a dense, glass-ceramic material.

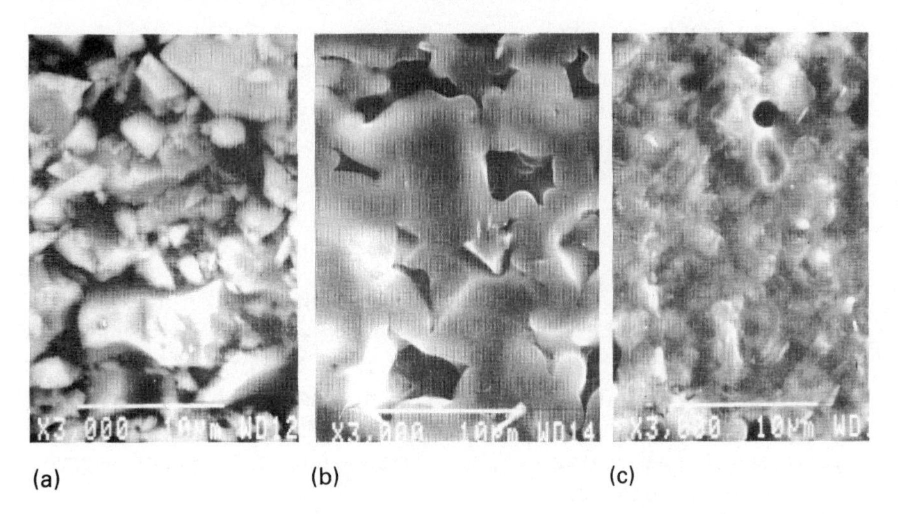

Fig. 7.18 SEM of fracture surfaces of clino-enstatite based glass powder compacts showing progressive sintering on heating to higher temperatures: (a) 800°C, (b) 850°C, (c) 900°C.

7.3.2 Compositions and properties

Typical examples of glass-ceramics which have been prepared in the GEC Engineering Research Centre, Stafford by the powder sintering process are described below.

(a) ZnO–Al_2O_3–B_2O_3–SiO_2 materials (British Patent 1 151 860)

These materials required firing temperatures in the temperature range 800°C to 1200°C depending on composition. Examples are given in Table 7.6. The materials made by powder sintering are microcrystalline with the principal crystal phases being zinc aluminate, zinc borate and quartz, these being similar to those developed in bulk crystallized glass-ceramics of this type. The surfaces of the fired materials are smooth and unblemished. Glass-ceramics of this type have linear expansion coefficients in the range 10–60×10^{-7} °C^{-1} with moduli of rupture up to 180 MPa and high volume resistivities.

(b) MgO–Al_2O_3–SiO_2 materials

The preparation via the powder route of glass-ceramic substrates and multilayer structures in the MgO–Al_2O_3–SiO_2 system has been studied in some detail by several workers. The emphasis has been placed on glass-ceramics in which the principal crystal phase is cordierite. This is because

Table 7.6 Examples of $ZnO–Al_2O_3–B_2O_3–SiO_2$ glass-ceramics prepared by firing glass powder compacts

	ZnO	*Al_2O_3*	*B_2O_3*	*SiO_2*	*Other additions*	*Firing conditions* (°C/h)	*Linear thermal expansion coefft.* $\times 10^7$ *20–500°C*	*Volume resistivity* (Ω cm at 500°C)	*Modulus of rupture* (MPa)
ZB1	20	20	30	20	–	800/½	41	–	180
ZB2	30	20	10	40	–	1200/¼	43	–	130
ZB3	30	17	15	30	$5CaO, 3P_2O_5$	900/1	48	–	150
ZB4	35	16	15	26	$5BaO, 3P_2O_5$	825/¼	49	–	115
ZB5	36	17	15	30	$4ZrO_2$	800/½	37	1.0×10^8	75

such glass-ceramics can possess similar linear thermal expansion characteristics to silicon, thereby offering the possibility of a matched expansion system. (It will be appreciated, from previous discussion in this chapter, that this possibility also exists for glass-ceramics from other systems, for example the $ZnO–Al_2O_3–SiO_2$ system.)

The cordierite compositions prepared by powder techniques can be sintered to near-theoretical density and crystallized to give the preferred α-cordierite phase at temperatures below 1000°C which is in the range required by thick film inks, which enables co-fired structures to be prepared.

An interesting recent development in glass-ceramics of this type is concerned with the materials, discussed earlier, in which the major crystalline phase is clino-enstatite when crystallized in bulk form (Budd and Partridge, 1985). When compositions of this type are processed via the

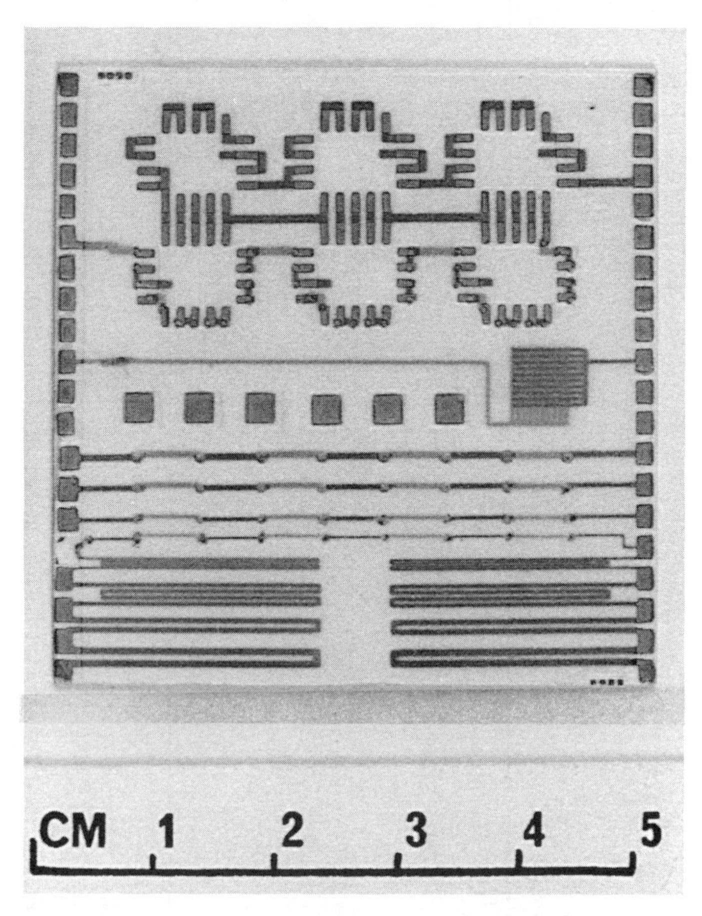

Fig. 7.19 Multilayer glass-ceramic/conductor substrate. (Reproduced by kind permission of Marconi Electronic Devices Ltd, Microsystems Division.)

powder route the clino-enstatite phase is not the major phase developed and lower expansion materials are produced which are compatible in expansion with silicon (Fig. 7.14). Firing temperatures are in the region of 900°C which is again compatible with the firing temperatures for thick film inks thereby allowing multilayer co-fired structures to be prepared. The dielectric constants of these materials at about 7.5 are higher than those of the cordierite glass-ceramic (5–6) but still significantly below those of alumina ceramics (approximately 10) thereby enabling potential use at higher frequencies to be considered.

7.3.3 Compatibility with thick film technology

As noted previously the glass-ceramics prepared via powder routes are compatible with thick film inks and firing technology. Conductor tracks have been in fact screen printed onto green state tape cast cordierite glass-ceramic, and layers prepared in this way have been stacked and co-fired to produce multilayer substrates on which silicon chips can be placed (Bridge and Holland, 1985). The glass-ceramic layers produced by screen printing or by tape casting can be prepared with through holes (vias) so that the conductor tracks at different levels can be interconnected. By successive application of dielectric and conductor layers a multilayer structure interconnected through the vias can be built up and co-fired. By this means the required circuitry is thus 'compressed' into a small volume.

Thick film inks which have been used successfully with glass-ceramics of the $MgO–Al_2O_3–SiO_2$ type in this manner have included gold, copper, and palladium–silver, although silver migration into the glass-ceramic may occur in some cases with the latter inks. Figure 7.19 shows a multilayer structure built up using screen printing technology.

7.4 GLASS-CERAMIC COATED METAL SUBSTRATES

Glass-ceramic coated metal substrates are exciting interest in a number of areas, principally because of the possibility they offer of good thermal dissipation in conjunction with ruggedness and large size.

Whilst bulk glass-ceramics, as discussed in the preceding sections, offer many excellent property characteristics, one problem is their comparatively low thermal conductivity compared with other ceramic substrate materials (e.g. alumina, beryllia and aluminium nitride) and, in common with these ceramics, there are limitations in the size of substrate which can realistically be made and processed. Current developments in hybrid circuitry are producing demands for greater density of packing of components and direct on-board mounting of discrete components such as resistors. This, in turn, provides a demand for increasing thermal dissipation which cannot be met by current epoxy based boards and may be difficult to meet with alumina

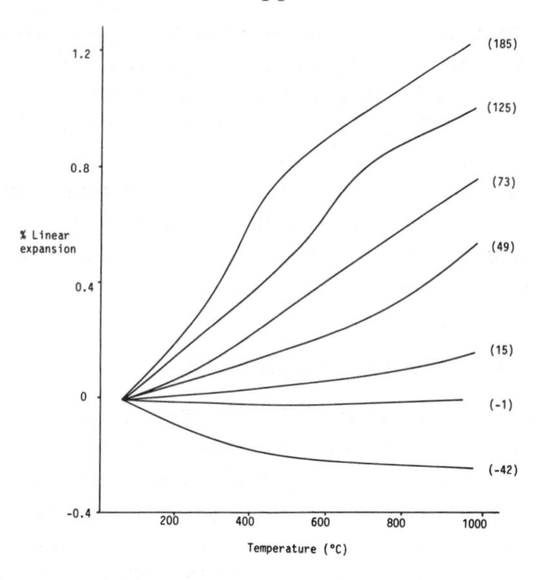

Fig. 7.20 Graph showing range of thermal expansion characteristics possible with glass-ceramics (20–500°C values quoted (10^{-7} °C^{-1})).

substrates. As noted previously, beryllia and aluminium nitride may provide the required thermal dissipation but each of these poses its own problems in terms of strength, surface finish, compatibility with thick film inks and ruggedness.

Recently, it has been proposed that vitreous enamels (also termed porcelain enamels) on suitable metal substrates such as steel sheet will overcome many of these problems and provide the desired substrate. However, the glasses used for enamelling purposes suffer from the disadvantage that they are insufficiently refractory to enable most commercially available thick film inks (which require firing temperatures in the region of 900°C) to be used satisfactorily. Although new, lower firing temperature inks are being developed, this still presents a considerable problem.

Glass-ceramics offer a solution to this problem. An important feature of glass-ceramics is the ability to produce materials covering a wide range of thermal expansion characteristics (Fig. 7.20), from negative, through zero, to highly positive expansions similar to those of, for example, copper. Out of this overall range of expansions, glass-ceramics can be prepared which are matched in expansion to many metals and alloys. Figures 7.21–7.24 show examples of the thermal expansion matching which can be achieved.

In addition to the ability to achieve thermal expansion matching, glass-ceramics are capable of being applied and bonded to many metals and alloys

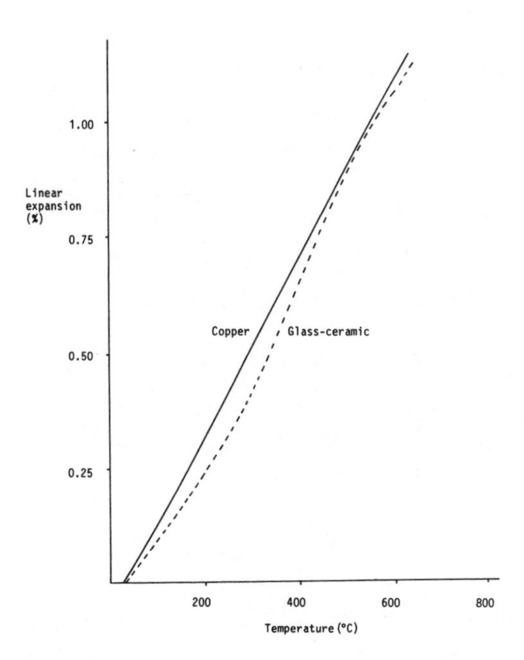

Fig. 7.21 Thermal expansion characteristics of copper and matching glass-ceramic.

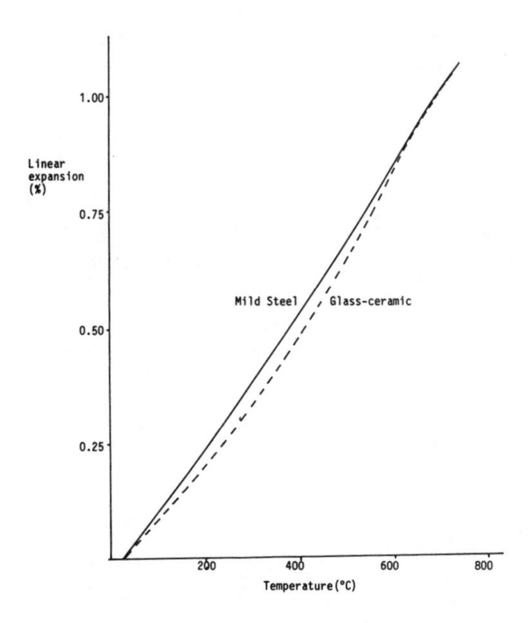

Fig. 7.22 Thermal expansion characteristics of mild steel and matching glass-ceramic.

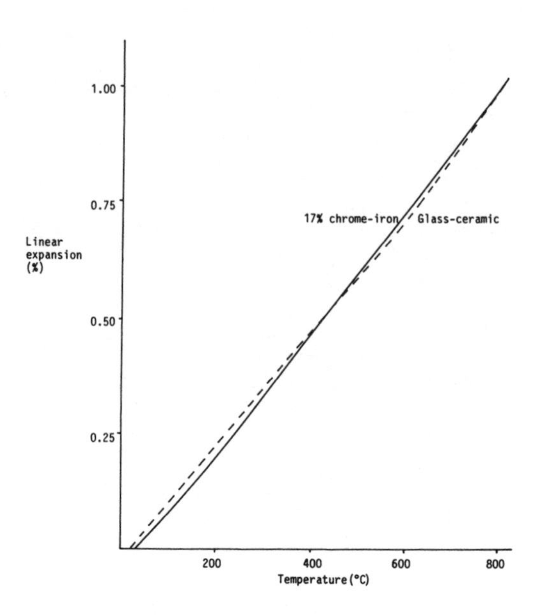

Fig. 7.23 Thermal expansion characteristics of 17% chrome–iron and matching glass-ceramic.

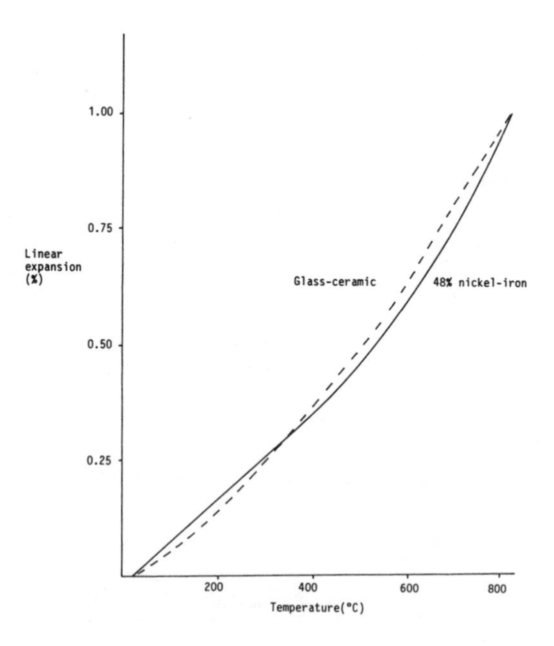

Fig. 7.24 Thermal expansion characteristics of 48% nickel–iron and matching glass-ceramic.

using essentially vitreous enamelling techniques but with the added step of crystallizing the enamel to provide the glass-ceramic material. This then provides the increased refractoriness to enable the coating to withstand the firing conditions required for thick film inks.

The principal requirements in the development of a glass-ceramic coating on a metal are that the coating shall wet and adhere to the metal and be reasonably matched in thermal expansion characteristics to the metal. This latter point is important in the context of the hybrid circuit/microelectronic substrate which is usually relatively thin (e.g. 0.6 to 1 mm) and where stressing arising from expansion mismatch will result in bowing. In other applications where the metal base is thicker then deliberate expansion mismatch can be introduced so that the coating is stressed in compression and mechanical properties are enhanced (Partridge, 1981). Where thin metal is required, both sides of the metal sheet may be coated to balance the stresses and provide the flat substrate which is required for ink printing, though this is not always convenient or desirable.

7.4.1 Metal selection and preparation

For substrate applications the metals and alloys which have received attention are many and diverse, including copper, iron–chrome alloys, iron–chrome–aluminium alloys, steel, nickel–iron alloys, molybdenum tungsten, copper–Invar–copper sandwiches and copper–tungsten composites. In terms of thermal conductivity copper offers the best heat dissipation characteristics, followed by molybdenum and tungsten but these are more expensive than steel. Further, copper permits use to higher frequencies than does steel, the losses generated in the steel itself limiting its use to frequencies less than about 400 MHz. Selection of the metal or alloy is thus governed by potential usage as well as cost. A further factor to be taken into account may be the necessity to match thermal expansion with circuit components such as alumina chip carriers, or even with the silicon chip itself.

In order to assist the development of a good adhesion between the glass-ceramic and the metal it is the usual practice firstly to roughen the metal surface, either by etching or grit blasting (although this latter method is frequently not practical with thin metal sheet owing to deformation problems), decarburize as necessary and suitably pre-oxidize. In the cases of copper and other metals and alloys which are highly reactive with oxygen at elevated temperatures, care has to be taken at this stage in order to obtain an adherent oxide and this usually means oxidizing at relatively low temperatures or carrying out the operation in a reduced oxygen content atmosphere (e.g. nitrogen or argon). In the case of alloys such as 17% chrome–iron, an oxide layer rich in chromium oxide is readily produced by firing in a wet hydrogen atmosphere (which also serves to decarburize the alloy). Iron–chromium–aluminium alloys can be preoxidized in air to provide an adherent layer of aluminium oxide.

7.4.2 Glass-ceramic application

It is important that the glass-ceramic coating on the particular metal or alloy is smooth, continuous (pin-hole free), and of sufficient thickness to provide the desired electrical insulation between the circuitry and the metal plate. Various techniques have been investigated and are employed for the application of the glass-ceramic coating to the metal. Many of these involve firstly preparation of the parent glass, reduction of this to powder form (as in the preceding section) followed by application of the glass powder to the prepared metal substrate. The methods which have been and are being used are listed in Table 7.7. Of these, the methods which are principally employed for the coating of metal substrates for electronics are as follows.

1. Screen printing: applying a suitable suspension via a screen on to the metal – good definition of areas to be coated can be achieved.
2. Electrophoretic deposition: application of an electric potential attracts charged glass particles in suspension to the substrate. This is a useful technique for the provision of insulated coatings through holes in the metal, or coating irregular shapes (Fig. 7.25).
3. Sol–gel: although strictly speaking this is a method of preparing material it offers a useful means of providing very thin coatings by direct application of the sol–gel to the metal followed by drying and firing (Fig. 7.26). Alternatively this process is a means of providing fine powders of closely controlled particle size which can be used in the above methods.

After the coating has been applied it is necessary to carry out a suitable firing cycle which will cause the powder to coalesce, wet and bond to the substrate. In some cases, the coating is produced in vitreous form and has to be subjected to a further firing cycle, essentially as shown in Fig. 7.1, to convert the glass to glass-ceramic. However, it has been shown that certain ranges of compositions can be fired to form a coherent coating and converted to glass-ceramic in the one firing cycle, and these are preferred from an economic point of view. Many firings are carried out under protective atmospheres (e.g. nitrogen) essentially to protect the metal from excessive oxidation.

Table 7.7 Methods for application of glass-ceramic coatings on to metal

Method
Dip coating
Wet spray
Screen printing
Electrophoretic deposition
Plasma spray
Sol–gel

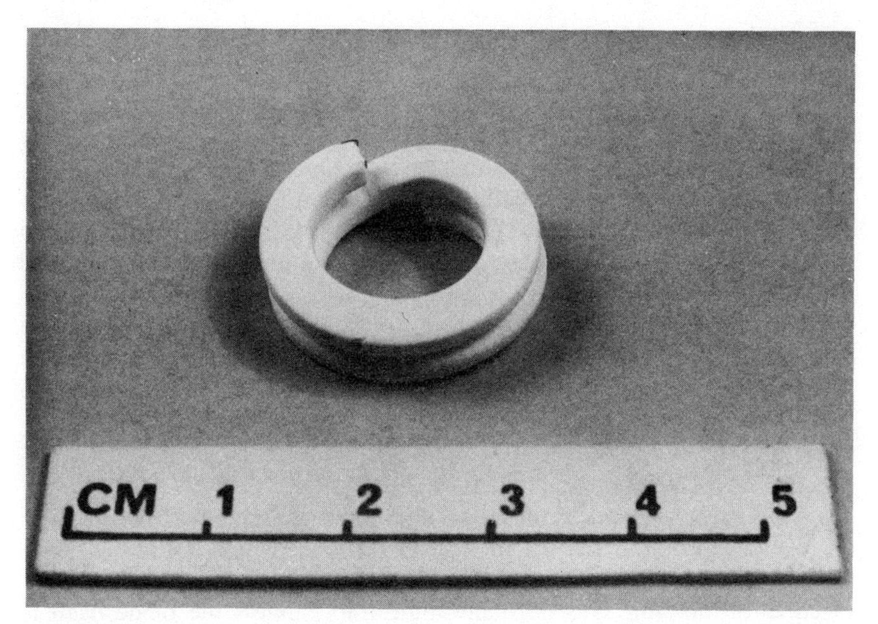

Fig. 7.25 Electrophoretically deposited coating showing ability to coat complex shapes.

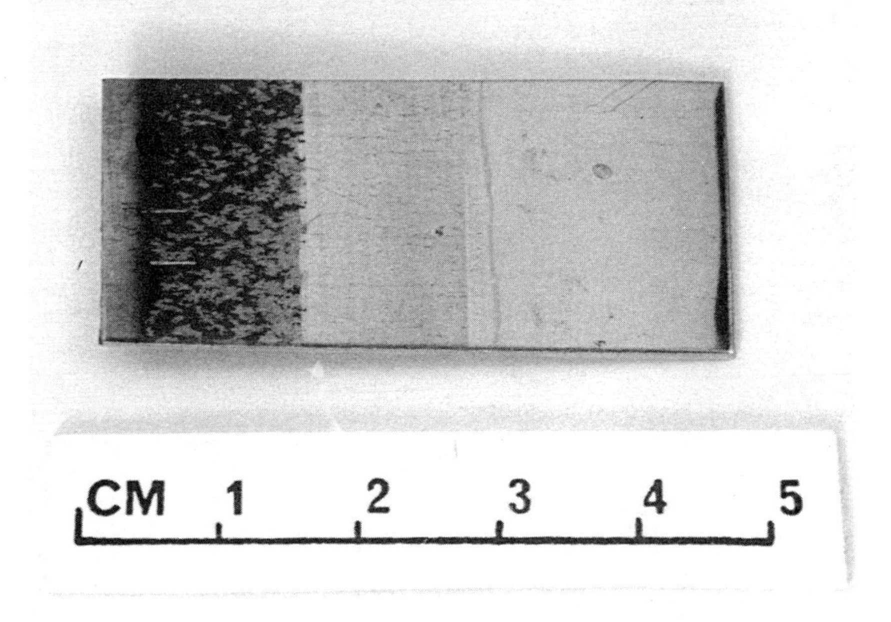

Fig. 7.26 Sol–gel coated metal (the coatings are too thin to be visible, but their presence is shown by oxidation of the sample). The protection afforded by regions of 0, 1 and 2 coatings, each $\sim 0.5\ \mu m$ thick, can be clearly seen.

7.4.3 Glass-ceramic to metal bond

In order to achieve the desired high bond strength between the glass-ceramic and the metal it is necessary that a chemical bond be attained through the interface. A chemical bond means that continuity of the electronic structure exists across the interface as well as a continuity of the atomic structure. To realize this structural continuity a transition layer or zone is needed that is compatible, or in equilibrium with both the metal and glass at the interface. In general this zone includes a layer of the oxide of the metal between the metal and glass which can only be retained when the metal and glass at the interface are saturated with the metal oxide. Figure 7.27 (Pask, 1971) shows schematic models of cross-sections across the glass–metal interface when an oxide layer is present and when absent. In the former case sharing of the oxygen ions between the metal and glass constituents, with bonding which has ionic/covalent characteristics, results in a cross-sectional continuity of the chemical bond across the interfacial zone. Although a discrete oxide layer as shown in Fig. 7.27(a) satisfies the requirement for a continuous zone of chemical bonding it is normally undesirable unless the oxide itself is strongly bonded to the metal (as in the case of chromic oxide on 17% chrome–iron). Figure 7.27(b) illustrates that only a single layer of the oxide

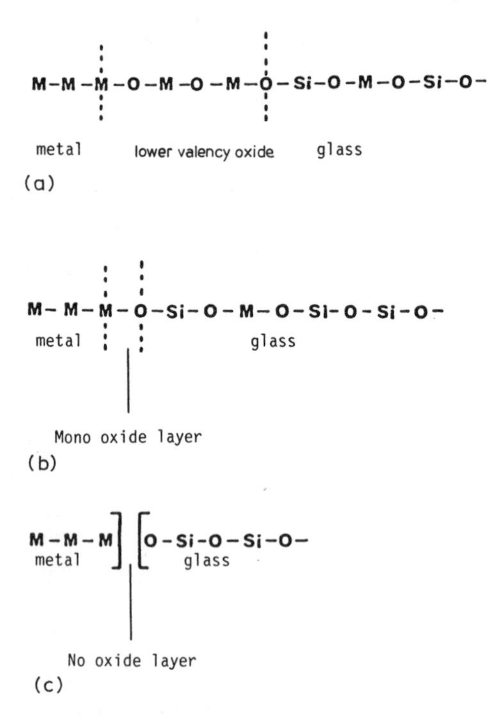

Fig. 7.27 Schematic development of glass to metal boundaries. (Reproduced by kind permission of the American Ceramic Society and Professor J. A. Pask.)

is necessary to realize chemical bonding across the interfacial zone. Redox conditions developing at the interface assist in ensuring that the interface remains saturated with the oxide, with reduction of multivalent cations in this interfacial region to a lower valency state and with retention of good adherence. When no oxide layer is present, as in Fig. 7.27(c), the glass comes into direct contact with the metal resulting in atomic continuity but not in electronic structure continuity. This van der Waals type of bonding is considerably weaker than the required chemical bond.

Whilst the above understanding has been developed largely from studies of glass–metal (particularly Fe) interfaces, similar considerations apply to the direct bonding of a glass-ceramic to a metal. In this latter case the situation is further complicated by the diffusion processes involved in the development of crystalline species interfering with and competing with the diffusion processes involved in bond development. Furthermore, the modification of the composition close to the metal surface is likely to result in changes in the crystallization of that region.

A number of studies have been carried out of interfacial areas of glass-ceramic to metal joins using microscopic sections, EPMA, EDAX and finite element analyses. Examples of typical micrographs together with

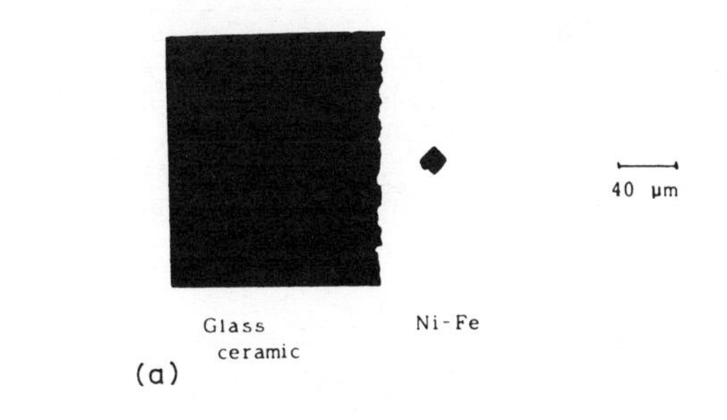

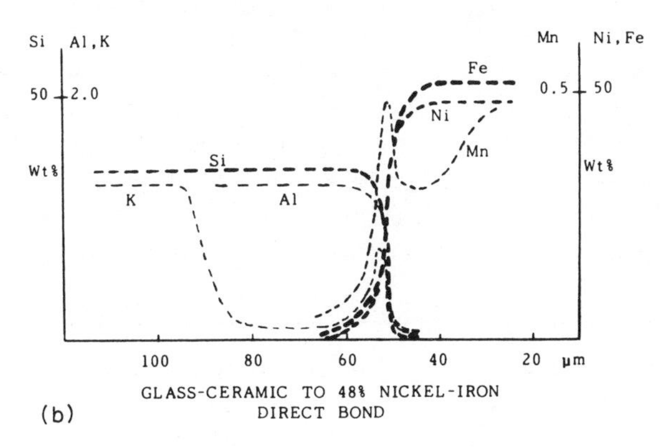

Fig. 7.28 Section (a) and EPMA scan (b) of glass-ceramic to 48% Ni/Fe join.

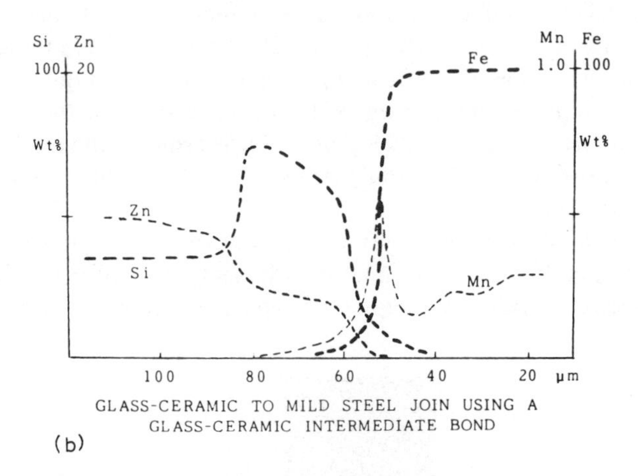

Fig. 7.29 Section (a) and EPMA scan (b) of glass-ceramic to mild steel join.

EPMA scans are given in Figs 7.28–7.34. Figure 7.28 illustrates a section and EPMA scan of a glass-ceramic bond to 48% nickel–iron alloy. Interesting features are movement of manganese from the alloy to the interface leaving a depleted layer of manganese, and the diffusion of potassium in the glass-ceramic to the interface (Partridge and Elyard, 1984).

Figure 7.29 shows a section and EPMA scan of a glass-ceramic bonded to mild steel via an intermediate glass-ceramic bond of different composition to the outer layer. The EPMA scan clearly shows the bond layer although this cannot readily be distinguished in the section. Diffusion of Fe, and to a lesser extent Cr, has taken place across the interface into the intermediate glass-ceramic but has not extended as far as the outer layer of glass-ceramic.

Fig. 7.30 Backscattered electron micrograph of a coating on mild steel rapidly cooled after firing.

Fig. 7.31 Backscattered electron micrograph of a coating on mild steel after heat-treatment.

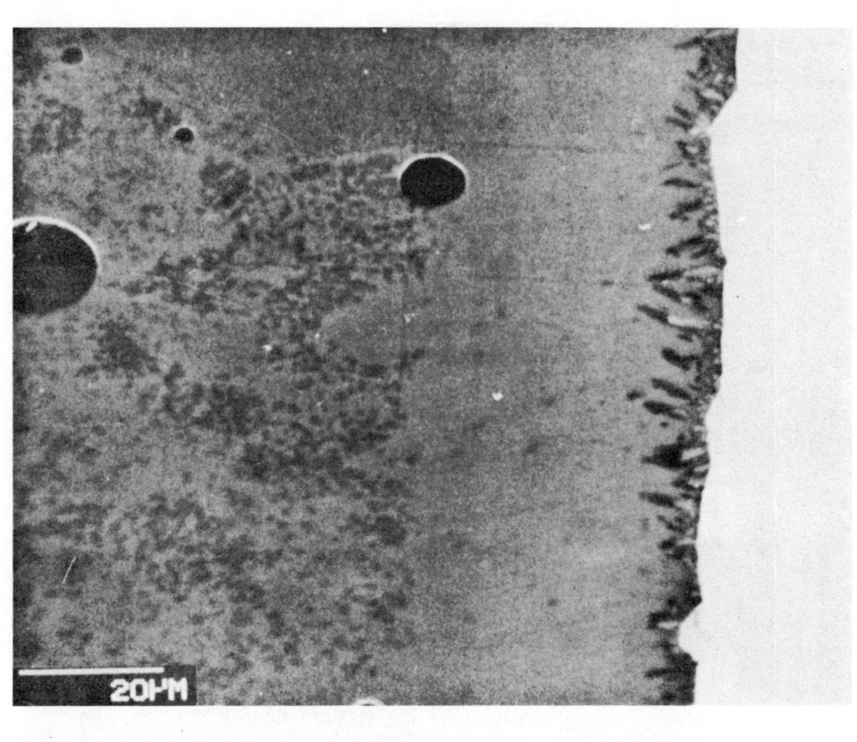
20µM

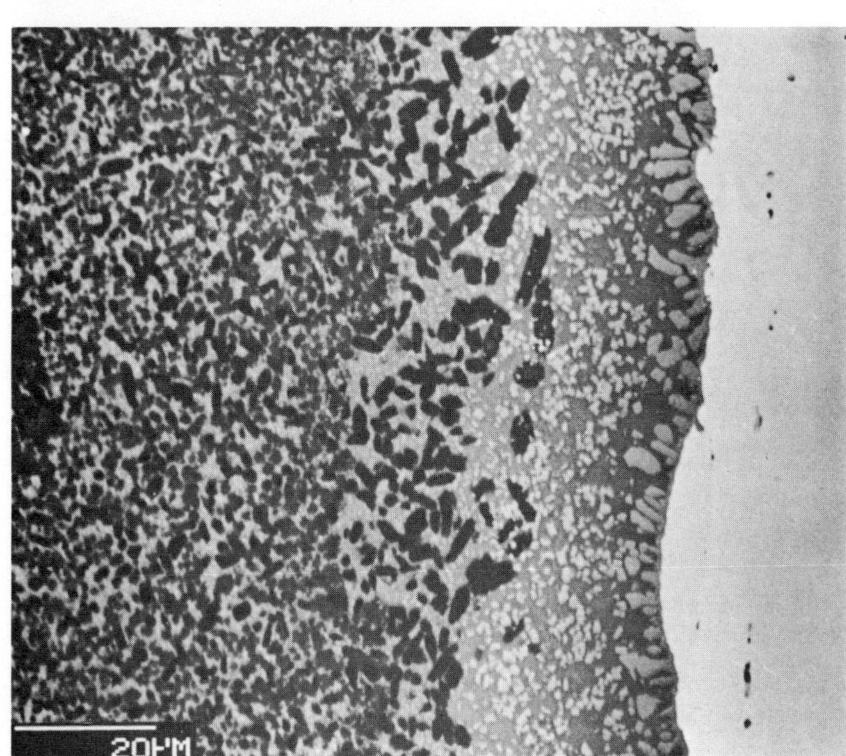
20µM

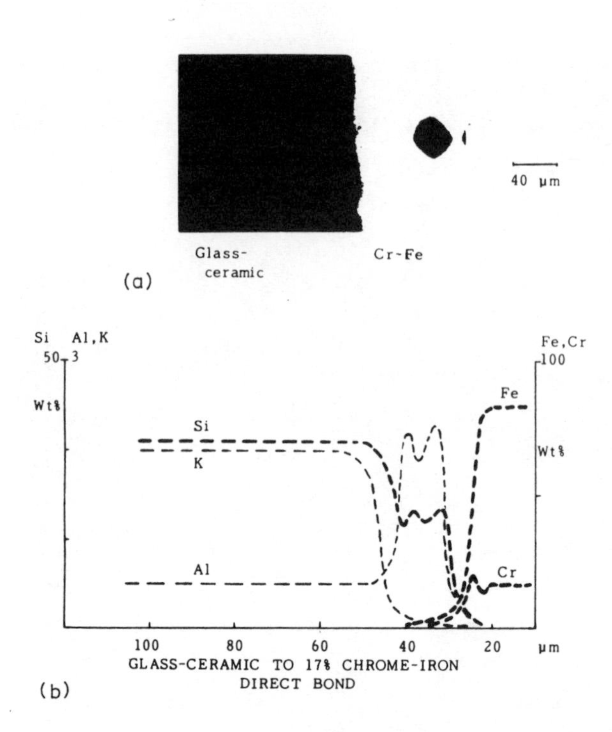

Fig. 7.32 Section (a) and EPMA scan (b) of glass-ceramic to 17% Cr/Fe join.

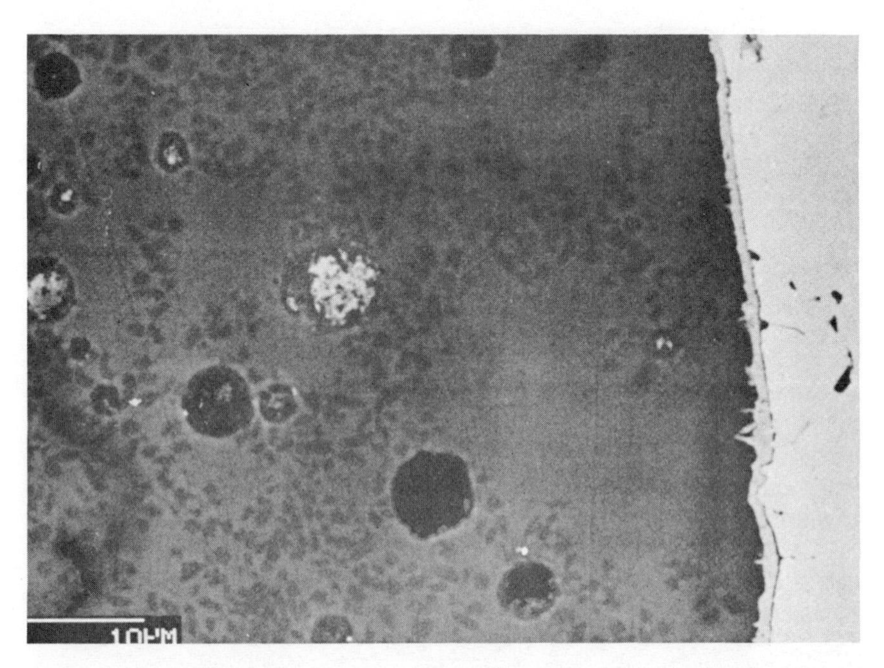

Fig. 7.33 Backscattered electron micrograph of a coating on 17% chrome–iron, cooled rapidly after firing.

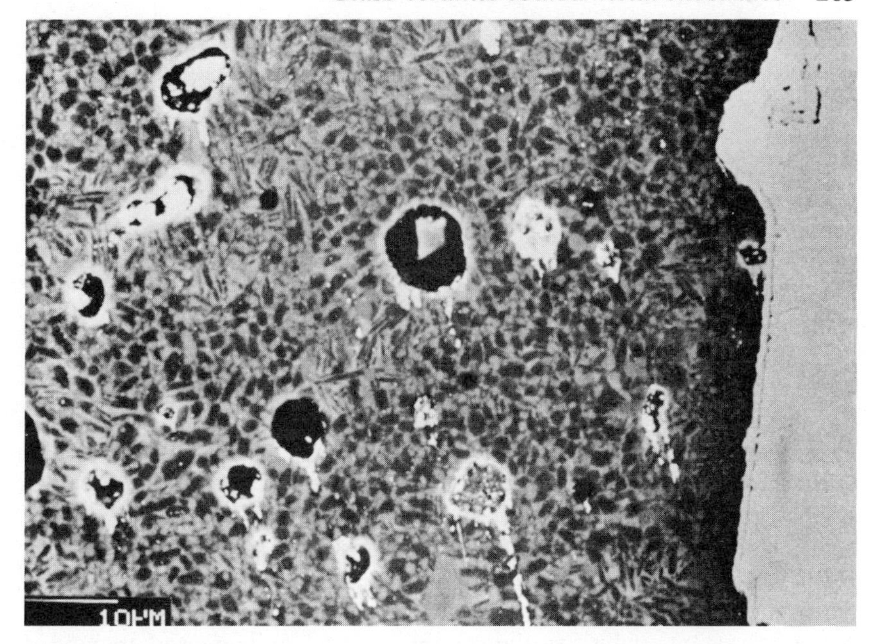

Fig. 7.34 Backscattered electron micrograph of a coating on 17% chrome–iron after heat-treatment.

Further work (Sturgeon *et al.*, 1986) to study the bonding of a Li_2O–ZnO–SiO_2 coating to mild steel has shown that the iron from the oxide layer (believed to be ferrous oxide) diffuses rapidly into the coating. This and the deep blue coloration of the coating after firing suggests the iron is diffusing predominantly as Fe^{2+}. The high iron concentration present in the coating adjacent to the interface has a noticeable effect on the microstructure, initially with dendritic growth from the interface (Fig. 7.30). EDAX analysis has indicated this phase to be a lithium iron silicate. Further heat-treatment results in the structure seen in Fig. 7.31. Well defined zones are now visible in the cross-section. At the interface the original dendritic growth has extended from a continuous layer of crystals attached to the metal substrate and residual oxide. Small crystals of the same composition form the band extending 35 μm from the interface. This then merges with a second zone consisting of large crystals of lithium disilicate, which extend from 25 to 40 μm from the interface, with the crystal size decreasing with distance. Beyond 40 μm both lithium disilicate and quartz, the normal crystalline species for this glass-ceramic, appear. The quality of the bond to the metal, however, depends on the compatibility of the lithium iron silicate phase with the metal substrate.

Distribution of the minor elements present in the coating is altered considerably by the subsequent heat-treatment. A loss of potassium occurs at the interface and a build-up of zinc is observed immediately adjacent to the interface. During heat-treatment the elements are redistributed by a build-up of both potassium and zinc in the interface region. Zinc is incorporated in the lithium iron silicate phase whilst the potassium remains in the residual glass.

Figure 7.32 shows a section and EPMA scan of a Li_2O–Al_2O_3–SiO_2 glass-ceramic bonded to 17% chrome–iron. This has also involved a layer of slightly different composition to the bulk. There are indications of a peak in the Cr profile corresponding to the oxide layer and other work using EDAX analyses has confirmed this.

Further work (Sturgeon *et al.*, 1986) to study the bonding of a Li_2O–Al_2O_3–SiO_2 glass-ceramic coating to 17% chrome–iron has shown clearly the presence of a discrete oxide layer at the interface with little tendency to form dendrites as occurred in the mild steel case (Fig. 7.33). The low rate of solution of the Cr^{3+} ions in the glass allows retention of the oxide layer during firing and heat-treatment and this oxide provides a transition layer between the base metal and the coating, resulting in excellent adhesion properties. In this case the same types of crystal which are present in the bulk of the glass-ceramic form right up to the oxide/glass-ceramic boundary. X-ray diffraction analysis of the coating indicates an increased amount of lithium disilicate, but little variation in the amount of quartz present. This phase is now clearly visible (Fig. 7.34) due to crystallization of the glassy regions and a coarsening process during the heat-treatment stage. Investigation of elemental distribution across the interface reveals that potassium variation is associated with the crystallization process during firing and heat-treatment and the potassium remains in the residual glassy regions. A higher potassium concentration observed adjacent to the interface before heat-treatment is associated with the absence of lithium disilicate. After heat-treatment, the peak adjacent to the interface remains, although not extending as far as the coating. It may be associated with a glassy layer at the boundary, or the presence of alkali ions may be a network stabilizing requirement necessitated by the presence of chromium ions in this region. Recent work suggests the former to be the more appropriate. It is to be noted, however, that it has been suggested (Jones, 1986), where similar potassium concentrations have been observed in a lithium zinc silicate glass-ceramic on an iron–chrome–aluminium alloy that, as the aluminium oxide layer initially formed is taken up into the glass and incorporated into the network structure forming AlO_4 tetrahedra, the charge balance is maintained by the increase of K^+ in this region.

7.4.4 Properties of the coatings

The properties of glass-ceramic coatings on metals have been reviewed in a number of publications (Partridge, 1981; Elyard and Partridge, 1983; Hang

Table 7.8 Properties of glass-ceramic coatings

	Glass-ceramic L6 on 17% Cr–Fe	*Glass-ceramic L7 on copper*	*Glass-ceramic coating on Fecralloy (ex Wade)*
Bulk resistivity (Ω cm)	5×10^{14}	10^{15}	2×10^{14}
Surface resistivity			
60% RH Ω/square	10^{15}	10^{15}	–
80% RH Ω/square	10^{12}	10^{15}	10^{14} (at 70% RH)
Dielectric strength			
(kV mm^{-1})	32		34
ε at 1 MHz	5–6	5–6	5
$\tan\delta$ at 1 MHz	4×10^{-3}	7×10^{-3}	4×10^{-3}

et al., 1981; Thaler *et al.*, 1981; Tsien *et al.*, 1981). Table 7.8 summarizes relevant properties for a number of glass-ceramic coated metal substrates.

A particular property of the glass-ceramic coated metal substrates is their ability to dissipate heat generated in the circuitry. A number of workers have carried out measurements of this parameter and the results are summarized below. It can be calculated that for a glass-ceramic coated steel substrate to possess better thermal dissipation characteristics than does an alumina ceramic the coating on the steel must be less than about 175 μm in thickness (Tsien *et al.*, 1981). Powder coating methods easily allow coatings less than this in thickness (thicknesses down to 75 μm are readily achievable). The sol–gel process allows coating thickness in the region of 0.5 to 5 μm to be achieved. Bulk glass-ceramics of the types reviewed exhibit a dielectric breakdown strength in the region of 40–50 kV mm^{-1}. Samples prepared by powder sintering either as monolithic materials or coatings on a metal substrate could be expected to have a dielectric breakdown strength of 20–26 kV mm^{-1} (or higher if the coating/firing technique enables closed pores to be virtually eliminated). For many applications voltage withstand capabilities of no more than 500 V are required and for microelectronic applications this might be as low as 50 V. A voltage withstand capability of 500 V requires, in the worst case, a glass-ceramic coating thickness of about 25 μm, and 50 V a coating thickness of 2.5 μm. It is clear that coating thicknesses of these values on steel will be expected to provide a substrate with significantly higher thermal dissipation characteristics than that of alumina. Such coatings on the higher thermal conductivity metals will provide even superior performances.

It is difficult to measure thermal conductivities of the composite structures and relate them in a general way to practical circuits, so it is useful to determine thermal dissipation characteristics. Measurements of temperature rise versus power input for a transistor mounted alternatively on either a coated steel substrate or an alumina board have shown a better

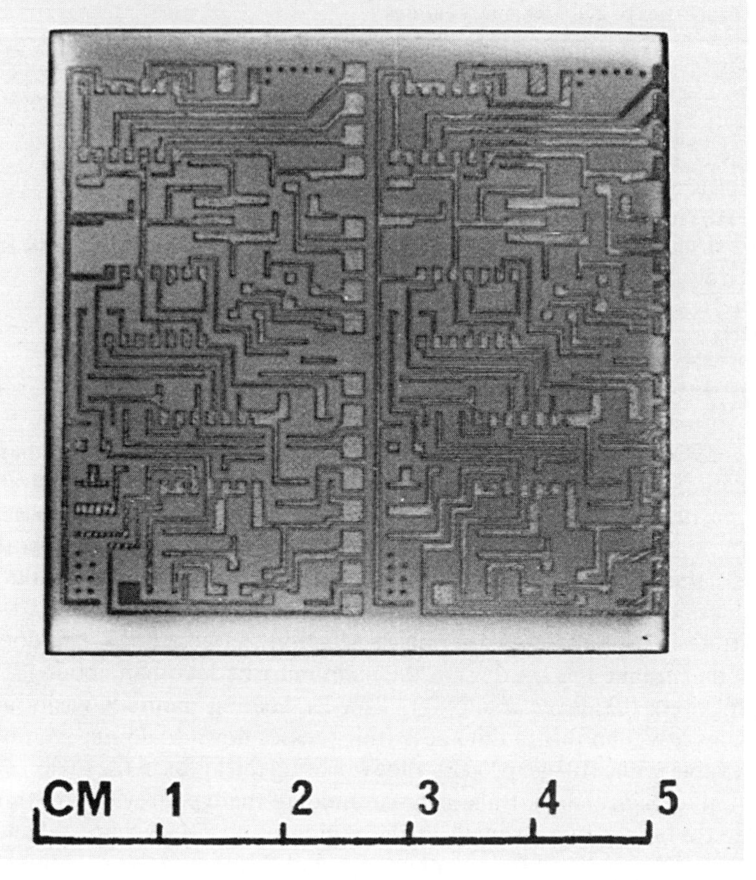

Fig. 7.35 Gold thick film conductor tracks on a glass-ceramic coated 17% chrome–iron substrate.

performance (i.e. lower temperature rise for a given power input) for the coated steel substrate (Tsien *et al.*, 1981). Similarly, a resistor mounted on coated steel remains about 50°C cooler than a similar one on an alumina substrate for 3.2 W cm^{-2} thermal loading (McCusker, 1981).

7.4.5 Compatibility with thick film inks

Glass-ceramic coatings have been shown to be compatible with a range of thick film conductor and resistor inks. The ability of the glass-ceramic coatings to withstand the firing temperatures required for the thick film inks (850–900°C) is of considerable importance in this respect and permits the

Fig. 7.36 Copper thick film conductor tracks on a glass-ceramic coated copper substrate. (Reproduced by kind permission of Marconi Electronic Devices Ltd, Microsystems Division.)

development of reproducible and stable circuitry on the coatings. It has been shown to be possible to provide gold, platinum–silver, palladium–silver and copper conductor tracks on the various coated metal substrates. Figures 7.35–7.37 show examples of thick film inks which have been applied to and fired onto glass-ceramic coated metal substrates illustrating the range of materials and capability of the process.

It is to be noted that in order to protect the metal of the substrates from oxidation during the firing process for the inks it is often necessary to carry out the firing procedures in an inert (e.g. nitrogen) atmosphere. This means that the inks must be capable of firing under these conditions. Fortunately suitable inks are readily available.

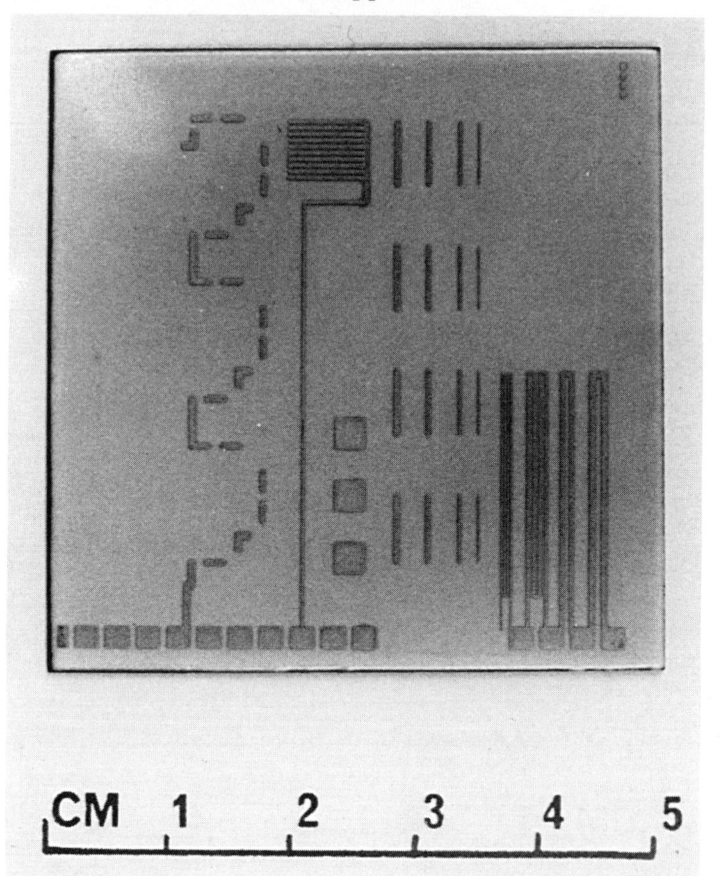

Fig. 7.37 Gold thick film conductor tracks on a glass-ceramic coated titanium substrate.

7.5 CONCLUSIONS

Work carried out by the authors and by other workers has demonstrated the potential of glass-ceramics both in bulk form and as coatings on metals for use as substrates in hybrid circuitry and microelectronics. The glass-ceramics are compatible with both thick and thin film circuitry and can be used in a variety of applications. The glass-ceramic coated metal substrates can provide high thermal dissipation characteristics together with ruggedness and large scale circuit boards.

ACKNOWLEDGEMENTS

The authors wish to thank GEC PLC for permission to publish the information given in this chapter.

REFERENCES

Barry, T. I., Cox, J. M. and Morrell, R. (1978) *J. Mater. Sci.*, **13**, 594–610.
Bridge, D. R. and Holland, D. (1985) *Glass Technol.*, **26**, 286–92.
Budd, M. I. and Partridge, G. (1985) High strength glass-ceramics in the MgO–Al_2O_3–SiO_2 system for substrate applications. Society of Glass Technology Symposium on the properties and applications of glasses in the modern world, Brunel University, 16 April, 1985.
Elyard, C. A. and Partridge, G. (1983) Glass-ceramics in substrate applications. Society of Glass Technology Symposium on the applications of glasses and glass-ceramics in the electrical and electronics industries, University of Reading, 23–24 March, 1983.
Hang, K. W., Andrus, J. and Anderson, W. M. (1981) *RCA Rev.*, **42**, 159–77.
Jones, R. W. (1986) *Glastech. Ber.*, **56**, 747–52.
McCusker, J. H. (1981) *RCA Rev.*, **42**, 281–97.
McMeeking, R. M. and Evans, A. G. (1982) *J. Am. Ceram. Soc.*, **65**, 242–6.
McMillan, P. W. (1979) *Glass-Ceramics*, 2nd edn, Academic Press, New York.
McMillan, P. W. and Partridge, G. (1972) *J. Mater. Sci.*, **7**, 847–55.
Partridge, G. (1981) *GEC J. Sci Technol.*, **47**, 87–94.
Partridge, G. (1982) *Glass Technol.*, **23**, 133–8.
Partridge, G., Phillips, S. V. and Riley, J. N. (1973) *Trans. J. Br. Ceram. Soc.*, **72**, 255–67.
Partridge, G. and Elyard, C. A. (1984) *Brit. Ceram. Proc.*, No. 34, 219–29.
Pask, J. (1971) *Proc. Porcelain Enamel Inst. Tech. Forum*, **33**, 1–16.
Sturgeon, A. J., Holland, D., Partridge, G. and Elyard, C. A. (1986) *Glass Technol.*, **27**, 102–7.
Thaler, B. J., McCusker, J. H. and Honore III, J. P. (1981) *RCA Rev.*, **42**, 198–209.
Tsien, W. H., McCusker, J. H. and Thaler, B. J. (1981) *RCA Rev.*, **42**, 210–20.

8

Glass-ceramics for piezoelectric and pyroelectric devices

Arvind Halliyal, Amar S. Bhalla, Robert E. Newnham and Leslie E. Cross

8.1 INTRODUCTION

The advantages of glass-ceramics for preparing large, complex, pore free bodies have been exploited in the electronics industry for a variety of applications. New glass-forming techniques which utilize fast quenching techniques have expanded the range of glass-forming materials and the number of their applications (McMillan, 1979).

Recently, mainly through the work in our laboratory, it has been realized that glasses with suitably selected composition can be crystallized into a polar glass-ceramic form which can then be used as a thermal or a pressure sensing element (Gardopee *et al.*, 1980, 1981). In these glass-ceramics the crystalline phase is polar and the crystallites are aligned in a polar parallel array. The macroscopic polarity thus developed gives rise to both pyroelectric and piezoelectric activity with characteristics markedly different from those which are realized in poled ferroelectric ceramics. Extensive investigations have been carried out to evaluate the properties of these glass-ceramics for pyroelectric and piezoelectric devices (Halliyal, 1984; Halliyal *et al.*, 1984) and they seem to be promising candidate materials for such devices as pyroelectric detectors, hydrophones and surface acoustic wave devices (Lee *et al.*, 1984). In this chapter the work done in the last decade on polar glass-ceramics is reviewed. Merits and demerits of ferroelectric and non-ferroelectric materials are briefly discussed in the next section. The earlier work done on ferroelectric glass-ceramics is also described. The advantages of preparing polar materials by the glass-ceramic route are described and guidelines for selection of glass compositions are presented. Later sections describe the preparation conditions, some useful glass compositions, heat treatment methods and microstructure. The dielectric and pyroelectric properties are described in sections 8.7 and 8.8. The piezoelectric properties of glass-ceramics and their evaluation for hydrophones are given in section 8.10. A connectivity model to predict and tailor the properties of polar glass-ceramics is discussed in section 8.11.

8.2 FERROELECTRIC AND NON-FERROELECTRIC MATERIALS

A large number of ferroelectric single crystals and ceramics are available at present for use in piezoelectric and pyroelectric devices (Herbert, 1982). In polycrystalline form, the polar axes in individual crystallites are normally oriented at random. An essential feature of ferroelectric ceramics which makes them usable for piezoelectric and pyroelectric applications is the ability to reorient the polar axes in individual crystallites under a strong electric field so as to impart long-range remnant polar order (the process called poling). In polar but non-ferroelectric ceramics this orientability is not possible, so that randomly axed ceramics of these materials do not exhibit piezoelectric or pyroelectric properties. In non-ferroelectric single crystals also, if there are polar twins, it is not possible to switch the twins to a single domain pattern and they cannot be used in devices. Quartz is used extensively in frequency control devices. A number of non-ferroelectric but polar single crystals ($Li_2B_4O_7$ (Shorrochs *et al.*, 1981), $Ba_2TiGe_2O_8$ (Kimura *et al.*, 1973), $Ba_2TiSi_2O_8$ (Ito *et al.*, 1981; Yamauchi, 1978; Kimura, 1977)) show excellent piezoelectric properties and look promising for surface acoustic wave (SAW) devices.

Although electrical polability makes the ferroelectric materials usable in ceramic form, it also carries several disadvantages. Since there is by definition a family of equivalent orientations in each crystallite, the material can also depole, either spontaneously with time, giving rise to the well-known ageing phenomenon or under high electrical or mechanical drive fields leading to instability in response and drift in sample dimensions. Engineers must necessarily put up with these inconveniences for the advantages of low expense, simple formability and large size which are available in conventional ceramics.

The present work on polar glass-ceramics was started with the intention of exploring the possibilities of preparing non-ferroelectric ceramics with polar properties. The glass-ceramic route provided an interesting possibility of developing polar order in a ceramic, as discussed in the next section.

8.2.1 Ferroelectric glass-ceramics

In the last two decades research workers at Corning, Bell Telephone Laboratory and in Japan have investigated a number of ferroelectric glass-ceramic systems in which ferroelectric crystalline phases were $BaTiO_3$,

$LiTaO_3$, $NaNbO_3$, $Pb_5Ge_3O_{11}$, $PbTiO_3$, etc. (Herczog, 1964, 1973; Borrelli, 1967; Borrelli and Layton, 1969, 1971; Layton and Herczog, 1969; Layton and Smith, 1975; Glass *et al.*, 1977a, 1977b; Takashige *et al.*, 1981). The main emphasis in this work was on the study of dielectric and electro-optic properties as a function of composition, heat treatment and grain size. By controlling the crystallite size to values below 0.2 μm, transparent ferroelectric glass-ceramics have been prepared which show a large electro-optic effect. Weak pyroelectric responses have been measured in glass-ceramics containing either $LiTaO_3$, $LiNbO_3$ (Glass *et al.*, 1977a) or $NaNbO_3$ crystalline phases. Takahashi and co-workers (Takahashi *et al.*, 1975, 1979) demonstrated a simple method of preparing $Pb_5Ge_3O_{11}$ single crystals from a glass of similar composition. They first prepared a glass of composition $Pb_5Ge_{3-x}Si_xO_{11}$ ($0 \leq x \leq 1.5$) by melting. A few pieces of glass were then remelted and annealed to obtain thin plate-like crystals with the polar *c*-axis oriented perpendicular to the surface. They showed that the pyroelectric figure of merit of these monocrystals was comparable to $LiTaO_3$. It should be noted, however, that in these studies the material obtained after crystallization was more like a single crystal than a glass-ceramic.

Apart from the papers referenced above there have been very few studies of the dielectric and electro-optic properties, and no detailed work seems to have been done on the piezoelectric and pyroelectric properties of ferroelectric glass-ceramics. One of the reasons for this is the difficulty in electrical poling of glass-ceramics. Very few ferroelectric materials are good glass formers. This necessitates the incorporation of a large percentage of a network former in the composition for glass formation. After crystallization, if there is a glassy phase of low dielectric constant between the crystallites, electrical poling is very difficult. For this reason, it is often much simpler to prepare ferroelectric ceramic meterials by powder processing and sintering methods.

8.2.2 Polar but non-ferroelectric glass-ceramics

In non-ferroelectric ceramics the domain configuration cannot be changed by an externally applied electric field. The polar order must be built in during the processing stage. Gardopee and co-workers (Gardopee *et al.*, 1980, 1981) demonstrated the possibility of preparing a polar but non-ferroelectric material through the glass-ceramic route. This was achieved by crystallizing a glass in such a manner that the crystallites had both crystallographic and polar orientation. The system chosen was $Li_2O \cdot 2SiO_2$ and the polar phase obtained after the crystallization was $Li_2Si_2O_5$ (orthorhombic point group mm2). It had previously been shown by Rindone (Rindone, 1962a, 1962b) that in lithium silicate glasses $Li_2Si_2O_5$ crystals grow with a high degree of orientation with respect to the surface of glass. Needle-like crystals forming a thin layer near the surface have their *c*-axes perpendicular to the sample surface. X-ray diffractometer patterns taken on surfaces of recrystallized samples confirmed a high degree of preferred orientation. If all or most of the $Li_2Si_2O_5$ crystallites in the glass-ceramic show similar polar orientation, it is obvious that we obtain a glass-ceramic which will exhibit the polar properties of single crystals. It was demonstrated by Gardopee (Gardopee *et al.*, 1981) that the degree of orientation of crystallites can be improved markedly if well-polished glass samples are crystallized in a large temperature gradient. These grain-oriented glass-ceramics showed substantial pyroelectric response, thus confirming the polar orientation of the crystallites. The piezoelectric properties of these glass-ceramics were, however, very weak. We call such a glass-ceramic with pyroelectric properties, a polar glass-ceramic. A schematic which defines a route to

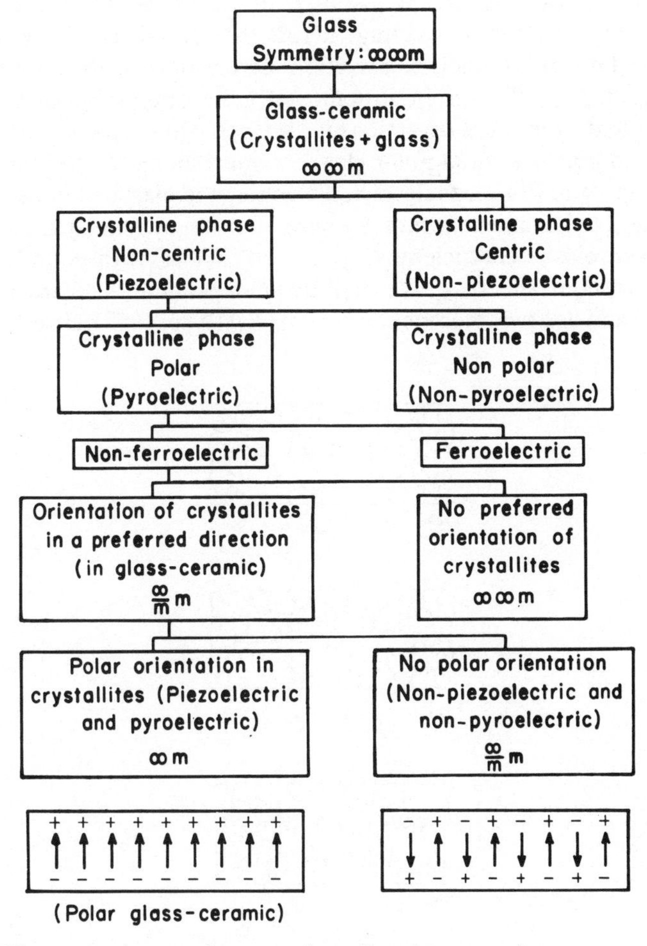

Fig. 8.1 Glass-ceramic route for preparing polar glass-ceramics.

obtain polar glass-ceramics is shown in Fig. 8.1. It is clear that the above procedure provides an interesting and novel technique to prepare grain-oriented polar but non-ferroelectric glass-ceramics. In this technique, the poling problems encountered in ferroelectric glass-ceramics are avoided. In addition the problems of depoling and ageing do not arise because polar glass-ceramics are non-ferroelectric and since they do not have a Curie temperature, they can also be used at high temperature. All the glass-ceramics described in this chapter are polar but not ferroelectric.

8.2.3 Tensor properties of grain-oriented polar glass-ceramics

The domains in an unpoled ferroelectric ceramic are randomly oriented with polarization vectors pointing in all directions as shown schematically in Fig. 8.2, where all the grains are assumed to be single domain grains for simplicity. Application of a voltage across the ceramic causes most of the domains to line up with their polarization vectors more nearly parallel to the field (Fig. 8.1(b)). The grain structure of grain-oriented glass-ceramic is similar to that of a poled ceramic (Fig. 8.1(c)). Thus a poled ferroelectric ceramic and grain-oriented polar glass-ceramic belong to the same conical point group, ∞m. For piezoelectric, dielectric, and elastic phonomena, this is equivalent to hexagonal symmetry 6mm. For this symmetry there are five independent elastic coefficients (s_{11}, s_{12}, s_{13}, s_{33}, s_{44}); three independent piezoelectric constants (d_{33}, d_{31}, d_{15}); two independent dielectric constants (K_{11}, K_{33}) and one pyroelectric coefficient (P_3) (Nye, 1957). By convention,

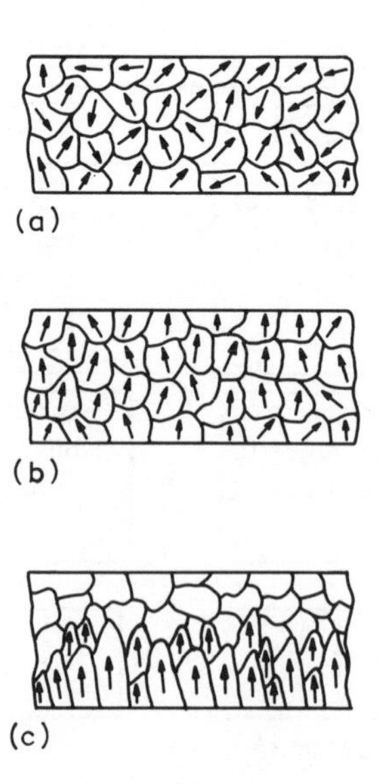

Fig. 8.2 Domain structure in (a) an unpoled ferroelectric ceramic; (b) poled ferroelectric ceramic and (c) polar glass-ceramic.

the X_3 direction is chosen to be the polar axis, normal to the surface of the crystallized polar glass-ceramic. Additional non-zero coefficients are related to these by symmetry. In order to evaluate the usefulness of polar glass-ceramics in devices, it is necessary to measure the above properties and compare them with other commonly used materials.

8.3 SELECTION OF GLASS COMPOSITIONS

Several factors had to be considered before making a proper choice of systems for the present study. The system chosen should satisfy the following criteria:

1. The crystalline phase must belong to one of the polar point groups;
2. It should form a glass easily;
3. The crystallites in the glass-ceramic should show both crystallographic and polar orientation after processing;
4. The glass-ceramic should have good mechanical strength.

Each of these four criteria is discussed in detail in the following sections.

8.3.1 Polar systems

Of the 32 crystallographic point groups, only 20 show piezoelectric properties and 10 of these 20 piezoelectric classes are pyroelectric (or polar). The ten polar point groups are 1, 2, 3, 4, 6, m, mm2, 3m, 4mm and 6mm (Nye, 1957). For the present study only crystalline phases belonging to one of the above polar point groups are considered so that both pyroelectric and piezoelectric properties could be studied.

8.3.2 Glass formation

For ease in processing, only compositions which form glasses readily by air quenching were considered. This places a major resriction on the number of systems that can be studied since the glass composition must then contain at least one glass former such as SiO_2, GeO_2, B_2O_3, or P_2O_5. If new fast-quenching methods like splat cooling or twin roller quenching are used for preparation of glasses, the number of polar systems that can be studied will be much larger.

8.3.3 Crystallographic and polar orientation

All the systems examined in the present work were non-ferroelectric. To achieve glass-ceramics with polar properties, the crystallization must be carried out in such a way that the individual crystallites possess both crystallographic and polar orientation. This requirement limits the number of systems that can be studied since many of the polar systems might form glasses easily but they may not yield glass-ceramics with crystallographic and polar orientation.

To satisfy this criterion a logical approach would be to select polar glass-forming systems which are known to recrystallize along a preferred crystallographic direction. Even if a glass crystallizes along a preferred direction, but fails to show any polar orientation, it will not exhibit either pyroelectric or piezoelectric properties. Since information concerning the crystallographic or polar orientation of polar glass-forming systems is lacking, a number of systems were tested in order to select a few which satisfy this requirement. If the crystal structure is highly anisotropic with molecular chains or layers, there is a strong possibility that the crystallites may grow along a preferred direction. But it is difficult to predict a priori whether the crystallites will show polar orientation or not. Polarity must be determined by experiment.

8.3.4 Physical properties of crystallized glass-ceramics

Once the chosen polar system satisfies the criteria for glass formation together with crystallographic and polar orientation, the last but equally important criterion is the crystallization behaviour of the system. If the glass-ceramic shows both crystallographic and polar orientation, but the individual crystallites are not bonded together by the glass matrix, as in the case of $Li_2Si_2O_5$ (Gardopee *et al.*, 1981), again the system is not of much use for practical devices. This is particularly true in the present study since the objective of the work was to prepare large area glass-ceramics with uniform properties. The glass-ceramics should have reasonably good mechanical strength to prepare samples for measurement.

8.3.5 Selection of polar glass-forming systems

Polar glass-forming systems that were tested during the course of the present work are listed in Table 8.1. The systems that were examined fall into four groups. The systems in group A did not form glass when stoichiometric compositions were melted. Group B systems formed glasses, but the crystallized glass-ceramics lacked good mechanical strength. Glasses of group C were hygroscopic. Systems of group D formed good glasses, but the crystallized glass-ceramics did not show crystallographic or polar orientation.

Hence the first task was to prepare optimized glass compositions which formed glasses easily and crystallized uniformly with maximum crystallographic and polar orientation. Among the systems listed in Table 8.1, the systems that partially satisfied the critieria are $Ba_2TiSi_2O_8$, $Ba_2TiGe_2O_8$, $Li_2Si_2O_5$, Li_2SiO_3, and $Li_2B_4O_7$. However, acceptable glass-ceramics were not obtained from glasses of stoichiometric compositions in any of these systems. Hence a wide range in compositions was investigated

Table 8.1 Polar glass-forming systems*

A	B
Li_2SiO_3	$Li_2Si_2O_5$
Li_2GeO_3	$Ba_2TiGe_2O_8$
$Li_2Ge_2O_5$	$Ba_2TiSi_2O_8$
$Li_2B_4O_7$	
Na_2SiO_3	
$Sr_2TiSi_2O_8$	
Li_3PO_4	
C	**D**
$Na_2Si_2O_5$	SrB_4O_7
ZnP_2O_6	PbB_4O_7

* All the systems are polar but non-ferroelectric.

Table 8.2 Crystal system and point group symmetry

Crystal	*Melting point (°C)*	*Crystal system*	*Point group symmetry*	*Polar axis*	*Reference*
$Ba_2TiSi_2O_8$	1440	Tetragonal	4mm	*c*	Moore and Louisnathan (1959)
$Ba_2TiGe_2O_8$	1260	Orthorhombic	mm2	*c*	Blasse (1968)
$Sr_2TiSi_2O_8$	1380	Tetragonal	4mm	*c*	Gabelica-Robert and Tarte (1981)
$Li_2Si_2O_5$	1200	Orthorhombic	mm2	*c*	Liebau (1961)
Li_2SiO_3	1201	Orthorhombic	mm2	*c*	Brun *et al.* (1979)
$Li_2B_4O_7$	917	Tetragonal	4mm	*c*	Krog-Moe (1962)

for each system to arrive at compositions which satisfied all four requirements. Further, various modifying oxides such as PbO, CaO, ZnO and SrO were added to optimize both the electrical and mechanical properties.

Data concerning crystal structure, melting point, point group symmetry and polar axes of the above materials are given in Table 8.2.

8.4 PREPARATION OF GLASS-CERAMICS

8.4.1 Glass melting

Reagent grade chemicals (silicic acid $SiO_2 \cdot nH_2O$, GeO_2, Li_2CO_3, $BaCO_3$, $CaCO_3$, TiO_2, H_3BO_3, PbO) were weighed and mixed by ball milling in alcohol for 8 to 10 h and then dried. Glass batches of approximately 50 g were melted in a platinum crucible in an open atmosphere using an electric globar furnace. Melting temperatures for different compositions were in the range 1200 to 1450°C. The melt was retained in the furnace for 4 to 8 h for fining and homogenization.

The fined glass melt was air quenched by pouring it into graphite molds to form cylinders of approximately 0.8 to 1 cm in diameter and 1 to 1.5 cm in length. The cylinders were cooled to room temperature and examined for cracking, devitrification and other defects using an optical microscope. If the glass samples contained small air bubbles, the batch was remelted and further fining was carried out to remove the bubbles from the melt. Thermal stresses were partially relieved by annealing the glasses for 12 h well below the nucleation temperature to avoid bulk nucleation in glasses. (Li_2O–SiO_2–B_2O_3 glasses were annealed at 400°C and BaO–TiO_2–GeO_2 and BaO–TiO_2–SiO_2 glasses were annealed at 500°C and 600°C respectively.)

For crystallization studies, glass samples were prepared by grinding and polishing one surface of the glass cylinders. The samples were ground using an abrasive diamond wheel, followed by polishing with 400 and 600 grit silicon carbide powder, and later with 12 and 3 μm alumina powder. The final polishing was done using 1 μm diamond polishing paste. The polished samples for crystallization studies were disks approximately 1 cm in diameter and 1 cm in length.

8.4.2 Crystallization of glasses

Differential thermal analysis (DTA) runs were made with glass powders to determine the crystallization temperature.

Two types of crystallization treatments were used in the present study. Heat treatments in which the entire sample was uniformly heated to the crystallization temperature are termed 'isothermal crystallization'. Heat treatments in which the glass samples were crystallized in a temperature gradient are called 'temperature gradient crystallization'. To obtain polar glass-ceramics with well orientated crystallites and good mechanical strength, many different heat treatment cycles were carried out. The effect of nucleation on oriented growth of crystallites was studied by crystallizing glasses after heating through a nucleation step.

Crystallization in a temperature gradient was carried out by positioning the polished glass samples on a microcope hot stage. The direction of crystallization was parallel to the direction of temperature gradient. The crystalline phases in the glass-ceramics were identified by powder X-ray diffraction (XRD) patterns. The degree of preferred orientation of crystallites in the glass-ceramic was evaluated from XRD patterns recorded on surfaces normal to the direction of temperature gradient. The surface XRD patterns were compared with powder diffraction patterns obtained from randomly oriented crystallites to determine the relative degree of orientation.

Table 8.3 Glass formation and recrystallization behaviour of various compositions of glasses

Composition	*Glass formation†*	*Mechanical strength of glass-ceramics‡*	p_3 ($\mu C\,m^{-2}\,K^{-1}$)	*Detailed study*
$Ba_2TiSi_2O_8$ system				
*$2BaO–2SiO_2–TiO_2$	hmp	p	–	no
$0.64SiO_2–0.36BaTiO_3$	g	s	2	no
$BaO–SiO_2–TiO_2$	s	s	2	no
$2BaO–4SiO_2–TiO_2$	g	p	–	no
$2BaO–2.5SiO_2–1.5TiO_2$	s	s	4	no
$2BaO–2.8SiO_2–1.3TiO_2$	s	p	–	no
$2.3BaO–3.1SiO_2–0.7TiO_2$	s	s	(0)	no
$2BaO–3SiO_2–TiO_2$	g	s	8	yes
$2BaO–0.15CaO–2.9SiO_2–TiO_2$	g	g	6	yes
$1.6BaO–0.4CaO–2.8SiO_2–TiO_2$	g	g	6	yes
$1.8BaO–0.2CaO–2SiO_2–TiO_2$	g	p	–	no
$1.7BaO–0.3SrO–3SiO_2–TiO_2$	g	p	–	no
$1.6BaO–0.4SrO–3SiO_3–TiO_2–0.2CaO$	g	g	5	yes
$1.9BaO–0.1PbO–3SiO_2–TiO_2$	g	g	8	yes
$1.6BaO–0.4PbO–3SiO_2–TiO_2$	g	p	–	no
$1.4BaO–0.4SrO–0.2PbO–3SiO_2–1.1TiO_2$	g	s	–	no
$2BaO–3SiO_2–TiO_2–0.2K_2O$	s	s	(0)	no
$2BaO–3SiO_2–ZrO_2$	hmp	–	–	no
$2BaO–3SiO_2–0.8TiO_2–0.2ZrO_2$	g	p	–	no
$2BaO–2.2SiO_2–TiO_2–0.2Al_2O_3$	s	s	(0)	no
$2BaO–3SiO_2–0.8TiO_2–0.2MnO_2$	g	s	(0)	no
$2BaO–3SiO_2–0.9TiO_2–0.1SnO_2$	s	p	–	no
$1.85BaO–0.15BaF_2–3SiO_2–TiO_2$	g	p	–	no

Table 8.3 *Continued*

Composition	*Glass formation†*	*Mechanical strength of glass-ceramics‡*	p_3 ($\mu C\,m^{-2}\,K^{-1}$)	*Detailed study*
$Ba_2TiGe_2O_8$ system				
*$2BaO–2GeO_2–TiO_2$	h	–	–	no
$2BaO–3GeO_2–TiO_2$	s	p	–	no
$BaO–GeO_2–TiO_2$	g	g	−2	yes
$BaO–GeO_2–TiO_2–0.1CaO$	g	g	−1	no
$BaO–GeO_2–TiO_2–0.2CaO$	g	p	–	no
$BaO–GeO_2–0.9TiO_2–0.1ZrO_2$	g	p	–	–
$BaO–GeO_2–0.9TiO_2–0.05Nb_2O_5$	p	p	–	–
$BaO–GeO_2–0.8TiO_2–0.2ZnO$	p	p	–	–
$0.8BaO–0.2SrO–GeO_2–TiO_2$	g	p	–	no
$Sr_2TiSi_2O_8$ system				
*$2SrO–2SiO_2–TiO_2$	ps	–	–	–
$2SrO–3SiO_2–TiO_2$	g	g	8	yes
$1.8SrO–0.2BaO–2.8SiO_2–0.1CaO–TiO_2$	g	g	2	yes
$Li_2B_4O_7$ system				
*$Li_2O–2B_2O_3$	r	p	–	no
$Li_2O–3B_2O_3$	s	s	–	no
$Li_2O–0.5B_2O_3–0.5SiO_2$	g	p	–	no

$Li_2Si_2O_5$ system				
*Li_2O–$2SiO_2$	g	p	–	no
Li_2O–$3SiO_2$	g	p	–	no
Li_2O–$2SiO_2$–$xZnO$ ($x = 0.1$ to 0.4)	g	s	−2	no
Li_2O–$1.5SiO_2$–$0.5GeO_2$	g	p	–	no
Li_2O–$1.8SiO_2$–$0.2CaO$	g	p	–	no
Li_2O–$1.9SiO_2$–$0.1MO$ ($MO = Ta_2O_5$, PbO, Al_2O_3, SnO_2, TiO_2, P_2O_5, Fe_2O_3)	g	p	–	–
($Li_2Si_2O_5 + Li_2B_4O_7$) system				
Li_2O–$(2-x)SiO_2$–xB_2O_3 ($x = 0.1$ to 0.7)	g	g	−6 to −11	yes
Li_2O–$1.5SiO_2$–$0.4B_2O_3$–$0.1Al_2O_3$	g	g	(0)	no
Li_2O–$1.6SiO_2$–$0.2B_2O_3$–$0.2WO_3$	g	s	(0)	no
Li_2O–$1.8SiO_2$–$0.1B_2O_3$–$0.1ZnO$	g	g	(0)	no
$0.85Li_2O$–$0.15Na_2O$–$1.8SiO_2$–$0.2B_2O_3$	g	p	–	no
$0.8Li_2O$–$0.2LiF$–$1.8SiO_2$–$0.2B_2O_3$	g	s	–	no
Li_2O–$1.8SiO_2$–$0.2B_2O_3$–$0.03Y_2O_3$	g	s	–	no
Li_2O–$1.8SiO_2$–$0.2B_2O_3$–$0.4P_2O_5$	g	s	(0)	no
Li_2O–$1.7SiO_2$–$0.2B_2O_3$–$0.06Si_3N_4$	g	p	–	no

* Stoichiometric compositions.
(0) Piezoelectric d_{33} coefficient, almost zero.
† Glass formation:
- g easy glass formation even by slow air quenching of the melt; no problem of recrystallization
- s glass formation possible by air quenching, but some of the samples crystallized partially
- p problem of recrystallization; fast quenching necessary
- r the melt crystallizes completely on air quenching
- ps the melt separates into two phases
- hmp high melting point (>1450°C); difficult to get rid of air bubbles from the melt.

‡ Mechanical strength of glass-ceramics: g glass-ceramics have good mechanical strength; possible to prepare thin sections (200 μm) by polishing; s glass-ceramics have reasonably good mechanical strength, but difficult to prepare thin sections by polishing; p the mechanical strength of glass-ceramics not enough to polish and prepare thin sections.

8.5 COMPOSITIONS OF GLASSES

The systems selected for detailed study were $Ba_2TiSi_2O_8$, $Ba_2TiGe_2O_8$, $Sr_2TiSi_2O_8$, $Li_2Si_2O_5$ and $Li_2B_4O_7$. A wide range in composition and additives were tested for each system to obtain glass-ceramics with optimized properties. A list of compositions tested is given in Table 8.3. The glass formation characteristics, mechanical strength, and approximate pyroelectric coefficients are also listed in the table. Out of several compositions tested, only those which yielded polar glass-ceramics with desired quality and substantial pyroelectric activity were selected for carrying out detailed study.

In the fresnoite system, for example, the melting point of a stoichiometric composition is very high (>1450°C) and hence it was not easy to prepare glass of this composition. In addition the crystallization of this glass was not uniform and the mechanical strength was poor. To overcome these problems, different ratios of BaO, SiO_2 and TiO_2 were tested to improve the crystallization. The composition which gave reasonably good glass-ceramics was $2BaO–3SiO_2–TiO_2$. However, for this composition also, it was necessary to improve further the physical properties. Several different substituents were tried for BaO and TiO_2 to improve the properties (Table 8.3). Addition of a small amount of PbO, CaO or SrO helped in obtaining glass-ceramics with good physical properties. With most of the other substituents, there was no problem with glass formation but there was no improvement in crystallization behaviour. Additions of Al_2O_3, K_2O or MnO_2 helped in obtaining well-crystallized glass-ceramics, but the measured d_{33} coefficient was zero for these glass ceramics, indicating very poor polar orientation of the crystallites.

Similar studies were carried out for other compositions listed in Table 8.3. Optimized compositions are listed in Table 8.4. The crystallization temperatures as determined by DTA runs and the crystalline phases identified from powder XRD patterns are also presented in the table.

8.6 HEAT-TREATMENT AND MICROSTRUCTURE

The ideal microstructure of polar glass-ceramics would consist of well-oriented crystallites extending between both sides of the glass-ceramic. Different heat-treatment schedules were tried to achieve this goal.

Generally nucleation and crystallization begins first at the surface of glasses because of the imperfect nature of the surface structure (McMillan, 1979). The bulk nucleation process does not begin until later. The ideal way to obtain polar glass-ceramics with oriented crystallites would be to start nucleation on the surface of the glass and allow the crystallites to grow into the bulk of the sample, thereby avoiding as far as possible bulk nucleation and the growth of spherulites.

Table 8.4 Compositions, crystallization temperature and crystalline pnases

Composition	*Glass melting temp (°C)*	*Crystallization temperature*	*Crystalline phases*
$Ba_2TiSi_2O_8$ system			
$2BaO–3SiO_2–TiO_2$	1450	930	$Ba_2TiSi_2O_8$
$1.9BaO–0.1PbO–3SiO_2–TiO_2$	1450	920	$Ba_2TiSi_2O_8$
$2BaO–0.15CaO–2.9SiO_2–TiO_2$	1450	920	$Ba_2TiSi_2O_8$
$1.6BaO–0.4CaO–2.8SiO_2–TiO_2$	1450	930	$Ba_2TiSi_2O_8$
$1.6BaO–0.4SrO–3SiO_2–TiO_2–0.2CaO$	1450	930	$Ba_2TiSi_2O_8$
$Ba_2TiGe_2O_8$ system			
$BaO–GeO_2–TiO_2$	1400	800	$Ba_2TiGe_2O_8$
$Sr_2TiSi_2O_8$ system			
$2SrO–3SiO_2–TiO_2$	1450	950	$Sr_2TiSi_2O_8$
$1.8SrO–0.2BaO–2.8SiO_2–0.1CaO–TiO_2$	1450	940	$Sr_2TiSi_2O_8$
$Li_2Si_2O_5$ system			
$Li_2O–2SiO_2$	1350	585	$Li_2Si_2O_5$
$Li_2O–2SiO_2–0.2ZnO$	1350	580	$Li_2Si_2O_5 + Li_2ZnSiO_4 + \alpha SiO_2$
$Li_2B_4O_7$ system			
$Li_2O–3B_2O_3$	1250	580	$Li_2B_4O_7 + Li_2B_6O_{10}$
$Li_2Si_2O_5 + Li_2B_4O_7$ system			
$Li_2O–1.8SiO_2–0.2B_2O_3$	1250	605–680	$Li_2Si_2O_5 + Li_2B_4O_7 + Li_2SiO_3$
$Li_2O–1.7SiO_2–0.3B_2O_3$	1250	595–670	$Li_2Si_2O_5 + Li_2B_4O_7 + Li_2SiO_3$
$Li_2O–1.6SiO_2–0.4B_2O_3$	1250	625–665	$Li_2Si_2O_5 + Li_2B_4O_7 + Li_2SiO_3$

It is clear that the following factors should be considered in order to obtain glass-ceramics with oriented crystallites.

1. The glasses should not contain nucleating agents which promote bulk nucleation;
2. The surface finish of glasses should be as fine as possible;
3. The heat-treatment process should be such that the crystallization takes place primarily by surface nucleation since this promotes oriented growth;
4. Bulk nucleation and crystallization produce randomly oriented spherulites and should be avoided; this may be possible if the bulk nucleation step is completely avoided.

It is obvious that if the glass is heat-treated under isothermal conditions, both surface and bulk crystallization can start simultaneously and hence the oriented growth of crystallites will stop when crystallites growing from the surface encounter randomly growing crystallites in the bulk of the sample.

It has been shown by several workers that directional crystallization is possible in amorphous materials if the crystallization is carried out in a temperature gradient. Gardopee (Gardopee *et al.*, 1980, 1981) crystallized Li_2O–$2SiO_2$ glasses in both isothermal and temperature gradient environments and showed that temperature gradient crystallization produces glass-ceramics in which the oriented region of crystallites is larger. Thermal gradient crystallization technique has been used to produce oriented crystallites of $CdGeAs_2$ (Sekhar and Risbud, 1981; Risbud, 1979), and Na_2SiO_3 (Melling and Duncan, 1980).

XRD patterns were taken on 'as-crystallized' surfaces of glass-ceramics listed in Table 8.4. The ratio of I_{002}/I_{211} (for fresnoite glass-ceramics) or I_{002}/I_{040} (for lithium silicate and lithium borosilicate glass-ceramics) was taken as a measure of the degree of preferred orientation of crystallites. The ratio was greater than 100 for all glass-ceramics indicating highly oriented growth of crystallites. In all cases, the polar *c* axis of the crystallites was perpendicular to the initial crystallizing surface.

Glass samples were crystallized in both isothermal and temperature gradient environments and X-ray diffraction and microstructure studies were carried out to assess the degree of preferred orientation of crystallites carried out as a function of depth from the initial crystallizing surface.

The glass composition selected for detailed study was Li_2O–$2SiO_2$–$0.2ZnO$, for which the nucleation and crystallization temperatures were 480 and 600°C, respectively. The isothermal heating cycles used are shown in Fig. 8.3. Heating cycles A, B and C did not include any nucleation step whereas heating cycle D included a nucleation step. The only difference between heating cycles A, B and C is the different upper cystallization temperature. In cycle E, the sample was heated slowly at a uniform rate up to the upper crystallization temperature. Glasses were crystallized by

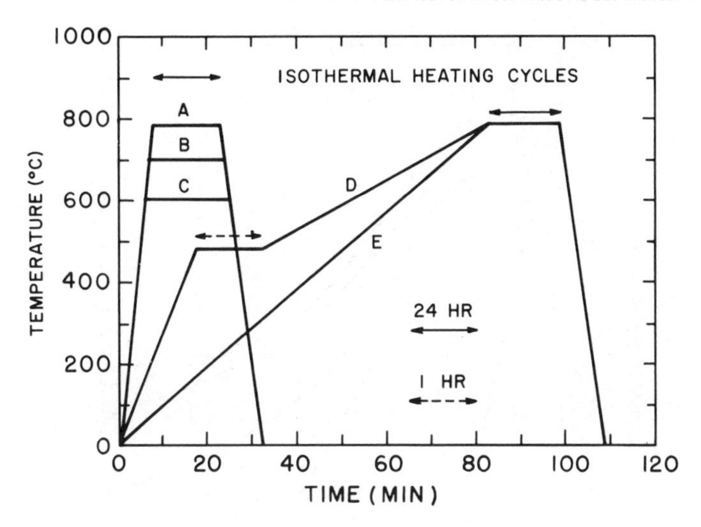

Fig. 8.3 Isothermal heat treatment cycles.

following one of the heating cycles similar to those shown in Fig. 8.3. The ratio of I_{002}/I_{040} was determined as a function of depth and plotted for different heat treatment cycles (Fig. 8.4). Several observations were made from this study.

1. The degree of orientation of crystallites was high ($I_{002}/I_{040} > 100$) up to a depth of about 50 μm for both temperature gradient and isothermal crystallization, for all heating cycles.
2. In the case of heating cycles in which there was enough time for nucleation (cycles D, E) the orientation of crystallites decreases very rapidly with depth from the initial crystallizing surface. Typically, the orientation becomes completely random at a depth of about 200 μm. The orientation of crystallites is worse for the isothermally crystallized samples. The reason for this is clear from the discussion given in the previous section. The nucleation step tends to promote bulk crystallization in the case of isothermally crystallized samples. The bulk crystallization is spherulitic and thus there is no preferred orientation of crystallites. The oriented surface crystallization will continue until the crystallites encounter spherulites at which point further growth of oriented crystallites stops.
3. In the case of heating cycles (both isothermal and temperature gradient) in which there was not enough time for nucleation (cycles A, B, C), the orientation of crystallites was reasonably good even at a depth of 300 μm.

For the present study, all the glass-ceramics were prepared in a temperature gradient, to obtain glass ceramics with well oriented crystallites.

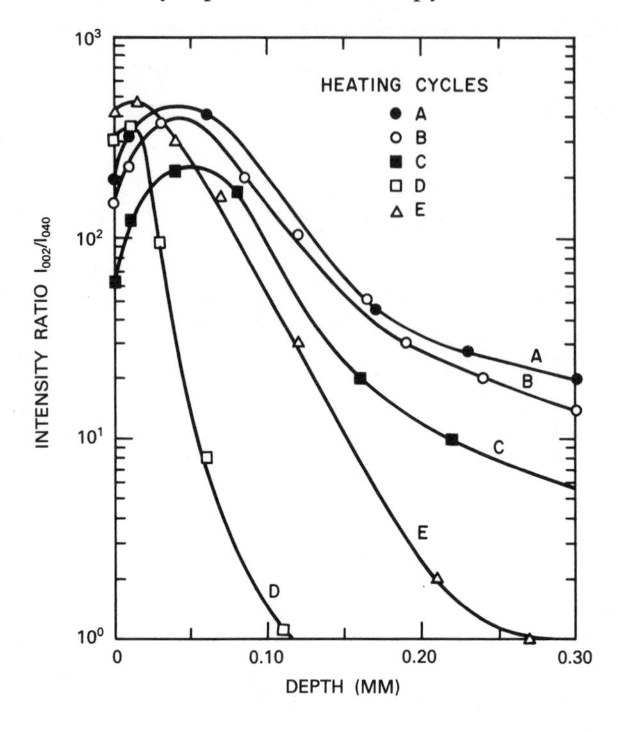

Fig. 8.4 The degree of preferred orientation as a function of depth for Li_2O–$2SiO_2$–$0.2ZnO$ glass-ceramic for different isothermal heating cycles.

The microstructures of glass-ceramics of selected compositions were examined using optical and electron microscopes. Typically the microstructure consisted of needle-like crystallites growing from the initial crystallizing surface into the bulk of the samples.

A vertical cross-sectional view (parallel to the temperature gradient) of Li_2O–$2SiO_2$–$0.2ZnO$ glass-ceramic is shown in Fig. 8.5(a), in which the oriented growth of $Li_2Si_2O_5$ crystallites from the surface and also spherulites resulting from bulk nucleation and crystallization can be seen. The oriented region extends to a depth varying from 200 to 500 μm. Figure 8.5(b) shows a horizontal cross-sectional view of the same sample cut perpendicular to temperature gradient. The diameter of the crystallites is generally in the range 1–3 μm.

8.7 DIELECTRIC PROPERTIES

For dielectric, pyroelectric and piezoelectric measurements, discs of approximately 1 cm diameter and 0.4 to 0.8 mm thick were prepared by sectioning the oriented region of the glass-ceramics near the surface. The

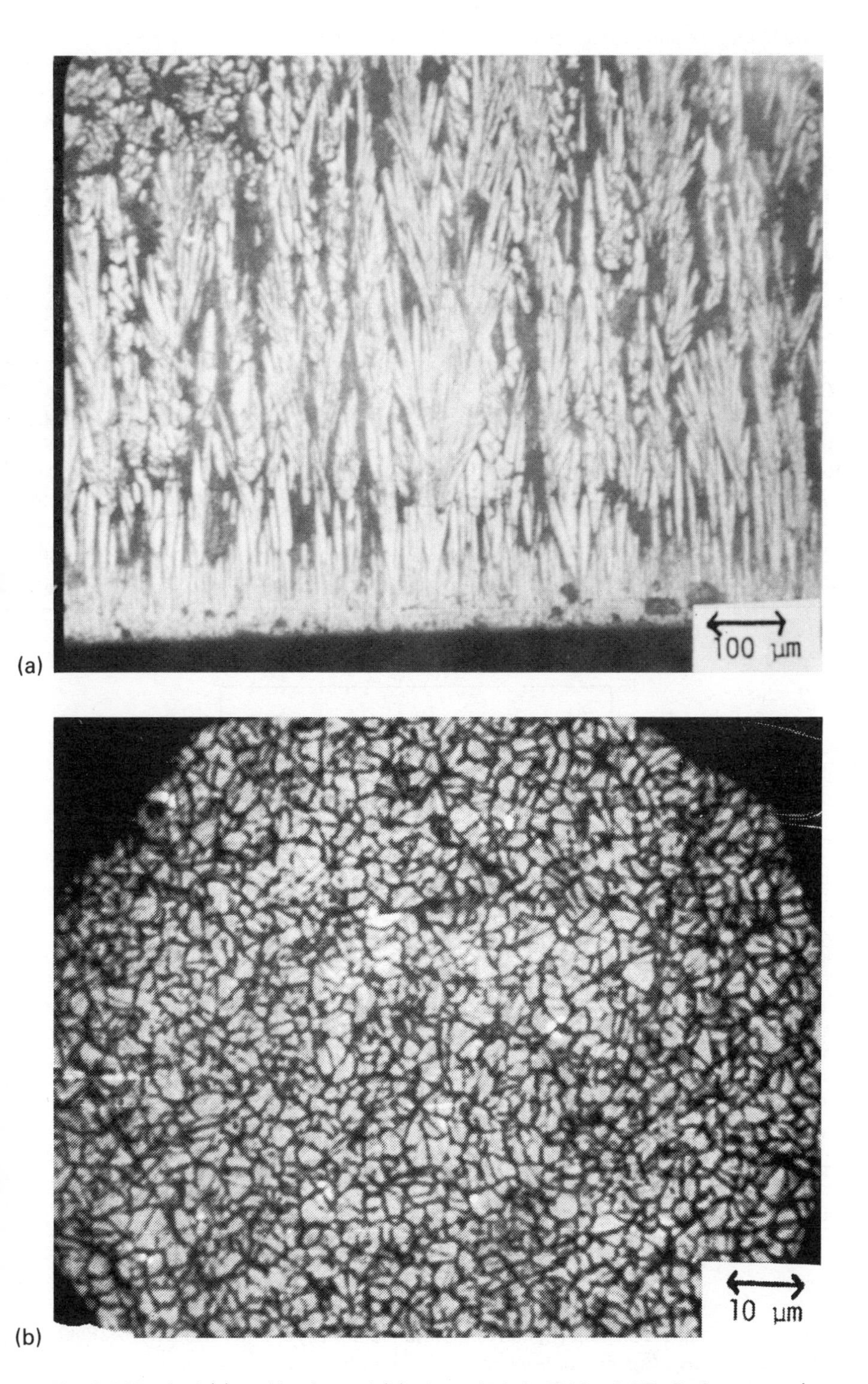

Fig. 8.5 Vertical (a) and horizontal (b) view of Li_2O–$2SiO_2$–$0.2ZnO$ glass-ceramic.

Table 8.5 Dielectric and physical properties of glass-ceramics

Composition	*Density* ($g\,cm^{-3}$)	*Specific* heat ($J\,cm^{-3}\,K^{-1}$)	K_3 (1 kHz)	*tan δ* (1 kHz)
Li_2O–$2SiO_2$–$0.2ZnO$	2.47	2.36	6.5	0.008
Li_2O–$1.8SiO_2$–$0.2B_2O_3$	2.34	2.39	7.0	0.05
Li_2O–$1.7SiO_2$–$0.3B_2O_3$	2.32	–	7.5	0.05
Li_2O–$1.6SiO_2$–$0.4B_2O_3$	2.35	–	7.0	0.05
BaO–GeO_2–TiO_2	4.78	–	15	–
$2BaO$–$3SiO_2$–TiO_2	4.01	2.04	9.0	0.001
$2BaO$–$0.15CaO$–$2.9SiO_2$–TiO_2	3.98	–	10	0.001
$1.6BaO$–$0.4CaO$–$2.8SiO_2$–TiO_2	–	–	10.5	0.001
$1.6BaO$–$0.4SrO$–$3SiO_2$–$0.2CaO$–TiO_2	3.87	–	9.8	0.001
$1.9BaO$–$0.1PbO$–$3SiO_2$–TiO_2	4.05	2.04	10.0	0.001
$2SrO$–$3SiO_2$–TiO_2	3.53	–	11.5	0.001
$1.8SrO$–$0.2BaO$–$2.8SiO_2$–$0.1CaO$–TiO_2	3.63	–	10.5	0.001

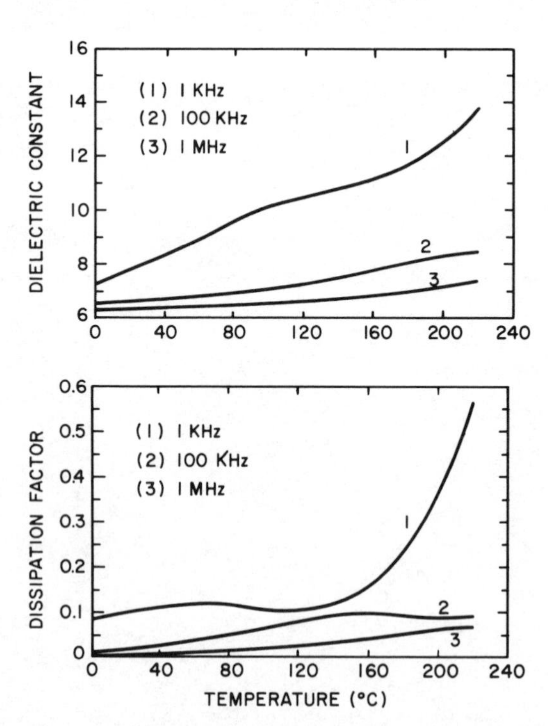

Fig. 8.6 Dielectric constant (a) and dissipation factor (b) of Li_2O–$1.8SiO_2$–$0.2B_2O_3$ glass-ceramic as a function of temperature and frequency.

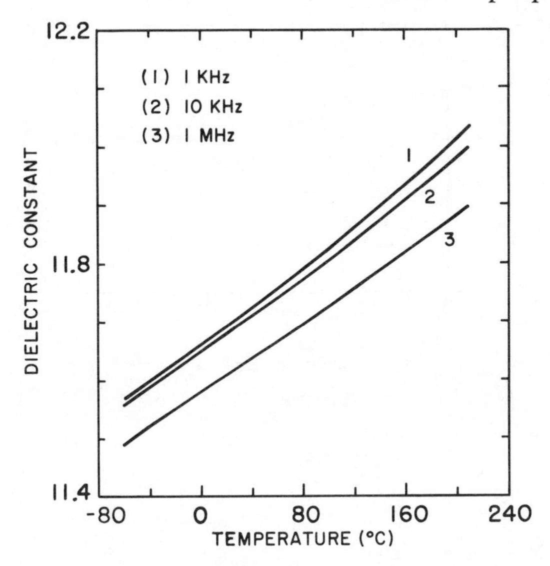

Fig. 8.7 Dielectric constant of $2BaO–3SiO_2–TiO_2$ glass as a function of temperature and frequency.

length of the crystallites was 200–500μm extending from the surface into the interior of the glass-ceramic. The remainder of the glass-ceramic consisted of randomly oriented crystallites. The samples were lightly polished with 3 μm alumina powder and electroded with sputtered gold films.

Dielectric constant and dissipation factors were measured over a temperature range from −150 to 200°C at frequencies from 10^2 to 10^7 Hz using a multifrequency LCR meter* and a desk top computer.† The physical and dielectric properties of glass-ceramics are listed in Table 8.5. Dielectric constants lie in the range 5–15, comparable to the values obtained in the corresponding single crystals. The variation of dielectric constant and dissipation factor with temperature is shown in Fig. 8.6 for $Li_2O–1.8SiO_2–0.2B_2O_3$ glass-ceramics. This is typical of all lithium borosilicate glass-ceramics. The general trend is that both the dielectric constant and losses increase as the temperature is raised, with the temperature dependence becoming less marked at higher frequencies. At lower frequencies the dielectric losses increase with temperature, due to the increasing contribution from ionic mobility. This trend is typical of alkali silicate glass-ceramics (McMillan, 1979).

The temperature variation of the dielectric constant of fresnoite glass ($2BaO–3SiO_2–TiO_2$), shown in Fig. 8.7 is featureless in the temperature

* Model 4274A and Model 4275A, Hewlett-Packard, Palo Alto, CA.
† Model 9825A, Hewlett-Packard, Palo Alto, CA.

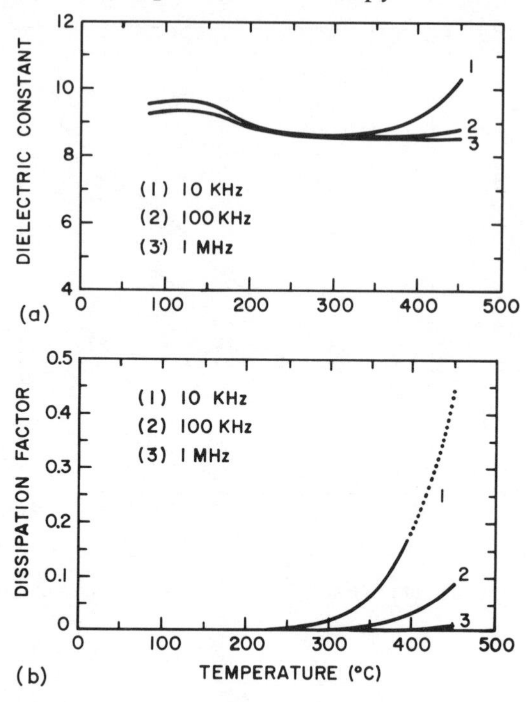

Fig. 8.8 Dielectric constant (a) and dissipation factor (b) of $2BaO–3SiO_2–TiO_2$ glass-ceramic as a function of temperature and frequency.

range −60°C to 220°C. The dielectric constant increases slightly with temperature. In the case of glass-ceramics of the same composition, a broad peak is observed in the dielectric constant curve at around 140°C (Fig. 8.8(a)). Both the dielectric constant and losses (Fig. 8.8(b)) begin increasing with temperature at about 300°C. The broad peak in the dielectric constant at 140°C was observed in all glass-ceramics in which $Ba_2TiSi_2O_8$ was the major phase. The cause of the dielectric maximum at 140°C in fresnoite based glass-ceramics has been discussed previously (Halliyal *et al.*, 1985c).

8.8 PYROELECTRIC PROPERTIES

A direct method (Byer and Roundy, 1972) was used for measuring pyroelectric coefficients. The measurement was carried out using a desk top computer with instrument control and data transfer provided by a multiprogrammer interface.* The computer controlled the heating or cooling rate through the multiprogrammer and the interface. Pyroelectric current was measured with a picoammeter.† The typical heating or cooling

* Model 59500A, Hewlett-Packard, Palo Alto, CA.
† Model 41140B, Hewlett-Packard, Palo Alto, CA.

rate was 3–4°C min^{-1}. Pyroelectric coefficients were calculated in the temperature range −150 to 220°C in both the heating and cooling cycles.

An evaluation of pyroelectric properties and pyroelectric figures of merit of glass-ceramics for different applications is given in this section.

Pyroelectric effect is quite frequently used for the detection of infrared radiation. Even though the detectivities (D^*) of pyroelectric materials are two to three orders of magnitude smaller than those of cooled photoconductive detectors, the advantage of operating at low temperature together with lower cost makes the pyroelectric detectors more versatile. This has led to an extensive use of pyroelectric detectors for military and commercial applications including intruder alarms, forest fire mapping and thermal imaging devices (vidicons) (Porter, 1981). Optimum property requirements for a pyroelectric target material have been reviewed in detail by a number of authors (Herbert, 1982; Putley, 1970; Liu, 1976, 1978; Whatmore *et al.*, 1980). Properties of both single crystals (such as triglycine sulphate, $LiTaO_3$, $Pb_5Ge_3O_{11}$, $Sr_{0.5}Ba_{0.5}Nb_2O_6$) and ceramics (such as $PbZr_{0.5}Ti_{0.5}O_3$, modified $PbTiO_3$ and $PbZrO_3$) have been evaluated for pyroelectric devices (Whatmore *et al.*, 1980).

It is desirable, when comparing the properties of pyroelectric materials, to define appropriate figures of merit for the particular application of interest. A few of the commonly used figures of merit are listed in Table 8.6 (Herbert, 1982; Whatmore *et al.*, 1980). In general to achieve high detectivity, it is desirable to use a material which has a high pyroelectric coefficient, low dielectric constant, low dissipation factor, low density, low specific heat and low thermal diffusivity.

Table 8.6 Pyroelectric figures of merit

Symbol	*Expression*	*Application*
M_1	p/K	For a quick evaluation of material
M_2	$p/\varrho c$	*Current mode*: a thin disc feeding current into a low impedance amplifier
M_3	$p/\varrho cK$	*Voltage mode*: a disc supplying voltage to a high impedance amplifier, the inherent noise of which limits the sensitivity of detector
M_4	$p/\varrho KD$	For vidicons
M_5	$p/\varrho cK^{1/2} \tan \delta^{1/2}$	*Voltage mode*: for a high impedance amplifier when the pyroelectric element is the main source of noise

p = pyroelectric coefficient; K = dielectric constant; $\tan \delta$ = dissipation factor; ϱ = density; c = specific heat per unit mass; c^1 = specific heat per unit volume (= ϱc); D = thermal diffusivity.

Currently triglycine sulphate [TGS = $(NH_2CH_2COOH)_3H_2SO_4$] or $LiTaO_3$ single crystals are used in pyroelectric devices. TGS has a high figure of merit but it is water soluble and hygroscopic. The lower Curie temperature of TGS (49°C) is also a disadvantage. The figures of merit of $LiTaO_3$ are lower than TGS but it has advantages of a high Curie temperature and insolubility in water.

The variation of pyroelectric coefficient p, with temperature for a few polar glass-ceramic compositions is shown in Figs 8.9 and 8.10. The following trends in the pyroelectric properties of glass-ceramics are evident.

1. The room temperature pyroelectric coefficients are in the range 5 to 10 $\mu C\ m^{-2}\ K^{-1}$;
2. For lithium borosilicate (Li_2O–B_2O_3–SiO_2) glass-ceramics, the sign of pyroelectric coefficient at room temperature is negative and remains practically constant up to 100°C. Above 100°C, the pyroelectric current is masked by large thermal currents;
3. For $Ba_2TiGe_2O_8$ glass-ceramics, the sign of p_3 is negative at room temperature and increases sharply at lower temperatures. The magnitude of p_3 is about $8\,\mu C\,m^{-2}\,K^{-1}$ at −140°C. There is a change

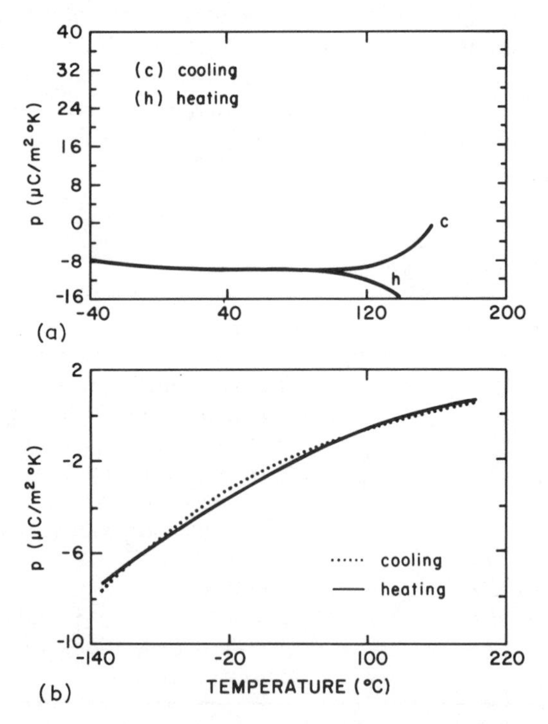

Fig. 8.9 Variation of pyroelectric coefficient with temperature for (a) Li_2O–$1.8SiO_2$–$0.2B_2O_3$ and (b) BaO–GeO_2–TiO_2 glass-ceramics.

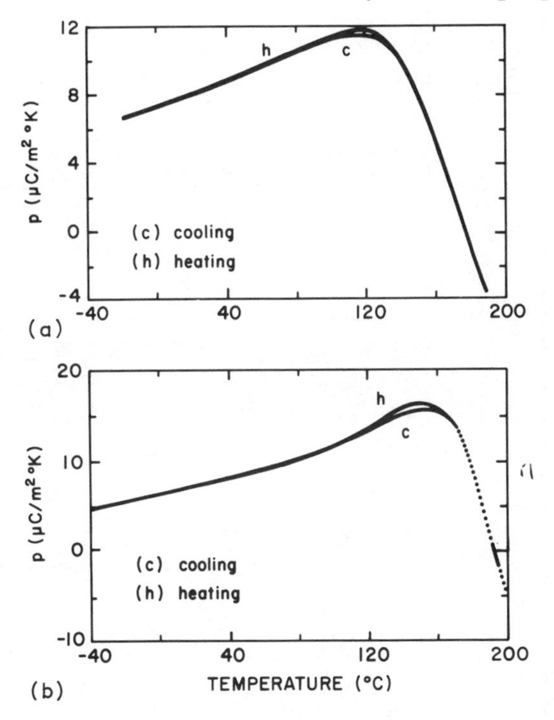

Fig. 8.10 Pyroelectric coefficient as a function of temperature for (a) $2BaO–3SiO_2–TiO_2$ and (b) $1.9BaO–0.1PbO–3SiO_2–TiO_2$ glass-ceramics.

in the sign of p_3 at about 130°C. Above 130°C, p_3 is positive and increases with temperature;

4. In glass-ceramics containing either $Ba_2TiSi_2O_8$ or $Sr_2TiSi_2O_8$ crystallites, the sign of p_3 is positive at room temperature and its magnitude increases slightly with increasing temperature. The sign of p_3 becomes negative at about 180°C, and its magnitude increases with temperature.

Pyroelectric properties of $Ba_2TiGe_2O_8$ (Halliyal *et al.*, 1985a), $Ba_2TiSi_2O_8$ (Halliyal *et al.*, 1985c) and $Li_2B_4O_7$ (Bhalla *et al.*, 1985) single crystals have been studied extensively. Room temperature pyroelectric properties of these single crystals are also listed in Table 8.7. It is difficult to grow defect free $Li_2Si_2O_5$ and $Sr_2TiSi_2O_8$ (Halliyal *et al.*, 1985b) single crystals and there are no reports about the pyroelectric properties of these materials. It is clear that the values of pyroelectric coefficient (p_3) of polar glass-ceramics are about 80 to 90% of their respective single crystal values.

The characteristic variation of p_3 with temperature and the reversal of its sign in $Ba_2TiGe_2O_8$ (at 130°C) (Halliyal *et al.*, 1985a) and in $Ba_2TiSi_2O_8$ (at 190°C) (Halliyal *et al.*, 1985c) single crystals has been attributed to the changing balance between primary and secondary pyroelectric effects with

Table 8.7 Room temperature pyroelectric figures of merit of glass-ceramics

Composition	p ($\mu C\,m^{-2}\,K^{-1}$)	p/K ($\mu C\,m^{-2}\,K^{-1}$) M_1	p/c^1 ($10^{-12}\,C\,m\,J^{-1}$) M_2	$p/c^1 K$ ($10^{-12}\,C\,m\,J^{-1}$) M_3	$p/c^1 (K \tan\delta)^{1/2}$ ($10^{-12}\,C\,m\,J^{-1}$) M_5
Li_2O–$2SiO_2$–$0.2ZnO$	−3.0	0.46	1.27	0.20	5.6
Li_2O–$1.8SiO_2$–$0.2B_2O_3$	−11.0	1.57	4.60	0.66	7.8
Li_2O–$1.7SiO_2$–$0.3B_2O_3$	−10.0	1.33	4.18	0.56	6.8
Li_2O–$1.6SiO_2$–$0.4B_2O_3$	−10.0	1.43	4.18	0.60	7.1
BaO–GeO_2–TiO_2	−2.0	0.13	0.98	0.07	8.0
$2BaO$–$3SiO_2$–TiO_2	+8.0	0.88	3.92	0.44	41.3
$1.9BaO$–$0.1PbO$–$3SiO_2$–TiO_2	+8.0	0.80	3.92	0.39	39.2
$2BaO$–$0.15CaO$–$2.9SiO_2$–TiO_2	+6.0	0.60	2.94	0.29	29.0
$1.6BaO$–$0.4CaO$–$2.8SiO_2$–TiO_2	+5.5	0.52	2.70	0.26	26.4
$1.6BaO$–$0.4SrO$–$3SiO_2$–TiO_2–$0.2CaO$	+4.5	0.46	2.21	0.23	22.3
$2SrO$–$3SiO_2$–TiO_2	+8.0	0.70	3.92	0.34	36.6
$1.8SrO$–$0.2BaO$–$2.8SiO_2$–$0.1CaO$–TiO_2	+7.0	0.66	3.43	0.32	33.3
$Li_2B_4O_7$ (single crystal)	−30.0	3.0	9.4	0.94	2.10
$Ba_2TiGe_2O_8$ (single crystal)	−4.0	0.35	1.97	0.17	18.4
$Ba_2TiSi_2O_8$ (single crystal)	+10.0	0.91	4.90	0.45	46.7

temperature in these crystals. The positive value of p_3 at room temperature in $Ba_2TiSi_2O_8$ and $Sr_2TiSi_2O_8$ may be due to the dominance of secondary pyroelectric effect. The above argument also applies to polar glass-ceramics since the pyroelectric properties of glass-ceramics are due to the oriented crystallites of these phases.

The pyroelectric figures of merit of polar glass-ceramics and single crystals are given in Table 8.7. The figure of merit M_4 was not calculated because data on thermal diffusivity are not available. A comparison of the figures of merit of glass-ceramics with those of commonly used pyroelectric materials is given below.

The figure of merit M_1 is highest for lithium borosilicate glass-ceramics and is about one-third of $LiTaO_3$. For glass-ceramics in the fresnoite family M_1 is about one-quarter of $LiTaO_3$. The figure of merit M_2 of glass-ceramics is much lower than that of $LiTaO_3$ and other pyroelectric materials, because the dielectric constant does not enter in the calculation of M_2. Lithium borosilicate glass-ceramics have higher M_3 (almost half of $LiTaO_3$) than fresnoite glass-ceramics because of their low density and dielectric constants. In the calculation of M_5, both the dielectric constant and $\tan \delta$ are considered. $\tan \delta$ values are very high for lithium borosilicate glass-ceramics (~ 0.05) whereas for fresnoite glass-ceramics $\tan \delta$ values are very small (<0.001). For calculation of M_5, the highest value of $\tan \delta$ (0.001) was taken for all fresnoite glass-ceramics. M_5 of lithium borosilicate glass-ceramics is very low because of high dielectric losses and those of fresnoite glass-ceramics are about 25% of $LiTaO_3$. Actual values of M_5 of fresnoite glass-ceramics will be slightly higher than the values listed in Table 8.7, if smaller $\tan \delta$ values are taken for the calculation.

The pyroelectric figures of merit of most of the glass-ceramics are in the range of 40 to 60% of those of $LiTaO_3$. Further improvement in the figures of merit of glass-ceramics can be achieved either by fine tuning the composition or by improved preparation techniques. Studies in our laboratory have shown that an oriented region up to 500 μm in length near the surface can be achieved even by isothermal recrystallization of glasses. This suggests the possibility of preparing large area targets by heating large glass sheets in a furnace. This technique can be applied only to materials which form glass easily. This, however, limits the choice of materials to silicates, germanates and borates if conventional glass melting techniques are followed. If alternative techniques for glass formation (e.g., sputtered glassy films, roller quenching, sol–gel method) are used, it may be possible to synthesize additional polar glass-ceramics.

8.9 PIEZOELECTRIC PROPERTIES

An evaluation of piezoelectric and electromechanical properties of polar glass-ceramics is given in this section, along with the temperature variation of the electromechanical properties of piezoelectric coefficients. Hydrostatic piezoelectric coefficients of glass-ceramics in the fresnoite family are presented and an evaluation of glass-ceramics for hydrophone applications is made by comparing their properties with those of lead zirconate titanate (PZT) and polyvinylidene fluoride (PVF_2).

Piezoelectric constant d_{33} parallel to the crystallization direction was measured using a Berlincourt d_{33} meter.* Resonance behaviour in the thickness mode and planar mode were investigated using a spectrum analyser.† The resonance frequency constants, electromechanical coupling factors and mechanical quality factor Q were determined by measuring the resonance and antiresonance frequencies. From these values, some of the elastic and piezoelectric coefficients were calculated. The temperature coefficient of resonance (TCR) was evaluated by measuring the resonance frequency as a function of temperature.

* Model CPDT 3300, Channel Products, Chesterland, OH.
† Model 3585A, Hewlett-Packard, Inc., Loveland, CO.

Table 8.8 Electromechanical and piezoelectric properties of glass-ceramics

Composition	N_p (m Hz)	k_p	N_t (m Hz)
$2BaO–3SiO_2–TiO_2$	3500	0.14	2250
$1.9BaO–0.1PbO–3SiO_2–TiO_2$	3450	0.14	2300
$2BaO–0.15CaO–2.9SiO_2–TiO_2$	–	–	–
$1.6BaO–0.4CaO–2.8SiO_2–TiO_2$	–	–	2500
$1.6BaO–0.4SrO–3SiO_2–TiO_2–0.2CaO$	3450	0.10	2350
$BaO–GeO_2–TiO_2$	3300	0.06	2500
$2SrO–3SiO_2–TiO_2$	3300	0.10	2550
$1.8SrO–0.2BaO–2.8SiO_2–0.1CaO–TiO_2$	–	–	2550
$Li_2O–2SiO_2–0.2ZnO$	–	–	–
$Li_2O–1.8SiO_2–0.2B_2O_3$	4500	0.15	3750
$Li_2O–1.7SiO_2–0.3B_2O_3$	4500	0.14	3700
$Li_2O–1.6SiO_2–0.4B_2O_3$	4500	0.12	3700
PZT 501A	2000	0.60	1860
$LiNbO_3$	–	0.048*	–
Quartz	–	0.0014*	–

* = Surface coupling coefficient k_s^2.
N_p, N_t = frequency constants for radial and thickness mode of resonance.
k_p, k_t = electromechanical coupling coefficients for radial and thickness mode of resonance.
Q = Mechanical quality factor; TCR = temperature coefficient of resonance ($(1/f)\,\partial f/\partial T$).

8.9.1 Electromechanical properties

The electromechanical properties of glass-ceramics resonating in the radial and thickness modes, the piezoelectric coefficients d_{33} (measured by d_{33}-meter) and d_{31} (measured by resonance technique) are listed in Table 8.8.

The coupling coefficients k_p and k_t of glass-ceramics range between 10 and 15% and 20 to 30%, respectively. In the temperature range -20 to 60°C, the temperature coefficient of resonance lie between 50 and 100 ppm °C^{-1}.

In most of the piezoelectric devices, high coupling coefficients and low TCR are desirable. For glass-ceramics, the coupling coefficients are small compared to commonly used piezoelectric materials (such as PZT or $LiNbO_3$) but the TCR of some compositions are low (50 to 60 ppm °C^{-1}). The strong dependence of TCR on composition for glass-ceramics suggests the possibility of obtaining temperature-compensated resonators by compositional modification.

8.9.2 Hydrostatic measurements

The hydrostatic voltage coefficient g_h was measured by the method described by Safari (Safari, 1984). The value of g_h for lithium borosilicate

Table 8.8 *Continued*

k_t	k_{31}	d_{33} (pC N^{-1})	d_{31} (pC N^{-1})	*Q* (*Radial*)	*TCR* (ppm °C^{-1})
0.25	–	7	–	800	100
0.20	–	7	–	800	120
–	–	6	–	–	–
0.15	0.04	6	1.8	–	100
0.22	–	6	–	500	–
0.08	0.04	6	1.2	2000	65
0.08	0.03	14	1.6	700	50
0.20	–	10	–	–	–
–	–	4	–	–	–
0.25	0.09	6	−2.4	800–1000	80–100
0.2	–	5	–	500–1000	80–100
low	–	5	–	500–1000	80–100
0.50	0.35	400	−175	80	250
–	0.02	6.0	0.9	–	95
–	–	$d_{11} = 2.31$ $d_{14} = 0.7$	–	–	0

Table 8.9 Hydrostatic measurements

Composition	K_{33}	d_{33} ($pC N^{-1}$)	g_{33} ($10^{-3} V m N^{-1}$)	g_h ($10^{-3} V m N^{-1}$)	d_h ($pC N^{-1}$)	$d_h g_h$ ($10^{-15} m^2 N^{-1}$)
$2BaO–3SiO_2–TiO_2$	9	7	88	110	8.8	970
$1.9BaO–0.1PbO–3SiO_2–TiO_2$	10	7	80	110	9.7	1070
$2BaO–0.15CaO–2.9SiO_2–TiO_2$	10	6	68	75	6.6	500
$1.6BaO–0.4CaO–2.8SiO_2–TiO_2$	10.5	6	65	85	7.9	670
$1.6BaO–0.4SrO–3SiO_2–TiO_2–0.2CaO$	9.8	6	70	100	8.7	870
$2SrO–3SiO_2–TiO_2$	11.5	14	138	85	8.7	740
$1.8SrO–0.2BaO–2.8SiO_2–0.1CaO–TiO_2$	10.6	10	107	100	9.4	940
$BaO–GeO_2–TiO_2$	15	6	45	70	9.3	650

Table 8.10 Comparison of hydrostatic properties

Property	*Glass-ceramics*	*PVF*$_2$	*PZT*
K	10	13	1800
$d_{33}(10^{-12}\,\mathrm{C\,N^{-1}})$	8–10	30	450
$d_{31}(10^{-12}\,\mathrm{C\,N^{-1}})$	+1.5	−18	−205
$d_h(10^{-12}\,\mathrm{C\,N^{-1}})$	8–10	10	40
$g_{33}(10^{-3}\,\mathrm{V\,m\,N^{-1}})$	100	250	28
$g_h(10^{-3}\,\mathrm{V\,m\,N^{-1}})$	100	100	2.5
$d_h g_h(10^{-15}\,\mathrm{m^2\,N^{-1}})$	1000	1000	100
Z^* (10^6 rayls)	15–25	2–3	30

* Z = Acoustic impedance (units of rayl: $\mathrm{kg\,m^{-2}\,s^{-1}}$).

glass-ceramics was almost zero whereas that of glass-ceramics in the fresnoite system was comparable to the g_h values of PVF_2. Values of dielectric constant K_{33}, d_{33} and g_h of glass-ceramics in the fresnoite family are listed in Table 8.9 along with the calculated values of g_{33}, d_h and $d_h g_h$. A comparison of the dielectric and hydrostatic properties of glass-ceramics is given in Table 8.10.

The values of g_h and $d_h g_h$ of glass-ceramics are comparable to PVF_2 and much higher than PZT. Although the values of d_{33} and d_h of glass-ceramics are comparatively low, the magnitudes of g_{33} and g_h of glass-ceramics are high because of their low dielectric constant. The unusually high values of g_h in fresnoite glass-ceramics can be attributed to the positive d_{31} coefficient in these materials. A composite model has been proposed (Halliyal *et al.*, 1984) to explain the positive sign of d_{31} in fresnoite based on the crystal structure of fresnoite and internal Poisson's ratio stress. A discussion of the advantages of these glass-ceramics in hydrophone applications will be given in the next section.

8.9.3 Application in hydrophones

A hydrophone is a passive device used as a hydrostatic pressure sensor (Safari, 1984). For hydrophone applications, the commonly used figures of merit are the hydrostatic voltage coefficient g_h and the product $d_h g_h$. Among the desirable properties of a transducer material used in a hydrophone are:

1. High d_h and g_h
2. A density suited for acoustic matching with the pressure transmitting medium, usually water
3. High compliance and flexibility such that the transducer can withstand mechanical shock, and can conform to any surface
4. No variation of g_h with static pressure

Acoustic impedance Z can be calculated from the relation $Z = \varrho c$ where ϱ is the density of the material and c is the velocity of sound in the medium. By measuring the thickness mode frequency constant N_t of the material, the velocity c can be calculated from the relation $c = 2N_t$. Accoustic impedances of glass-ceramics, PZT and PVF_2 are listed in Table 8.10.

PZT ceramics are used extensively as piezoelectric transducers despite having several disadvantages. The values of g_h and $d_h g_h$ of PZT are low because of its high dielectric constant (~ 1800). In addition the high density of PZT (~ 7900 kg m^{-3}) makes it difficult to obtain a good impedance match with water. PZT is also a brittle ceramic, whereas for hydrophone applications, a more compliant material with better shock resistance would be desirable (Safari, 1984). PVF_2 offers several advantages over PZT ceramic for hydrophone applications. It has low density (1760 kg m^{-3}) and is a flexible material, and although it has low d_{33} and d_h, the dielectric constant of PVF_2 is low enough that large values of piezoelectric voltage coefficients g_{33} and g_h are possible. Overall, this combination of properties is very attractive and PVF_2 transducers are under intensive development (Sessler, 1981). However, a major problem in the use of PVF_2 is the difficulty in poling PVF_2 sheets. A very high voltage is necessary to pole PVF_2 (about 10 to 100 MV m^{-1}) and this places a limitation on the thickness of PVF_2 transducers.

To overcome these problems, a number of composites of PZT and polymer have been studied in recent years. A detailed description of different kinds of composites and the principles involved can be found elsewhere (Newnham *et al.*, 1980; Safari *et al.*, 1986). In a composite the polymer phase lowers density and dielectric constant and increases elastic compliance. Very high values of g_h and $d_h g_h$ have been achieved by fabricating composites with different connectivity patterns.

From Table 8.10 it is clear that the piezoelectric coefficients d_{33} and d_h of polar glass-ceramics are comparable to PVF_2, but much lower than that of PZT. However, because of the low dielectric constant of glass-ceramics, the g_{33} and g_h values of glass-ceramics are considerably higher than PZT. Hence, polar glass-ceramics will be useful in passive devices such as hydrostatic pressure sensors where g_h is more important than d_h. The variation of g_h with pressure was measured for glass-ceramics up to 8 MPa. There was no significant variation of g_h with pressure (Ting *et al.*, 1984, 1986).

In practice polar glass-ceramics offer several possible advantages over PVF_2 and other ferroelectric materials for application in piezoelectric devices. Since the polar glass-ceramics are non-ferroelectric, no poling step is involved, which is a major problem with PVF_2. Thus there will be no problem of depoling or ageing which are encountered in many ferroelectric materials. Hence, polar glass-ceramics can be used in devices operating at high temperatures. Large area devices can be prepared by routine glass preparation techniques, which reduces the cost of the device significantly.

Since acoustic impedances of glass-ceramics are in the range 18–20 × 10^6 rayls, good acoustic matching can be obtained with metals like aluminium. Non-destructive testing of aircraft metals over a wide temperature range is a possibility. These glass-ceramics also look attractive for use in devices in which glass fibres are used because of the good impedance matching.

8.10 SURFACE ACOUSTIC WAVE (SAW) PROPERTIES

In recent years piezoelectric materials have become of increasing importance for SAW devices (Whatmore, 1980; Mathews, 1978). High electromechanical coupling coefficient (k_s^2) and low temperature coefficient of surface wave velocity are desirable material properties for SAW devices. Only a few piezoelectric materials are useful for SAW applications because an undesirable trade off between temperature coefficient of delay (TCD) and k_s^2 is observed in most of the materials presently used for SAW devices (Whatmore, 1980). For example, ST-cut quartz exhibits zero TCD but shows only weak electromechanical coupling (Table 8.11). On the other hand, Y-cut Z-propagating $LiNbO_3$ shows strong electromechanical coupling, but exhibits a high TCD (94 ppm °C^{-1}). Other piezoelectric materials which exhibit good piezoelectric properties for SAW devices present severe crystal growth problems (Whatmore, 1980).

Attempts have been made to use piezoelectric ceramic materials for SAW devices (Ito *et al.*, 1981a, 1981c; Jyomura *et al.*, 1981; Tamura and Yonezawa, 1974). Unfortunately, in ceramic materials the SAW propagation losses are very high, presenting a major problem in device fabrication (Tamura and Yonezawa, 1974). Lead titanate ceramics modified with additives of Nd_2O_3, MnO_2 and In_2O_3 have been developed which have zero TCD, low propagation losses and an electromechanical coupling of

Table 8.11 Saw properties

Property	*Material*				
	Fresnoite glass ceramic	*ZX-fresnoite single crystal*	*YZ-LiNbO$_3$*	*ST-quartz*	*Lead titanate*
Coupling coefficient k_s^2	0.008 to 0.012	0.016 ± 0.003	0.0482	0.00116	0.022
Velocity (m s^{-1})	2600	2678 ± 3	3488	3158	2610
TCD (ppm °C^{-1})	10 to 60	51	94	0	<1

about 2% (Ito *et al.*, 1981a). However, variation in the properties of ceramics makes it difficult to use lead titanate ceramics for SAW devices.

Since surface wave propagation in SAW devices is generally limited to within a few wavelengths of the surface it is possible that polar glass-ceramics with oriented crystallites near the surface might exhibit useful SAW properties. To test this possibility, polar glass-ceramics of modified fresnoite compositions were prepared and their SAW properties were evaluated on polished surfaces perpendicular to the oriented crystallites.

For SAW measurements a number of uniform overlapping interdigital transducer pairs, in either single or double electrode format, were fabricated on polished surfaces of glass-ceramics. The transducer structure had the following characteristics: periods = 19.5, aperture = 1 mm and wavelength = 30 μm. With this arrangement several odd harmonics were detectable. At higher frequencies SAW response was masked by excessive electromagnetic feed through.

Figure 8.11 shows the SAW response for fresnoite glass-ceramics having single finger electrodes of 30 μm wavelength. Essential features of this response are the clean $(\sin x/x)^2$ response with very little interference from bulk wave coupling. It was shown that bulk wave coupling was almost

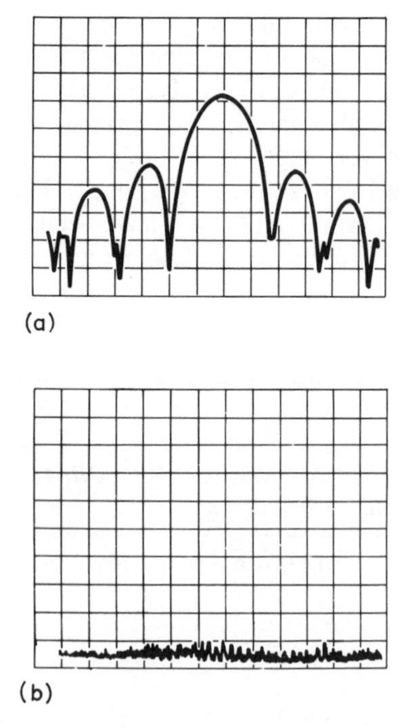

Fig. 8.11 Relative amplitude against frequency response of (a) surface wave and (b) bulk wave. Vertical scale 10 dB div^{-1}; horizontal = 72–203 MHz.

entirely absent within the SAW passband. For the response shown in Fig. 8.11, the insertion loss was 30 dB and the measured centre frequency was 87.96 MHz.

For modified fresnoite glass-ceramics the SAW coupling was about 0.8 to 1.2% and the temperature coefficient of delay was in the range 10 to 60 ppm $°C^{-1}$. However, propagation losses were high. Work is now in progress to prepare glass-ceramics with optimized composition and microstructure to reduce the SAW propagation loss (Ylo, 1986).

Based on preliminary results, polar glass-ceramics look promising for SAW devices. A particularly attractive feature of the glass-ceramic processing route is the ease of fabrication of large area piezoelectric substrates. Although large area lithium niobate single crystals are available, they are expensive. In principle, the glass-ceramic process could be used to advantage for planar processing on a much larger scale than is possible for lithium niobate SAW fabrication. This might also help in fabricating SAW devices with more reproducible properties than ceramics.

A further advantage of the glass-ceramic processing is that piezoelectric properties of glass-ceramics can be tailored by modification of the parent glass-composition. The possibility of further reducing temperature coefficient of delay and SAW attenuation in fresnoite glass-ceramics by strontium doping is being explored (Ylo, 1986).

8.11 CONNECTIVITY MODEL FOR PIEZOELECTRIC AND PYROELECTRIC PROPERTIES OF POLAR GLASS-CERAMICS AND TAILORING THE PROPERTIES

Polar glass-ceramics consist of needle-like crystallites oriented along the *c*-axis in a glassy matrix phase. The excellent piezoelectric and pyroelectric properties of glass-ceramics (approaching the single crystal properties) indicate that polar glass-ceramics possess a high degree of both crystallographic and polar orientation. In this section, the growth behaviour of crystallites in polar glass-ceramics of different composition will be discussed. An attempt will be made to analyse the properties of these glass-ceramics by considering it as a composite of a crystalline phase and a glass phase. Several interesting combinations of piezoelectric and pyroelectric properties can be achieved by adjusting the composition and degree of crystallinity.

8.11.1 Growth behaviour of crystallites in glass-ceramics

All the glass-ceramics studied in the present work were non-ferroelectric. Hence, unlike ferroelectric materials, the electrical twin configuration of the non-ferroelectric crystalline phases in these glass-ceramics cannot be switched by an externally applied electric field. Thus both crystallographic and polar orientation must be achieved during the mucleation stage of the recrystallization process, and must remain unaltered during subsequent growth of crystallites. Since the recrystallization process takes place through surface nucleation and crystallization, the initial stage of recrystallization at the surface and its influence on the crystallographic and polar orientation of crystallites during subsequent growth are extremely important. It is clear that the initial crystallographic and polar orientation are retained as the crystallization progresses from the surface into the bulk.

During the present work an interesting effect occurred in the polar orientation of different crystalline phases in glass-ceramics. It was observed that the sign of d_{33} and the polar orientation of crystalline phases were dependent on the composition of the parent glass matrix (Halliyal *et al.*, 1983). To clarify the discussion, the following conventions will be followed for the signs of piezoelectric coefficient d_{33} and pyroelectric coefficient p_3. The growth behaviour of the crystalline phase is defined on the basis of the sign of d_{33} measured on the initial crystallizing surface (the sign of d_{33} defined in this way may differ from the actual sign of d_{33} of the particular crystalline phase defined according to the IRE 1949 standard on piezoelectricity, 1949). According to this convention, the same sign of d_{33} for two phases means that the orientation of dipoles is similar in both phases. In other words, the two phases exhibit identical growth habit. The accepted definition of the sign of pyroelectric coefficient is as follows. The sign of pyroelectric coefficient is positive if on heating a pyroelectric material,

Table 8.12 Sign and magnitude of d_{33} and p_3

Composition	d_{33} (10^{-12} pC N^{-1})	p_3 (μC m^{-2} K^{-1})
$2BaO–3SiO_2–TiO_2$	+7	+8
$1.9BaO–0.1PbO–3SiO_2–TiO_2$	+7	+8
$2BaO–0.15CaO–2.9SiO_2–TiO_2$	+6	+6
$1.6BaO–0.4CaO–2.8SiO_2–TiO_2$	+6	+5.5
$1.6BaO–0.4SrO–3SiO_2–TiO_2–0.2CaO$	+6	+4.5
$BaO–GeO_2–TiO_2$	+6	−2
$2SrO–3SiO_2–TiO_2$	+14	+8
$1.8SrO–0.2BaO–2.8SiO_2–0.1CaO–TiO_2$	+10	+7
$Li_2O–2SiO_2$	−1	–
$Li_2O–2SiO_2–0.2ZnO$	−4	−3.0
$Li_2O–1.95SiO_2–0.05Fe_2O_3$	−3	–
$Li_2O–3B_2O_3$	+3	–
$Li_2O–1.8SiO_2–0.2B_2O_3$	+6	−11.0
$Li_2O–1.7SiO_2–0.3B_2O_3$	+5	−10.0
$Li_2O–1.6SiO_2–0.4B_2O_3$	+5	−10.0
$Li_2O–1.8SiO_2–0.1(ZnO, B_2O_3)$	−2	–

positive charges develop on the surface towards positive d_{33} of the crystallites. This convention for the signs of piezoelectric and pyroelectric coefficients will be followed throughout the discussion.

The sign and magnitude of d_{33} measured on the surface from which crystallization commenced (which is towards the high temperature end of the temperature gradient axis) are given in Table 8.12. The sign and magnitude of pyroelectric coefficient p_3 are also listed in the table. Based on the above sign convention for d_{33}, the growth behaviour of all the systems listed in the table can be divided into two groups as shown in Fig. 8.12. In the case of glass-ceramics belonging to the systems $Ba_2TiSi_2O_8$, $Ba_2TiGe_2O_8$, $Sr_2TiSi_2O_8$, Li_2O–B_2O_3 and Li_2O–SiO_2–B_2O_3 the sign of d_{33} is positive, indicating that the positive end of the dipoles point towards the high temperature end of the sample. For Li_2O–$2SiO_2$, Li_2O–SiO_2–ZnO and Li_2O–SiO_2–Fe_2O_3 glass-ceramics, the sign of d_{33} is negative, indicating that the negative end of the dipoles point towards the high temperature end. There are several possible reasons for this characteristic growth behaviour in the above systems.

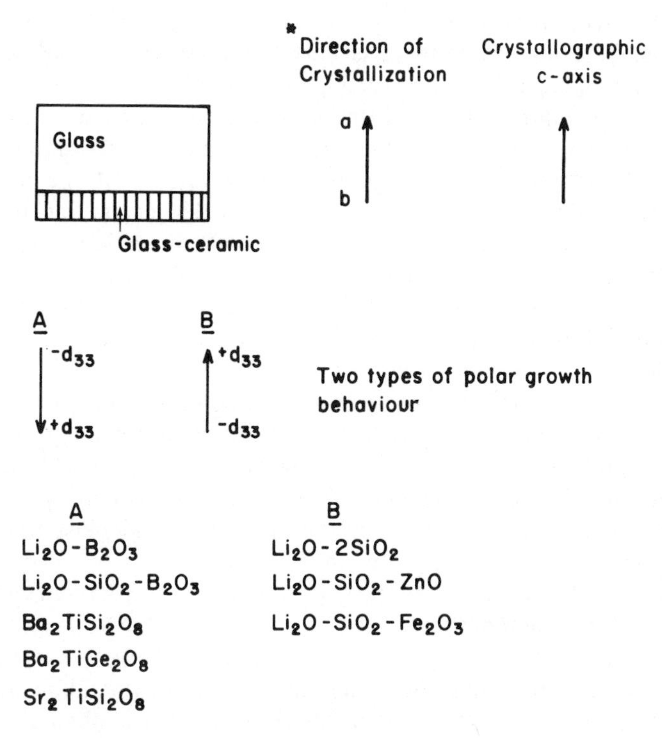

Fig. 8.12 Polar orientation – two types of growth behaviour.

For glass-ceramics in the fresnoite system ($Ba_2TiSi_2O_8$, $Ba_2TiGe_2O_8$ and $Sr_2TiSi_2O_8$) there is only one crystalline phase present in the glass-ceramic, and d_{33} is positive for all the phases. Hence this growth behaviour or polar orientation (positive d_{33}) seems to be characteristic of fresnoite glass-ceramics. On the other hand, for glass-ceramics in the system Li_2O–$2SiO_2$, Li_2O–SiO_2–ZnO, Li_2O–SiO_2–Fe_2O_3 and Li_2O–SiO_2–B_2O_3 the major phase is $Li_2Si_2O_5$. Minor amounts of Li_2ZnSiO_4 and quartz were present in Li_2O–SiO_2–ZnO system and $Li_2B_4O_7$ and Li_2SiO_3 in Li_2O–SiO_2–B_2O_3 system. The presence of ZnO, B_2O_3 or Fe_2O_3 appears to influence the growth habit of $Li_2Si_2O_5$ crystallites, possibly by controlling the surface nucleation sites. The presence of Li_2SiO_3 or $Li_2B_4O_7$ as separate phases, which are also polar phases, may be influencing the growth habit of $Li_2Si_2O_5$. In any case the present study indicates clearly that the polar orientation of crystallites and the resulting piezoelectric and pyroelectric properties depend on the composition of the initial glass (Halliyal *et al.*, 1983).

8.11.2 Diphasic glass-ceramic composites

A polar glass-ceramic is essentially a composite of a glassy phase and one or more crystalline phases. If there is only one crystalline phase, it can be considered a diphasic system consisting of a crystalline phase embedded in an amorphous matrix. X-ray diffraction and microstructure studies of polar glass-ceramics indicate that the oriented region of crystallites extends deep into the sample and then tapers off. However, not all the crystallites extend throughout the thickness of the samples. The crystalline and glassy phases are partly in series and partly in parallel connection. As the grain orientation is improved, the length of the crystallites increases and the composite approaches pure parallel connectivity. In the present section expressions will be derived for piezoelectric and pyroelectric properties of a diphasic glass-ceramic composite.

For a composite consisting of two phases, one-dimensional solutions for dielectric, piezoelectric and pyroelectric properties have been presented for both series and parallel connectivity (Newnham *et al.*, 1978; Skinner *et al.*, 1978). If the dielectric constant, piezoelectric coefficient, pyroelectric coefficient and volume fraction of the crystalline and glass phases are designated as ${}^1K_3$, ${}^1d_{33}$, ${}^1p_3$, ${}^1v$ and ${}^2K_3$, ${}^2d_{33}$, ${}^2p_3$, ${}^2v$, respectively, expressions for K_3, d_{33} and p_3 of the glass-ceramic can be derived by making certain simplifying assumptions.

For the glass phase, ${}^2d_{33}$ and ${}^2p_3$ are equal to zero. In many non-ferroelectric glass-ceramics, the dielectric constants of crystalline and glass phases are roughly the same. With this assumption it is obvious that the dielectric constant of the composite is the same for both series and parallel connections.

Further, if we assume that the elastic compliance of crystalline and glassy phases are the same (${}^{1}s_{33} = {}^{2}s_{33}$), the expressions for d_{33} and p_3 reduce to the following expressions for both series and parallel connectivity (Halliyal *et al.*, 1983)

$$d_{33} = {}^{1}d_{33}\,{}^{1}v \tag{8.1}$$

and

$$p_3 = {}^{1}p_3\,{}^{1}v \tag{8.2}$$

The magnitudes of ${}^{1}d_{33}$ and ${}^{1}p_3$ of the crystalline phase itself depend upon the degree of both crystallographic and polar orientation of the crystallites, which is not accounted for in (8.1) and (8.2). We conclude that in the ideal case of complete crystallographic and polar orientation of the crystallites, the magnitudes of d_{33} and p_3 of the composite are directly proportional to the percentage of crystalline phase, irrespective of whether the crystallites are connected in series or parallel. In addition, for this ideal case, if the glass-ceramics contain a very high percentage of crystalline phase, the values of d_{33} and p_3 approach the single crystal values of the crystalline phase. The pyroelectric coefficient p_3 of fresnoite ($Ba_2TiSi_2O_8$) and $Ba_2TiGe_2O_8$ glass-ceramics are 75 to 80% of the single crystal values. The lower value of p_3 of polar glass-ceramics can be accounted for by the lack of perfect orientation of the crystallites. The piezoelectric d_{33} coefficient of the glass-ceramics is also about 75 to 80% of the single crystal values.

8.11.3 Multicomponent glass-ceramic composites

Most of the compositions listed in Table 8.12 result in polar glass-ceramics with several crystalline phases after recrystallization. In such cases, the polar vectors in different crystalline phases may be oriented parallel (or antiparallel) to the temperature gradient. If we assume that the crystallites of two or more phases preserve their characteristic growth habit in relation to the direction of temperature gradient, then the piezoelectric and pyroelectric properties of polar glass-ceramic containing several phases can be predicted by the analysis discussed below. For simplicity, an analysis for glass-ceramic composites containing only two crystalline phases will be given here, recognizing, however, that the analysis can be extended to more than two phases. For the signs of d_{33} and p_3, the same convention will be followed as defined in the previous section.

Consider a polar glass-ceramic containing two crystalline phases. Let the dielectric constant, piezoelectric coefficient, and the volume fraction of the two phases be ${}^{1}K_3$, ${}^{1}d_{33}$, ${}^{1}p_3$, ${}^{1}v$ and ${}^{2}K_3$, ${}^{2}d_{33}$, ${}^{2}p_3$, ${}^{2}v$, respectively. Let us assume that both the phases have similar elastic compliances (${}^{1}s_{33} = {}^{2}s_{33}$) and thermal expansion coefficients, thereby eliminating any secondary contributions to the pyroelectric effect. To simplify the analysis further, let us assume that the properties of glass-ceramic are not influenced by the glassy phase (${}^3v = 0$). Irrespective of whether the crystalline phases are connected in series or parallel, we get the following equations for the properties of the composite (Halliyal *et al.*, 1983).

$$K_3 = {}^1v\,{}^1K_3 + {}^2v\,{}^2K_3$$
$$d_{33} = {}^1v\,{}^1d_{33} \pm {}^2v\,{}^2d_{33}$$
$$p_3 = {}^1v\,{}^1p_3 \pm {}^2v\,{}^2p_3$$

The resultant values of d_{33} and p_3 of the composite depend upon the sign and magnitude of the properties of individual phases, and also upon the volume fraction of the individual phases. Since ${}^1d_{33}$ and ${}^2d_{33}$ can be positive or negative at the growth end and ${}^1p_3$ and ${}^2p_3$ can be positive or negative, there are 16 possible cases of d_{33} and p_3, eight of which are listed in Table 8.13. If we consider the resulting piezoelectric and pyroelectric properties of the glass-ceramic, these 16 cases can be grouped into four classes.

The crystalline phases have the same growth behaviour in the case of glass-ceramics belonging to class (a) and (b) and opposite growth behaviour in cases (c) and (d). Depending on the signs of pyroelectric coefficients ${}^1p_3$ and ${}^2p_3$ of the two phases, we obtain glass-ceramic composites with completely different piezoelectric and pyroelectric properties. Glass-ceramics of class (a) are fully piezoelectric as well as pyroelectric and in the ideal case their properties can be expected to approach those of single domain single crystal materials. Glass-ceramics of class (b) are fully piezoelectric but only partially pyroelectric (or non-pyroelectric) because of the opposite signs of ${}^1p_3$ and ${}^2p_3$. Glass-ceramics of class (c) are both non-piezoelectric and non-pyroelectric and hence are not of much interest for devices. Class (d) glass-ceramics are fully pyroelectric but non-piezoelectric. Glass-ceramics of classes (b) and (d) are interesting for piezoelectric and pyroelectric device applications where the interference of the two properties may be a problem, as will be discussed below.

8.11.4 Examples of the four classes

The growth habit and the signs of pyroelectric coefficients of the crystalline phases can be exploited to design compositions which give glass-ceramics with properties of any one of the four classes just described.

Glass-ceramics in the system Li_2O–SiO_2–ZnO, Li_2O–SiO_2–Fe_2O_3, and Li_2O–SiO_2–B_2O_3 have negative pyroelectric coefficient, but exhibit opposite growth behaviour (case c). Compositions in the quaternary system Li_2O–SiO_2–ZnO–B_2O_3 illustrate the reduced pyroelectric and piezoelectric effect characteristic of glass-ceramics of class (c). $Ba_2TiSi_2O_8$ and $Ba_2TiGe_2O_8$ phases crystallized from glasses of composition $2BaO$–$3SiO_2$–TiO_2 and BaO–GeO_2–TiO_2 respectively showed similar growth habit, but opposite signs for pyroelectric coefficients. As glass-ceramic of composition

Table 8.13 Eight combinations of piezoelectric and pyroelectric properties

Case	*Growth habit* sign of d_{33}*		*Sign of d_{33}*	*Sign of p_3*		*d_{33} of composite*	*p_3 of composite*	*Resultant properties*
	Phase I	*Phase II*		*Phase I*	*Phase II*			
(a)	+	+	+	+	+	${}^1v^1d_{33} + {}^2v^2d_{33}$	$+({}^1v^1p_3 + {}^2v^2p_3)$	piezo and pyro
	+	+		−	−	${}^1v^1d_{33} + {}^2v^2d_{33}$	$-({}^1v^1p_3 + {}^2v^2p_3)$	
(b)	+	+	+	+	−	${}^1v^1d_{33} + {}^2v^2d_{33}$	$+{}^1v^1p_3 - {}^2v^2p_3$	piezo but non-pyro
	+	+		−	+	${}^1v^1d_{33} + {}^2v^2d_{33}$	$-{}^1v^1p_3 + {}^2v^2p_3$	
(c)	+	−	+	+	+	${}^1v^1d_{33} - {}^2v^2d_{33}$	$+{}^1v^1p_3 - {}^2v^2p_3$	non-piezo and non-pyro
	+	−		−	−	${}^1v^1d_{33} - {}^2v^2d_{33}$	$-{}^1v^1p_3 + {}^2v^2p_3$	
(d)	+	−	+	+	−	${}^1v^1d_{33} - {}^2v^2d_{33}$	$+({}^1v^1p_3 + {}^2v^2p_3)$	pyro but non-piezo
	+	−		−	+	${}^1v^1d_{33} - {}^2v^2d_{33}$	$-({}^1v^1p_3 + {}^2v^2p_3)$	

* Sign of d_{33} measured on the initial crystallizing surface. Eight more combinations can be obtained with a negative d_{33} sign for phase I.

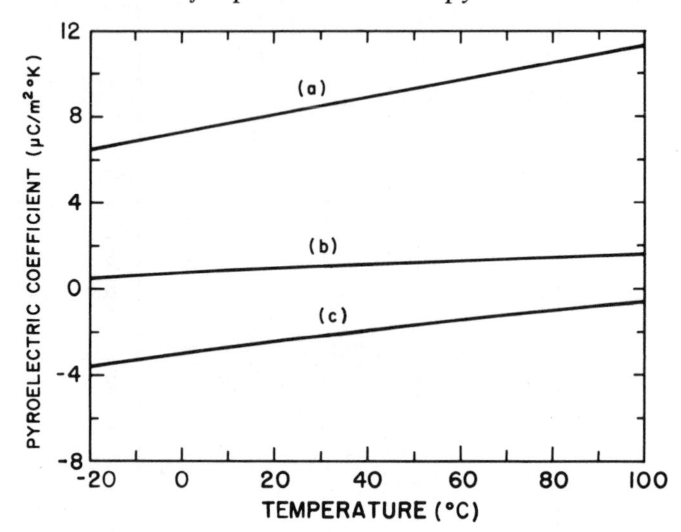

Fig. 8.13 Pyroelectric coefficients of glass-ceramics in the fresnoite system: (a) $2BaO–3SiO_2–TiO_2$; (b) $2BaO–2SiO_2–GeO_2–TiO_2$; (c) $BaO–GeO_2–TiO_2$.

$2BaO–2SiO_2–GeO_2–TiO_2$ which gave a solid solution phase after recrystallization had almost zero pyroelectric coefficient, but piezoelectric properties remained the same (Fig. 8.13). In this case, there is no appreciable change in the piezoelectric properties, whereas the pyroelectric property reduced substantially over a wide temperature range.

8.11.5 Optimizing the piezoelectric and pyroelectric properties

From the examples just presented, it is clear that glass-ceramics with tailored piezoelectric or pyroelectric properties can be prepared by crystallizing multicomponent glasses of suitable composition. The growth habit of the crystalline phases provides a means to tailor the piezoelectric and pyroelectric properties of the final glass-ceramic composite. Two of the cases look especially interesting for device applications; piezoelectric glass-ceramics which are non-pyroelectric (case b) and pyroelectric glass-ceramics which are non-piezoelectric (case d).

In piezoelectric devices used in ambient conditions, pyroelectric noise is usually undesirable. In hydrophone elements fabricated from PZT–polymer composites, for instance, pyroelectric noise can be reduced only within a relatively narrow working temperature range (Lynn, 1982).

In a similar way, piezoelectric noise interferes with the high frequency infrared pyroelectric signals. In some cases alternate methods are used to eliminate the interfering piezoelectric voltages developed in such devices. In the case of $LiTaO_3$ detectors, piezoelectric oscillations are damped by embedding the pyroelectric element in epoxy (Glass and Abrams, 1970); and in PLZT detectors, piezoelectric noise is eliminated by modifying the electronics (Simhony and Bass, 1979). Piezoelectric oscillations superimposed on the pyroelectric response have been shown to have an incubation time of 40–60 ns, and occur after the onset of pyroelectric signals (<40 ns). Thus the piezoelectric oscillations occur in the tail of the pyroelectric response, and can be eliminated with suitable electronics.

Piezoelectric glass-ceramics which are non-pyroelectric should satisfy the relation $\pm\, {}^1v^1p_3 + {}^2v^2p_3 = 0$ (case b). Referring to Table 8.13 candidate compositions can be found in the solid solution of $Ba_2TiSi_2O_8$ and $Ba_2TiGe_2O_8$. Piezoelectric glass-ceramics which are not pyroelectric would be useful as pressure sensors that are insensitive to temperature changes. (Quartz is piezoelectric and non-pyroelectric, but cannot be used to measure pressure changes because $d_{11} + d_{22} + d_{33} = 0$. This is generally true for all piezoelectric crystals belonging to non-pyroelectric point groups.)

The second case of interest is pyroelectric glass-ceramics which are not piezoelectric (case d). Here the requirement is that ${}^1v^1d_{33} - {}^2v^2d_{33} = 0$ should be satisfied. Candidate compositions may exist in the Li_2O–B_2O_3–SiO_2–ZnO quaternary system containing different amounts of ZnO. Such pyroelectric materials will not be sensitive to mechanical oscillations.

8.12 SUMMARY

The preparation and properties of polar glass-ceramics have been described in this chapter. Polar glass-ceramics consist of well-oriented crystallites of a polar but non-ferroelectric crystalline phase.

Glasses were prepared from Li_2O–SiO_2–B_2O_3, BaO–GeO_2–TiO_2, BaO–SiO_2–TiO_2 and SrO–SiO_2–TiO_2 systems and were recrystallized to obtain polar glass-ceramics with crystallites of $Li_2Si_2O_5$, Li_2SiO_3, $Li_2B_4O_7$, $Ba_2TiGe_2O_8$, $Ba_2TiSi_2O_8$ or $Sr_2TiSi_2O_8$, oriented along their polar *c*-axes. The oriented region of the crystallites was 200–500 μm from the surface of glass-ceramics. Dielectric, piezoelectric and pyroelectric properties of glass-ceramics are in close agreement with the respective single crystal properties. Glass-ceramics in the fresnoite family are excellent candidate materials for hydrophones. Magnitudes of hydrostatic piezoelectric coefficient $d_h(\sim 100 \times 10^{-3}\,\mathrm{V\,m\,N^{-1}})$ and dielectric constant (~10) of fresnoite glass-ceramics are comparable to the corresponding values of polyvinylidene fluoride. The electromechanical coupling coefficients k_p and k_t were in the range 15 to 20%. Several glass-ceramics showed pyroelectric figures of merit up to 60% of those obtained on $LiTaO_3$ single crystal. Surface acoustic wave properties of glass-ceramics in the fresnoite family are promising.

The polar growth behaviour of crystallites from glass matrix depends on the original composition of glass. This property can be exploited to tailor the piezoelectric and pyroelectric properties of polar glass-ceramics. A connectivity model, based on the principles of series and parallel connectivity models has been described to predict the piezoelectric and pyroelectric properties of polar glass-ceramics.

ACKNOWLEDGEMENTS

It is a pleasure to thank our colleagues at the Materials Research Laboratory for their help in this work. This work was supported by the Army Research Office through contract DAAG29-83-K-0062 and the National Science Foundation through contract DMR-8010811.

REFERENCES

Bhalla, A. S., Cross, L. E. and Whatmore, R. W. (1985) *J. Phys. Soc. Jpn*, **24**, 727.
Blasse, G. (1968) *J. Inorg. Nucl. Chem.*, **30**, 2283.
Borrelli, N. F. (1967) *J. Appl. Phys.* **38**, 4243.
Borrelli, N. F. and Layton, M. M. (1969), *IEEE Trans. Electron Devices*, **ED-16**, 19.
Borrelli, N. F. and Layton, M. M. (1971) *J. Non-Cryst. Solids*, **6**, 37.
Brun, M. K., Bhalla, A. S., Spear, K. E. *et al.* (1979) *J. Cryst. Growth*, **47**, 335.
Byer, R. L. and Roundy, C. B. (1972) *Ferroelectrics*, **3**, 333.
Gabelica-Robert, M. and Tarte, P., (1981) *Phys. Chem. Minerals*, **26**, 26.
Gardopee, G., Newnham, R. E. and Bhalla, A. S. (1981) *Ferroelectrics*, **33**, 155.
Gardopee, G., Newnham, R. E., Halliyal, A. and Bhalla, A. S. (1980) *Appl. Phys. Lett.*, **36**, 817.
Glass, A. M., and Abrams, R. L. (1970) *J. Appl. Phys.*, **41**, 4455.
Glass, A. M., Lines, M. E., Nassau, K. and Shiver, J. W. (1977a) *Appl. Phys. Lett.*, **31**, 249.
Glass, A. M., Nassau, K. and Shiever, J. W. (1977b) *J. Appl. Phys.*, **48**, 5213.
Halliyal, A. (1984) PhD thesis, The Pennsylvania State University.
Halliyal, A., Bhalla, A. S. and Cross, L. E. (1985a) *Ferroelectrics*, **62**, 3.
Halliyal, A., Bhalla, A. S., Cross, L. E. and Newnham, R. E. (1985b) *J. Mater. Sci.*, **20**, 3745.
Halliyal, A., Bhalla, A., Markgraf, S. A. *et al.* (1985c) *Ferroelectrics*, **62**, 27.
Halliyal, A., Bhalla, A. S. and Newnham, R. E. (1983) *Mater. Res. Bull.*, **18**, 1007.
Halliyal, A., Safari, A., Bhalla, A. S. *et al.* (1984) *J. Am. Ceram. Soc.*, **67**, 331.
Herczog, A. (1964) *J. Am. Ceram. Soc.* **47**, 107.
Herczog, A. (1973) *IEEE Trans. Parts, Hybrids Packag.*, **PHP-9**, 247.
Herbert, J. M. (1982) *Ferroelectric Transducers and Sensors*, Gordon and Breach, New York.
Institute of Radio Engineers (1949) IRE Standards on Piezoelectric Crystals, *Proc. IRE*, **37**, 1378.
Ito, Y., Nagatsuma, K., and Ashida, S. (1981a) *Jpn. J. Appl. Phys.*, **20**, Suppl. 20–4, 163.
Ito, Y., Nagatsuma, K., Takeuchi, H. and Jyomura S. (1981b) *J. Appl. Phys.*, **52**, 4479.
Ito, Y., Takeuchi, H., Nagatsuma, K. *et al.* (1981c) *J. Appl. Phys.*, **52**, 3223.
Jyomura, S., Nagatsuma, K. and Takeuchi, H. (1981) *J. Appl. Phys.*, **52**, 4472.
Kimura, M. (1977) *J. Appl. Phys.*, **48**, 2850.

Kimura, M., Doi, K., Nanamatsu, S., and Kawamura, T. (1973) *Appl. Phys. Lett.*, **23**, 531.
Krog-Moe, J. (1962) *Acta Crystallogr.* **15**, 190.
Layton, M. M. and Herczog, A. (1969) *Glass Technol.*, **10**, 50.
Layton, M. M. and Smith, J. W. (1975) *J. Am. Ceram. Soc.*, **58**, 435.
Lee, C. W., Bowen, L. J., Browne, J. M. *et al.* (1984) IEEE Ultrasonic Symposium, p. 285.
Liebau, V. F. (1961) *Acta Crystallogr.*, **14**, 389.
Liu, S. T. (1976) *Ferroelectrics*, **10**, 83.
Liu, S. T. (1978) *Proc. IEEE*, **66**(1), 14.
Lynn, S. Y. (1982) MS thesis, The Pennsylvania State University.
Matthews, H. (ed.) (1978) *Surface Wave Filters*, John Wiley, New York.
McMillan, P. W. (1979) *Glass-Ceramics*, Academic Press, New York.
Melling, P. J. and Duncan, J. F. (1980) *J. Am. Ceram. Soc.*, **63**, 264.
Moore, P. B. and Louisnathan, J. (1969) *Z. Kristallogr.*, **130**, 438.
Newnham, R. E., Bowen, L. J., Klicker, K. A. and Cross, L. E. (1980) *Mater. Eng.*, **2**, 92.
Newnham, R. E., Skinner, D. P. and Cross, L. E. (1978) *Mater. Res. Bull.*, **13**, 525.
Nye, J. F. (1957) *Physical Properties of Crystals*, Oxford University Press, London.
Porter, S. G. (1981) *Ferroelectrics*, **33**, 193.
Putley, E. H. (1970) *Semiconductors and Semimetals,* Vol. 5, Academic Press, New York, p. 259.
Rindone, G. E. (1962a) in *Proc. of the Symposium of Nucleation and Crystallization in Glasses and Melts* (eds M. K. Roser, G. Smith and H. Insley), American Ceramic Society, Columbus, Ohio.
Rindone, G. E. (1962b) *J. Am. Ceram. Soc.*, **45**, 7.
Risbud, S. H. (1979) *Metall. Trans.*, **10A**, 1953.
Safari, A. (1984) PhD thesis, The Pennsylvania State University.
Safari, A., Sa-gong, G., Giniewicz, J. and Newnham, R. E. (1986) Materials Science Research, Vol. 20 (eds R. E. Tressler, G. L. Messing, C. G. Pantano and R. E. Newnham), p. 445. Plenum Press, New York.
Sekhar, J. A., and Risbud, S. H. (1981) *Mater. Res. Bull.*, **16**, 681.
Sessler, G. M. (1981) *J. Acoust. Soc. Am.*, **70**, 1596.
Shorrocks, N. M., Whatmore, R. W., Ainger, F. W. and Young, I. M. (1981) IEEE Ultrasonic Symposium, p. 337.
Simhony, M. and Bass, M. (1979) *Appl. Phys. Lett.*, **34**, 426.
Skinner, D. P., Newnham, R. E. and Cross, L. E. (1978) *Mater. Res. Bull.*, **13**, 599.
Takashige, M., Mitsiu, T., Nakamura, T. *et al.* (1981) *Jpn J. Appl. Phys.*, **20**, L159.
Takahashi, K., Cross, L. E. and Newnham, R. E. (1975) *Mater. Res. Bull.*, **10**, 599.
Takahashi, K., Hardy, L. H., Newnham, R. E. and Cross, L. E. (1979) Proc. 2nd Meeting of International Symposium on Application of Ferroelectrics, p. 257.
Tamura, M. and Yonezawa, M. (1974) *Proc. IEEE* **62**, 416.
Ting, R. Y., Halliyal, A. and Bhalla, A. S. (1984) *Appl. Phys. Lett.*, **44**, 852.
Ting, R. Y., Halliyal, A. and Bhalla, A. S. (1986) *Jpn J. Appl. Phys.*, **24**, 982.
Whatmore, R. W. (1980) *J. Cryst. Growth*, **48**, 530.
Whatmore, R. W., Herbert, J. M. and Ainger, F. W. (1980) *Phys. Status Solidi A*, **61**, 73.
Yamauchi, H. (1978) *J. Appl. Phys.*, **49**, 6162.
Ylo, E. (1986) Private communication.

9

Interfacial electrochemical aspects of glass in solid state ion-selective electrodes

R. E. Belford and A. E. Owen

9.1 INTRODUCTION

The growth in industrial automated systems for quality control and product analysis has caused an upsurge in the use of ion-selective electrodes (ISEs). Continuous quality testing and feedback information has progressed from wet chemical analysis to electrical on-line systems which can be computer controlled and directly linked to input data. For example, in biological systems and pharmaceutical preparations, the pH of the solutions during reaction needs careful monitoring; this must be done under sterile conditions or sometimes at elevated temperatures. These conditions preclude wet methods or conventional pH electrode use; development of the appropriate robust solid state ISEs has therefore received a great deal of attention. Added advantages are (a) low-voltage output, achieved by adding *in-situ* FETs, and (b) the small size of the device, which makes *in-vivo* monitoring possible.

This chapter is concerned with the fundamental underlying differences between solid state electrodes and their membrane precursors. The areas of contention are the interfaces of the ion-selective glass medium. Both the solution and electrical contact interfaces are considered with respect to their physical mechanisms and electrochemical characteristics.

Different glasses are used as the sensing part of a number of ion-selective electrodes (ISEs) (Eisenman, 1967). These sensors use glass as a selective membrane in a bulb form, and the electrical back contact is via an internal filling solution. This is illustrated in Fig. 9.1. Much work has been done to produce more robust solid structures which dispense with this fragile arrangement (Cattrall and Hamilton, 1984). More pertinently most of the recent work has been on planar structures which are or can be interfaced with silicon integrated circuits giving numerous advantages, especially that

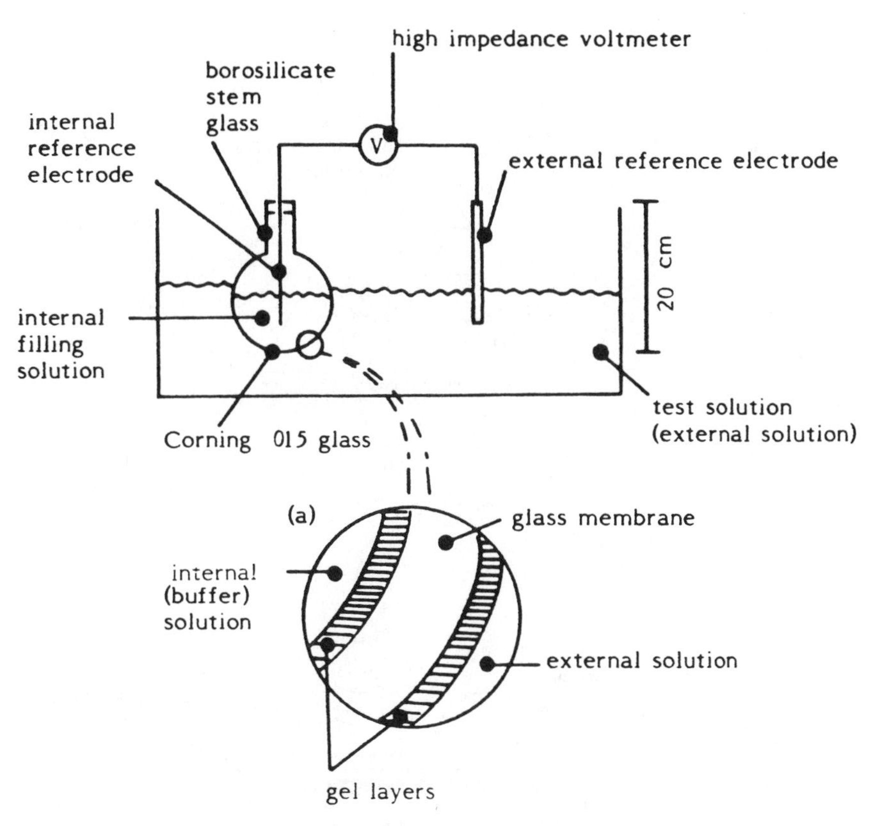

Fig. 9.1 General arrangement of the pH glass electrode in operation.

of low output impedance (for a comprehensive collection of titles see Moody and Thomas, 1984). There have been a number of fundamental problems in producing such structures (Nikolskii and Materova, 1985). Research in our own laboratories has been centred on thick-film pH electrodes. This particular mode of construction allows experimental variation of the glass–metal contact interfaces as shown in Fig. 9.2 (Belford, 1985). A number of fundamental and some subtle chemical and electrochemical considerations have been revealed which are basic to all forms of ISE fabrication, but which require to be resolved for each individual sensing material. The background collated here is drawn from a wide range of disciplines ranging from colloid chemistry to that of glass–metal seals.

The structures shown in Fig. 9.2 are simple in concept and basically achievable using established microelectronic technology. There are, however, fundamental and practical problems, particularly with the interfaces on either side of the glass 'sensing' layer, i.e. with both the glass–solid-conductor interface and the glass–solution interface.

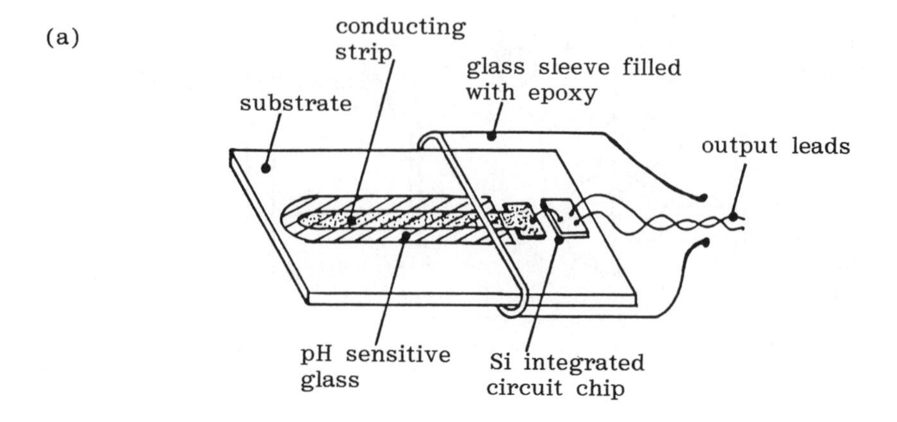

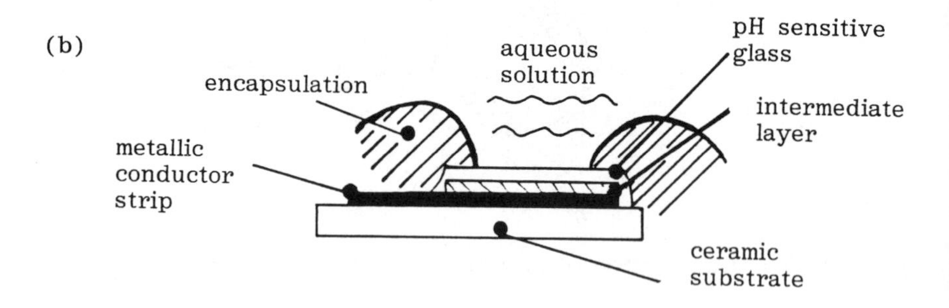

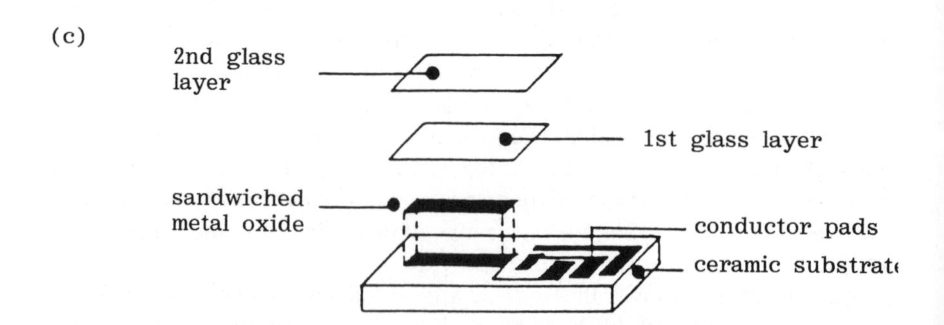

Fig. 9.2 The hybrid pH electrode: (a) a possible probe structure; (b) a section view of an actual thick film electrode and its experimental arrangement; (c) the internal layered structure formed using standard thick film printing techniques.

9.1.1 The glass–solid-conductor interface

The essential difference between conventional membrane-type and solid-state electrodes is shown by comparing Figs 9.1 and 9.2. Ion-selective glasses are mostly ionic conductors and hence the important question is: can the direct contact between an electronic conductor (metal) and the glass perform the same function as the combination of electrically reversible phases in the membrane ISE (Fig. 9.1)?

The term 'reversible' needs some definition because, as pointed out by Bergveld, there are several subtleties in its usage. Electrodes by definition are reversible systems thermodynamically, electrochemically and chemically. For practical purposes the extent of reversibility required of any sensor is that it should function stably and reproducibly. The processes at the back contact should facilitate charge transfer in both directions and in doing so must not alter the morphology or chemical condition of that boundary in such a way as to impair its function. Smith *et al.* (1973) remarked that for an ion-selective electrode to function as a thermodynamically defined reversible electrode the system requires an internal reference solution and an outer ion-selective medium. The internal reference electrode must contain a reversible couple to allow ionic to electronic charge interconversion. The internal solution must contain the ion of interest and a counter ion which is reversible to the internal electrode. Attempts have been made to satisfy these criteria using solid electrolyte systems. In order to achieve a stable reproducible response, some reversible mechanism must operate at the ion selective material–metal back contact, i.e. ionic–electronic interconversion of current. Many solid-contacted electrodes have performed less ideally than their membrane counterparts (Bousse, 1985; Nicolskii and Materova, 1985) in which electrochemical potentials can equilibrate with ease at both liquid–glass interfaces and a quasi-equilibrium results which leads to a maximum response given by the Nernst (1889) equation

$$E = E_0 + \frac{RT}{nF} \ln \frac{a_i}{a_j}$$

where $E =$ is the e.m.f. of the half cell, E_0 is the standard potential of the glass electrode, $R =$ is the gas constant, T is the temperature in kelvins, F is Faraday's constant, n is the number of moles involved, and a_i and a_j are the activities of products and reactants. The quantity $(RT/nF) = 59\,\mathrm{mV}$ for each decade change of reactant concentration assuming product concentration is kept constant (at STP).

Less than ideal responses infer a less than true electrochemical potential equilibrium at the glass–metal interface. The quest for a reversible contact has led to the use of intermediate materials (Fjeldy *et al.*, 1979; Shults *et al.*, 1981) as illustrated schematically in Fig. 9.3. Direct simulation of the

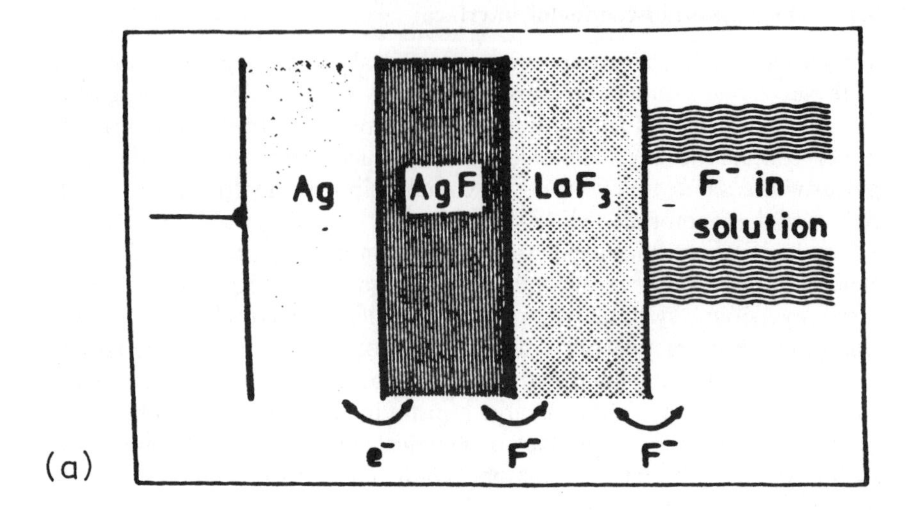

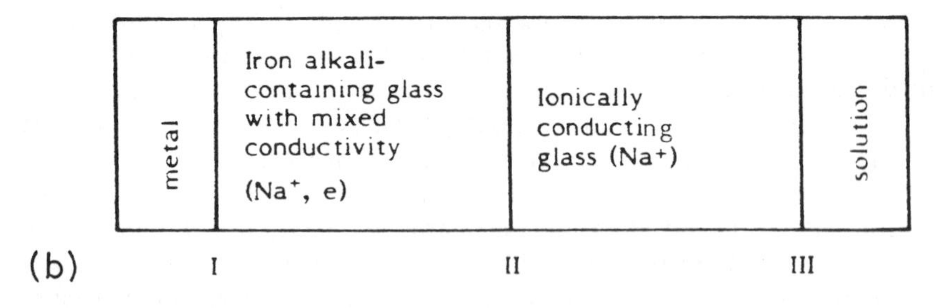

Fig. 9.3 (a) Structure for the reversible solid state contact on LaF_3; (b) iron alkali-containing glass layer to give a reversible contact.

membrane electrodes using gels has also been attempted and is reported in the review by Cattrall and Hamilton (1984). Back contacts with O^{2-} as the counter ion have been reported by Niedrach (1982), Boyce and Huberman (1979) and Subbarao (1980), and with particular reference to Cu/Cu_2O contacted systems by Niedrach (1980). Oxygen is also thought to participate in the workings of metal evaporated contacts on glass (Tomozawa, 1977).

9.1.2 The glass–solution interface

Superficially, the glass–solution interface, in the case of a solid-state ISE, is no different from the situation when the same glass is used in a membrane-type ISE. Fabrication of solid devices requires thermal cycling and this can affect the sensing surface.

The purpose of this chapter is to consider, in detail, the two interfaces and the way in which they influence the performance of solid state ISEs. The discussion will be focused specifically on the authors' recent experience in the development of a thick-film hybrid pH electrode based on *Corning 015 glass as the sensing material. The fabrication processes involve temperature treatment of the glass up to 1000°C, and these are also discussed. Much of this work is generally applicable to most potentiometric devices.

9.2 THE GLASS–METAL INTERFACE

9.2.1 The chemistry

The chemistry of the glass–metal interface is important for both mechanistic and practical reasons. Chemical processes at this interface have a major role in determining the electrode reaction at the back contact and will also influence factors such as adhesion and the general compatability of materials in thick film fabrication. The particular metals used as back contacts were Fe, Cu and Pt, applied at high temperatures, and thus incorporating their oxides, and low temperature evaporated contacts of Au, Ag and Cu.

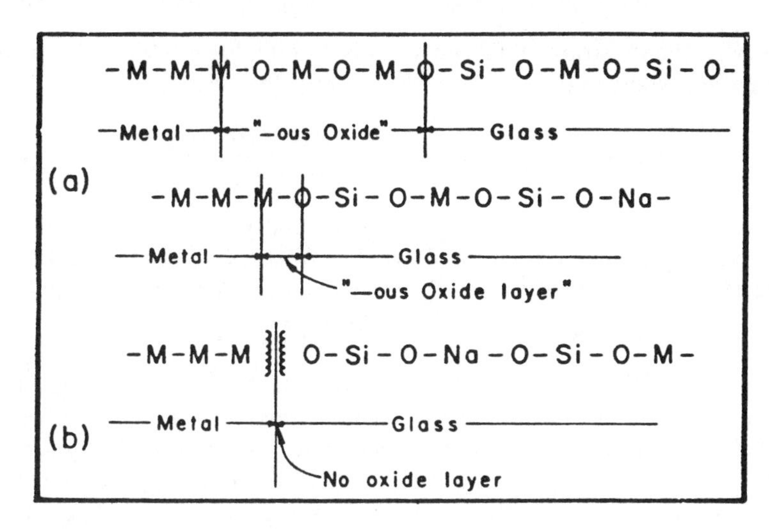

Fig. 9.4 Schematic representation of transition zones through metal–glass interfaces, showing conditions for chemical and van der Waals types of bonding: (a) chemical bonding with and without discrete intermediate oxide layer, and (b) van der Waals bonding.

* Corning 015 is a well-documented pH sensitive glass.

Most of the work on glass–metal interfaces has been studied in the fields of glass-to-metal seals and in the forming of glass by metal moulds. Notable contributions were made by Pask and co-workers, and according to Pask (1971) in order to realize continuity of atomic and electronic structure a transition zone is required, compatible and in equilibrium with both metal and glass at the interfaces. In general it must contain at least a monomolecular layer of the metal oxide, and both metal and glass must be saturated with the metal oxide. Figure 9.4 shows schematically a sodium silicate glass and metal M interface. The chemical bonds in Fig. 9.4(a) are considerably stronger than the weak van der Waals forces in Fig. 9.4(b). When the glass dissolves the metal oxide layer and comes into contact with the metal, there is a driving force for oxidation of the metal; this is important for formation of the interfacial zone. As the sodium ion is the principal charge carrier in silicate glasses its interaction at the interface with the contacting metal is of prime significance. The most extensive reactions involve the reduction of interstitial cations to their metallic states, e.g. the reaction of iron and a disilicate glass:

$$xFe^0 + Na_2Si_2O_5 \rightleftharpoons Na_{2-2x}Fe_xSi_2O_5 + 2xNa^0 \qquad (9.1)$$

The oxidation mechanism of the substrate metal, in this case Fe, can be represented generally by the following:

$$M_I^0 + M_{II}^{2+}O\,(\text{glass}) \rightleftharpoons M_I^{2+}O(\text{glass}) + M_{II}^0 \qquad (9.2)$$

where M_I is the substrate metal (Fe in the last example) and M_{II} is the alkali ion present initially in the glass. The free energy for this reaction is

$$\Delta G = \Delta G^0 + RT \ln \frac{a(M_IO\,(\text{glass}))\,a(M_{II})}{a(M_{II}O\,(\text{glass}))\,a(M_I)} \qquad (9.3)$$

The conditions for (9.3) to occur spontaneously, given M_I = Fe and M_{II} = Na, require a temperature above the boiling point of sodium, i.e. to reduce the activity quotient to much less than unity. Consider the reaction:

$$Fe + Na_2O(\text{glass}) \rightarrow FeO(\text{glass}) + 2Na^0\uparrow \qquad (9.4)$$

At 1000°C, $\Delta G^0 = 2.91$ kJ mol^{-1} (Tomsia and Pask, 1981). Thus the activity quotient must be sufficiently small for the reaction to occur, i.e. if removal of Na(g) is efficient. It has been shown that just such a reaction occurs between molten glass and iron. Ferrous iron is also produced when Fe^{3+} comes into contact with the metal surface and reacts with Fe^0. Our devices which were made with iron contacts were preoxidized prior to coating with glass, but it has been shown (Pask, 1971) that on subsequent heating the iron oxide is dissolved by the glass. After formation of iron oxide, the mixing process is:

$$xFeO(\text{interface}) + Na_2Si_2O_{5-x} \rightleftharpoons Na_{2-2x}Fe_xSi_2O_5 + 2xNa \qquad (9.5)$$

or

$$x\text{FeO(interface)} \rightleftharpoons x\text{FeO(glass)}$$

This occurs spontaneously, i.e. ΔG^0 is negative.

Several important observations were made in sessile drop experiments where iron was used (amongst other metals) as a substrate. Change in colour of the sessile drop from bluish to greenish-blue indicates an oxidation reaction involving $Fe^{2+} \rightleftharpoons Fe^{3+}$. Ferrous glasses are blue whereas ferric are yellow-brown; (Fe^{2+}/Fe^{3+}) in combination give a greenish colour (Weyl, 1951). There are several different mechanisms possible for this oxidation, most involving the evolution of sodium gas. The changes in colour and adherence properties with glass composition are shown in Table 9.1 (Pask, 1964; Tomsia and Pask, 1981). The amount of sodium in a glass will affect the ratio of (Fe^{3+}/Fe^{2+}) present in the interfacial region between the metallic iron contact and the glass bulk (Paul, 1982; Tomsia and Pask, 1981). Studies of glass containing both sodium and iron have included spectroscopic evidence for the existence of two oxidation states of iron (Fe^{3+} and Fe^{2+}) and they occupy specific sites (Fe^{2+} in octahedral sites and Fe^{3+} in two types of fourfold co-ordinated environments). Small variations in Na^+ content produce shifts in their optical absorption as the spatial environments around the iron ions are altered (Fox *et al.*, 1982). Reaction sequences involving Fe^0 oxidation by Na^+ are energetically unfavourable. Charge transfer at ambient temperatures is more likely to involve the (Fe^{3+}/Fe^{2+}) couple, as the presence of Na^+ ions can alter this ratio and thus transfer charge from an ionic to an electronic mechanism. The formation of an interfacial zone (Beckman Instruments, 1972; Fjeldy *et al.*, 1979) may be fundamental to electrode response and in the case of an iron metallic contact an intermediate zone gives very good results, but as yet no mechanism has been postulated for its function. In the case of Shults electrode (Fig. 9.3(b)) it was recognized that electrochemical potentials must equilibrate at each boundary for the half cell to function reversibly. The following equilibrium is required:

$$Na^{\oplus}(\text{in the ISM}) + e^{\ominus}(\text{in transition layer}) \rightleftharpoons Na^{\oplus}(\text{soln}) + e^{\ominus}(\text{metal})$$

Table 9.1 Contact angles, colours and adherences of sodium silicate glasses on iron at 1000°C and pressures of 10^{-5} mm Hg

Glass composition (mol %)		*Colour*	*Adherence*	*Contact angle (degrees)*
Na_2O	SiO_2			
0.20	0.80	Blue-green	Good	62
0.25	0.75	Green	Good	59
0.33	0.67	Green	Poor	59

This electrode gave stable results when Na ISMs were used. The nature of the interfaces was studied using thermal and current pulses; these are claimed to confirm charge transfer reversibility.

The conduction mechanisms attributed to the combined effects of Fe_2O_3 and Na_2O in lead silicate glass provided interesting information. For small concentrations of iron oxide (2%) there is a rise in the activation energy for conduction. Further addition of Fe_2O_3 results in a sharp drop in conductivity. According to Grechanik *et al.* (1962) the initial increase is attributed to the co-existence of both ionic and electronic conduction. It is also of interest to note that electronic conduction is present at concentrations of Fe_2O_3 as low as 2 mol %, i.e. where the iron ions are well separated. It is not inappropriate to picture a situation in which there are local regions of predominantly electronic conduction, separated by regions of ionic conduction. This would involve many (Na^+/Fe^{2+}) interactions which could account for the increase in activation energies observed at these concentrations.

Apart from iron, other metals studied by the authors include Au, Cu and Pt, which were fused to glass at high temperature (800°C). Low temperature evaporation techniques were used to deposit Cu, Ag and Au. The literature concerning these metals with respect to soda lime silicate glass–metal interfaces is fragmented and this section is a collation of relevant data on soda lime silicate and similar glasses. In their study of glass to metal seals, Nagesh *et al.* (1983) investigated the adherence properties using sessile drop experiments. Molten lead borosilicate glass was dropped at 700°C on to Au, Ag and Pt substrates in air, vacuum and He(g) atmospheres. Wetting occurred in all cases with strong adherence in air due to reaction with the metal oxides. Strong adherence to Ag in vacuum was shown to result from the following redox reactions:

$$2\text{Ag(substrate)} + \text{PbO(glass)} \rightleftharpoons \text{Ag}_2\text{O(interface)} + \text{Pb(g)}$$

$$\Delta G = \Delta G^0 + RT \ln \frac{a_{Ag_2O} P_{Pb}(g)}{a_{PbO}} \tag{9.6}$$

The standard free energy ΔG^0 is positive as the oxidation potential of silver is lower than that of lead. Thus the reaction can only proceed if removal of Pb gas is possible because the activity of Ag_2O is low due to P_{O_2} being less than the dissociation pressure for silver oxide. This was observed experimentally and as can be seen from Table 9.2, poor adherence in $P_{He} = 1$ atm. indicates that redox reaction (9.6) occurs. In all cases adherence is attributed to saturation of the interface zone with the metal oxide. The contact angle of gold is 18°, which is very high compared to Ag (0°) and Pt (2–3°). A low contact angle indicates preferential adherence of glass to metal or metal oxide formation, the two being synonymous. The smaller the

Table 9.2 Sessile drop measurements of lead borosilicate glass on silver, gold and platinum

Metal	*Atmosphere*	*Temperature* (°C)	*Contact angle* *(degrees)*	*Adherence* *(qualitative)*
Ag	Air	600	0	Very strong
	Vacuum	700	2–12	Good
	He	700	5–48	Poor
Au	Air	700	18	Strong
	Vacuum	700	7	Poor
Pt	Air	700	2–3	Strong
	Vacuum (no carbon)	700	58	Poor
	Vacuum (with carbon)	700	73	Poor

Vacuum at 2.6×10^{-4} Pa; helium atmosphere about 1 atm. From Nagesh *et al.*, 1983.

contact angle the better the wetting. Combining information from Tables 9.2 and 9.3 gives an insight into the character of the glass–metal regions.

For the metals of interest, i.e. Au, Ag, Cu, Pt and Fe, gold adheres the least and silver the best under a variety of conditions. Confusion with regard to the mechanism of adherence may arise due to the diffusion capabilities of Ag in glass. Platinum is of interest as the presence of its oxide makes such a marked difference to the bond strength with the glass (Table 9.2). It is also interesting to note that no sodium loss in observed (Table 9.3). Preoxidized Pt has been used as a back contact for a dipped electrode (Nicols, 1982). Platinum is not the inert material that it is often considered to be. Over 70

Table 9.3 Reactions of sodium silicate glass with a number of metals

Metal	*Graphite furnace* $P_{O_2} = 10^{-20}$ *atm*		*Alumina furnace* $P_{O_2} = 10^{-10}$ *atm*	
	Contact angle (degrees)	*Weight loss of Na/total Na* (mass %)	*Contact angle* (degrees)	*Weight loss of Na/total Na* (mass %)
Au	56	0.0	60	0.0
Pt	15	0.0	16	0.0
Ni	29	10.5	16	4.0
Co	13	42.0	20	15.7
Fe	0	95.4	10	57.3

From Pask and Tomsia, 1981.

redox and decomposition reactions are catalysed by platinum metal, or more pertinently by platinum metal and its oxides. Copper (I) is similar to Pt in that it takes part in reversible redox reactions as a catalyst. It is also recognized as a fast ion conductor in materials such as CuI, CuS and CuSe (Boyce and Huberman, 1979; Subbarao, 1980). Application to a pH sensor was first reported in 1980 by Niedrach; Cu/Cu_2O was used as a solid alternative to the internal filling solution in the production of a high temperature pH sensor. This has since been favourably reported on by different authors (Hettiarachchi and Macdonald, 1984). There was only one attempt at a mechanistic view and thermodynamic balancing of the interface potential between phases. The net charge is equal to that of the simple Cu/Cu_2O couple;

$$2Cu^0 + H_2O \rightleftharpoons Cu_2O + 2H^{\oplus} + 2e^{\ominus} \tag{9.7}$$

which was related directly to the standard hydrogen electrode.

$$H_2 \rightarrow 2H^{\oplus} + 2e^{\ominus} \tag{9.8}$$

As an analogy in the conventional glass electrode Cl^- is present as the counter ion in the internal reference solution; in the Cu/Cu_2O contact, O^{2-} is the counter ion between the membrane (ZrO_2 in this particular case) and Cu_2O (Niedrach, 1980, 1982).

9.2.2 Mixed conduction zones

Many back contacts which employ a definable region of mixed ionic and electronic conduction are reported to exhibit good reversible characteristics (Shults, 1981; Nicols, 1982; Pfaudler, 1981; Fjeldy *et al.*, 1979). Within this region, however, redox reactions must be responsible for charge transfer and it is possible that by providing a sufficiently large region for these reactions to occur, reversibility can be more complete. The case of an electrode with such a region involving metallic iron has been considered and it is believed that a transition layer of mixed ion–electronic conduction exists, and it is recognized that electronic conduction does occur in these glasses (Owen, 1977). It is implicit that reversible charge transfer occurs between Na^+ and Fe^{2+} and subsequent charge transfer from Fe^{2+} to the metallic contact occurs via the mixed valence mechanism. Another metal and its oxide (Cu/Cu_2O) has been used as a back contact in a high temperature pH sensor, employing a ZrO_2 membrane, and cuprous oxide is known to conduct electronically by holes (Kroger, 1964). The solid state fluoride sensor investigated by Nagy can also be regarded as having an intermediate layer of mixed ionic–electronic conduction.

Platinum contacts are more subtle in this sense but preoxidation can cause an intermediate layer of mixed conductivity to form. Several oxides exist but Pt_3O_4 is of most interest (Salmeron *et al.*, 1981). This oxide is very difficult to produce free from its sodium analogue $Na_xPt_3O_4$ (where $0 < x < 1$). Historically this sodium analogue was studied first (Waser and McClanahan, 1951) and initially thought to be an electronic conductor which is consistent with the short Pt–Pt distance 2.8 Å (Cohen *et al.*, 1974). These platinum bronzes are of interest due to their role in the Adams catalyst (Cohen and Ibers, 1973). Further studies of sodium–platinum oxide have shown conflicting evidence for metallic and ionic (Na^+) conduction (Shannon *et al.*, 1982). It is feasible that both conduction processes are possible under the appropriate conditions. This would suggest that in those pH sensors in which a pre-oxidized Pt wire has been used as the back contact (Nicols, 1982) an intermediate zone of mixed conductivity is present. Finally, in this context a conventional internal reference electrode has a substantial layer of a metal salt such as AgCl which is potentially a mixed conductor.

9.2.3 Summary of the metal–glass interface

The information collected so far can be generalized to provide a model of the charge transfer at various metal interfaces with a Na^+ ion conducting glass. It is assumed that:

1. An oxide of the metal is present as an intermediate zone between the metal and glass;
2. In the case of Ag metal the corresponding Ag^+ ions are mobile within this zone;
3. In the case of Pt, Fe and Cu, Na ions are mobile within this zone;
4. A reversible flux of Na ions/electrons directed from the glass to the metal contact, or vice versa, is possible.

Consider each metal case separately:

(a) Ag contacts

The Ag ions will act similarly to Na^+ ions in the glass, i.e. they will drift in a concentration gradient or applied electric field. There will therefore be a change in flux from purely Na^+ flow to purely Ag^+ through the interface zone and on reaching the Ag metal contact the Ag ions discharge. Silver contacts were not made at high temperatures, i.e. molten glass temperatures, as silver migration through the glass was so rapid at these temperatures that the glass ion selectivity was affected by its presence. Silver contacts were applied by the relatively lower temperature method of vaccuum evaporation.

(b) Cu contacts

The interaction of the Cu–Cu_2O–glass system is not well documented. Cuprous oxide is known to be a p-type semiconductor and it is also known that the Cu ion is not mobile in either soda–lime–silicate glasses or in its own oxide. A possible mechanism for charge transfer could therefore be:

$$2Na^{\oplus} + Cu_2O \rightleftharpoons Na_2O + 2Cu^{\oplus} \quad (9.9)$$

$$Cu^{\oplus} \rightarrow Cu^0 + (h)^{\oplus} \quad (9.10)$$

The Cu^+ ions generated in (9.9) would effectively cause the Cu_2O phase to become more p-type (i.e. more holes are generated–reaction (9.10)). Holes would drift toward the Cu metal electrode and the discharge process could be regarded as equivalent to Cu^+ discharging at the Cu^0 contacts.

(c) Pt contacts

In this case the Na^+ ion is mobile within the platinum oxide. The transfer of charge may be:

$$3Na^{\oplus} + 2Pt_3O_4^{\ominus} \xrightarrow{e} Na_2O + 3PtO + NaPt_3O_4 \quad (9.11)$$

Thus altering the equilibrium ratio of Pt^{III}–Pt^{II},

$$Pt^{3+} + e^{\ominus} \rightarrow Pt^{2+}$$

(d) Fe contacts

The Fe contact is similar to the Pt case in that both develop mixed ionic/electronic conduction regions, and charge transfer occurs via a change in the equilibrium ratio of Fe^{3+}/Fe^{2+}.

Examples of stable, long lived electrodes exhibiting good pH responses were produced with each of the different metal contacts discussed here, and so each must be reversible to some extent under the prevailing operating conditions.

9.3 THE GLASS–SOLUTION INTERFACE

9.3.1 Hydration and chemical aspects of the glass surface

After a few hours' exposure to aqueous media the glass surface becomes swollen and hydrated. Mild dissolution of the soda lime silicate results in the formation of the so-called gel layer. It is within this region that ion-exchange equilibrium, association and dissociation occurs. This is the sensing part of the glass electrode, where the chemical potential of H^+ ions in solution equilibrates with those ions already present.

The extent of this layer has been studied for some pH sensitive glasses including Corning 015 where a gel layer thickness of 2800 Å was estimated (Wikby and Karlberg, 1974). This seems consistent with the 500 ionic layers quoted earlier by Isard (1967). The mechanism of this dissolution has been attributed to the attack of bridging oxygens by either hydrogen ions, hydroxyl ions or both (Boksay and Bouquet, 1980). In weakly alkaline or acid solution the glass surface loses nearly all of its Na ions near the surface and takes up hydrogen ions to form Si–OH groups as follows:

$$\equiv\text{Si–O–Si}\equiv \; + \; \text{H}^{\oplus} \rightarrow \left[\begin{array}{c}\equiv\text{Si–O–Si}\equiv \\ | \\ \text{H}\end{array}\right]^{\oplus} \quad \begin{array}{c}\text{intermediate} \\ \text{(A)}\end{array} \tag{9.12}$$

$$\downarrow + \text{H}_2\text{O}$$

$$2(\equiv\text{Si–OH}) \; + \; \text{H}^{\oplus}$$

The formation of the intermediate (A) is rate determining and is quickly followed by hydrolysis. During the decomposition molecules of silicic acid and perhaps its oligomers are formed. For basic solutions the following reaction sequence is proposed:

$$\equiv\text{Si–O–Si}\equiv \; + \; \overset{\ominus}{\text{OH}} \rightarrow \left[\begin{array}{c}\text{Si–O–Si}\equiv \\ \quad\;\; | \\ \quad\;\; \text{OH}\end{array}\right]^{\ominus} \tag{9.13}$$

$$\downarrow$$

$$\equiv\text{SiOH} + \equiv\text{Si–O}^{\ominus}$$

The initial leaching of sodium from the glass has been attributed to the action of molecular water (Smets and Lommen, 1983), i.e.:

$$\equiv\text{Si–O}^{\ominus}\text{–Na}^{\oplus} \; + \; \text{H}_2\text{O} \rightarrow \; \equiv\text{SiOH} \; + \; \text{Na}^{\oplus}\text{OH}^{\ominus} \tag{9.14}$$

Correlation between the gel layer properties and electrochemical behaviour was studied by Wikby and Karlberg (1974). Several pH-sensitive glasses were characterized with respect to pH response, layer thickness, durability and ion-exchange capacity. In general, glasses of low durability have a thick gel layer (13000 Å) and a high exchange capacity compared to high durable glasses. It was also concluded that the thinner gel layer, the more ideal the ion-selective electrode function. This is in keeping with a site binding mechanism dicussed later (Section 9.3.2). The glass compositions studied by Wikby and Karlberg included lithia glass compositions which were shown to exhibit the best results in terms of durability, and have a small gel layer thickness. This is an agreement with earlier studies (Perley, 1941) of glass composition as a function of pH response, i.e. these lithia glasses also gave the best pH response. Generally alkali metal ion leaching was shown to be complete after 90 min in 0.02 M $HClO_4$ solution: the gel layer formed has a looser structure and an electrical resistivity 5–10 times lower than the bulk. Unless both sides of a glass membrane are hydrated synchronously an asymmetric potential develops. The phase boundary between the dry glass and the gel layer is thought to be sharp and the drastic changes in chemical and electrical properties indicate that the gel layer is a separate phase (Wikby, 1974). Only a small part of the gel layer is active with respect to the electrode function. The thicker the gel layer the greater the response time and the larger the acid and alkaline errors. This is obvious when considering the steric hindrance of transport to and from active exchange sites, making equilibrium very slow and diffusion controlled. A thick gel layer and low

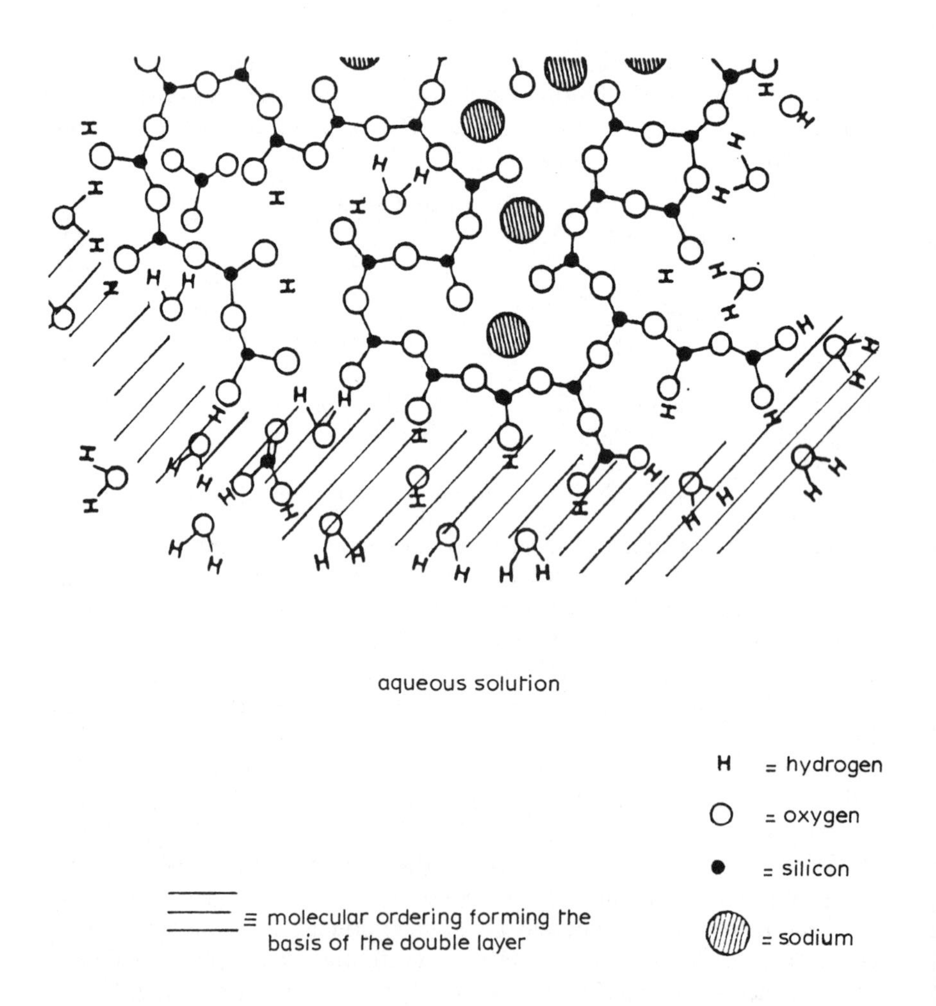

Fig. 9.5 Speculative two-dimensional representation of the gel layer.

durability may be regarded as a direct consequence of aqueous corrosion or slow dissolution of the silica framework. Diffusion coefficients were found to be high within a thick gel layer. Also contributing to the problems caused by a thick layer is that with the much greater surface area available, problems of adhesion, adsorption and steric retention of active species are more evident.

A recent study of surface –OH groups proved interesting in understanding the gel layer and the pH responses of glasses with different thermal histories. Kanazawa *et al.* (1984) reported three different types of –OH groups on porous Corning 7930 glass. The three types are: isolated –OH groups, gem type –OH groups (i.e. two –OH groups on the same Si atom) and thirdly H-bonded OH groups. Water is adsorped more strongly on gem type of H-bonded –OH groups which are more readily removed than isolated groups at temperatures around 600°C and are completely absent at 800°C. Glass heated at 800°C for 4 h was shown to contain only isolated groups, and of those present originally only 70% remained.

Figure 9.5 is a speculative two-dimensional representation of the gel layer. It is based on the concepts discussed (Belford, 1985) and includes the double layer containing interfacial ion-pairs (Yates *et al.*, 1974). With regard to pH response, if prehydrated electrodes are baked at 400°C to 500°C, the surface layer is irreversibly dehydrated to an impervious high silica, low alkali structure, and the pH response is lost (Isard, 1967). Etching with HF can restore the response. Annealing pH electrodes in the same temperature range prior to hydration does not affect the pH response. The reaction is thought to be:

$$\equiv SiO^{\ominus}H^{\oplus} + H^{\oplus}O^{\ominus}\text{–}Si\equiv \xrightarrow{500°C} \equiv Si\text{–}O\text{–}Si\equiv + H_2O \quad (9.15)$$

Effects such as these become important when glass is processed prior to high temperature treatment. For example in fabricating thick-films electrodes the glass may be ground in aqeuous media prior to processing.

9.3.2 The site binding model

The ion-exchange model of processes occurring at the glass–solution interface of a pH glass electrode predicts a strictly nernstian response for dilute solutions. Indeed under controlled conditions conventional glass electrodes generally obey the Nernst equation, which predicts a slope of $59\,mV\,pH^{-1}$. However, the solid state versions often give responses which are sub-nernstian and a typical value of the slope is $40\,mV\,pH^{-1}$. It is therefore necessary and relevant to consider the site binding model which predicts a continuously variable response with the nernstian slope of $59\,mV\,pH^{-1}$ as an upper limit. The site binding model was originally developed by Yates *et al.* (1974) in studying the surface chemistry of oxides such as SiO_2, and Al_2O_3 which have amphoteric surface sites. The model has been applied to

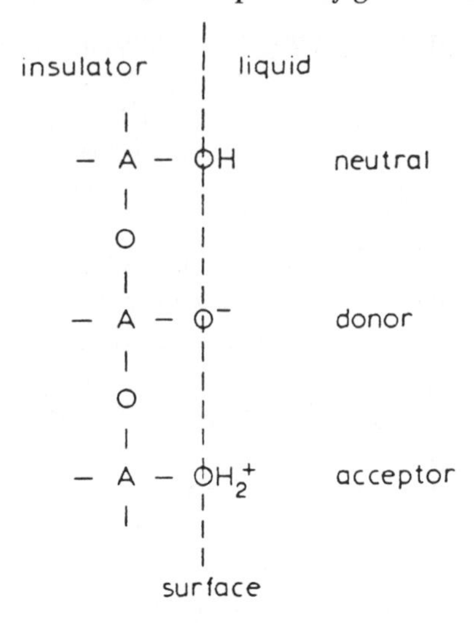

Fig. 9.6 The site binding surface model: representation of the three possible forms a surface site may take in the simple model of amphoteric site dissociation, without counter-ion adsorption.

integrated solid state pH electrodes which utilize similar oxides as their active material. Figure 9.6 represents such a surface (Bousse, 1982). The site binding theory has its origin in colloidal chemistry where the extremely large surface area of a powder in colloidal dispersion gives a large surface to volume ratio. The surface reactions have a measurable effect on the electrolyte in which the colloid is dispersed. There are two measurable quantities, surface charge (σ_0) and the zeta potential. Surface charge is measured by titration with acid or base and the zeta potential by moving the electrolyte relative to the oxide surface which produces a potential equal to that at the plane of shear (this is the limit of the water molecules which cannot move). Although a useful parameter, the physical meaning of the zeta potential is ill-defined. The bases for the site binding model are the acid–base reactions at the amphoteric surface sites (AOH) which are:

$$\mathrm{SOH_2^{\oplus}} \rightleftharpoons \mathrm{SOH} + \mathrm{H^{\oplus}_{(s)}} \quad K = \frac{[\mathrm{H^{\oplus}}][\mathrm{SOH}]}{[\mathrm{SOH_2^{\oplus}}]} \tag{9.16}$$

$$\mathrm{SOH} \rightleftharpoons \mathrm{SO^{\ominus}} + \mathrm{H^{\oplus}_{(s)}} \quad K = \frac{[\mathrm{H^{\oplus}}][\mathrm{SO_2^{\ominus}}]}{[\mathrm{SOH}]} \tag{9.17}$$

Viewing the surface as having a double layer capacitance and taking into account that saturation is not possible in high site density surfaces, e.g. oxides, Yates *et al.* (1974) and Healy *et al.* (1977) deduced a theoretical expression with terms which are consistent with other theoretical reports (Hunter and Wright, 1971). A major variable within this expression is β^0 characterizing the sensitivity of a particular surface to change in pH. For large β^0 the Nernst slope is observed and this is a maximum value. When β^0 is low, much lower theoretical slopes are predicted. Optimizing β^0 is important for nernstian behaviour. This can be done by achieving a high surface site density (optimum 2×10^{15} sites cm^{-2}) and optimizing the site dissociation constant, as neutral surface sites are characterized by (ψ_0/pH) being non-linear, where ψ_0 is the potential difference H^+ experiences when moving from the bulk electrolyte to a surface at which it reacts.

9.3.3 Summary of the glass–solution interface

The surface structure of glass in contact with aqueous solution has a bearing on the models used for the origin of the glass potential. The formation of the gel layer is essential for the sensing function of the glass (Isard, 1967). Of particular interest are the different surface sites available for potential determining reactions (Kanazawa *et al.*, 1984). The hydrogen bonded sites and gem sites would intuitively appear to be responsible for the sensitivity associated with pH electrodes, as glasses with only isolated –OH groups are relatively insensitive in their response. The presence of the gel layer does not preclude any particular potential determining reaction whether it be ion-exchange, association–dissociation or acid–base in nature. In reality a combination of all these reactions, along with simple hydrogen bonding, are probable, to varying extents, at the glass–solution interface. Adhesion, association and exchange of ions must occur to enable the surface to equilibrate (in terms of electrochemical potential) with the ions in solution to which it responds. The question of the origin of the potential is obscured by the different approaches to the problem. The classical treatment of the glass response based on ion-exchange reactions, although not without its inconsistencies, was and is acceptable when referring to membrane glass electrodes. The development of pH electrodes with solid back contacts has produced sensors of various pH responses. Only fairly recently has any theoretical reasons for sub-nernstian response been proposed (Yates *et al.*, 1974; Hunter and Wright, 1971). Baucke (1985) attempted to solve some of the inconsistencies of the classical approach by introducing association and dissociation mechanisms. If one or other of these predominates a potential difference is established. This approach is more relevant to the origin of the potential than the exchange of unit charge which is inherent in the ion-exchange mechanism. The kinetic model, which deals with a single glass–solution interface, is also relevant and does enable variations of response when parameters such as surface characteristics are non-ideal. The accounts given by Bousse (1982) and Healy *et al.* (1977) approach this single interface from the double layer physics viewpoint and potentials are predicted for different parameters relevant to different surface sites and solvent characters.

9.4 CONCLUSIONS

A variety of metal contacts were studied (section 9.2.3) and for each metal, examples of satisfactory electrodes were obtained. Although insufficient evidence exists for quantitative description of the contact mechanism, some fundamental aspects of its function have been considered and qualitative models proposed (see section 9.2.3).

The pH-sensing performance of the authors' electrodes was found to be more dependent on the glass-solution interface than on the type of metal contact. Devices subjected to temperatures of over 850°C for more than 10 min were insensitive to pH.

The recipe for an ideally active surface (maximizing β^0) would be to limit fabrication heat treatment to 700°C and certainly if higher temperatures are required, they should be limited to short time periods. The retention of gem- and H-bonded sites seems paramount to the electrode sensitivity.

The major difference between solid and membrane-type ISEs, not yet appraised in the literature, is that the processes occurring at the surface of a solid state pH electrode are governed by 'electrodics'. The surface of the glass is in this case the electrode surface and nernstian response may be observed only if this surface acts as a perfect electrode sensitive to $[H^+]$. This situation is different from the membrane case where a pH sensitive glass is interposed between two solutions, and can be regarded as a selective membrane.

REFERENCES

Baucke, F. G. K. (1985) *Glass – Current Issues* (eds A. F. Wright and J. Dupuy), NATO ASI Series, Martinus Nijhoff, Dordrecht.

Beckman Instruments (1972) US Patent 1 260 065.

Belford, R. E. (1985) *Principles and practice of hybrid pH sensors*, PhD thesis, Edinburgh University.

Bergveld, P. and Saamon, A. A. (1985) *Sensors and Actuators*, **7**, 69–71.

Boksay, Z. and Bouquet, G. (1980) *Phys. Chem. Glasses*, **21**, 110–3.

Bousse, J. L. (1982) Doctorate thesis, Technische Hogeschool Twente, The Netherlands.

Bousse, J. L. (1985) Progress in ion-sensitive FETs. *Modelling and Materials*, **4**.

Boyce, J. B. and Huberman, B. A. (1979) *Phys. Rep.* (Review section of *Phys. Lett.*), **51**, 189–265.

Cattrall, R. W. and Hamilton, I.C. (1984) *Ion-Selective Electrode Reviews*, **6**, 125–72.

Cohen, D. and Ibers, J. A. (1973) *J. Catal.*, **31**, 369–71.

Cohen, D., Ibers, J. A. and Wagner, J. B. (1974) *Inorg. Chem.*, **13**, 1377–88.

Eisenman, G. (ed.) (1967) *Glass Electrodes for Hydrogen and other Cations*, Marcel Dekker, New York.
Fjeldy, T. A., Nagy, K. and Johannesen, I. S. (1979) *J. Electrochem. Soc.*, **26**, 793–5.
Fox, K. E., Furukawa, T. and White, W. B. (1982) *Phys. Chem. Glasses*, **23**, 169–78.
Grechanik, L. A., Fainberg, E. A. and Zertsalova, I. N. (1962) *Sov. Phys. – Solid State*, **4**, 331–3.
Healy, T. N. *et al.* (1977) *J. Electroanal. Chem.* **80**, 57–66.
Hettiarachchi, S. and Macdonald, D. D. (1984) *J. Electrochem. Soc.*, **131**, 2206–7.
Hunter, R. J. and Wright, H. J. L. (1971) *J. Colloid Interface Sci.*, **37**, 564–80.
Isard, J. O. (1967) *Glass Electrodes for Hydrogen and other Cations*, Marcel Dekker, New York, Ch. 3.
Kanazawa, T. *et al.* (1984) *J. Ceram. Soc. Jpn*, **92**, 654–59.
Kroger, F. A. (1964) *Chemistry of Imperfect Crystals*, North Holland, Amsterdam.
Moody, G. J. and Thomas, J. D. R. (1984) *Ion Selective Electrode Reviews*, **6**, 209–63.
Nagesh, V. K., Tomsia, A. P. and Pask, J. A. (1983) *J. Mater. Sci.*, **18**, 2174.
Nernst, W. (1889), *Z. Phys. Chemis.*, **4**, 129.
Nicols, M. F. (1982) US Patent 4 312 734.
Niedrach, L. W. (1980) *J. Electrochem. Sci. Technol.*, **127**, 2122–30.
Niedrach, L. W. (1982) *J. Electrochem. Sci. Technol.*, **129**, 1445–9.
Nikolskii, B. P. and Materova, E. A. (1985) *Ion Selective Reviews*, **7**, 3–39.
Owen, A. E. (1977) *J. Non-Cryst. Solids*, **25**, 372.
Pask, J. A. (1964) *Modern Aspects of the Vitreous State*, Vol. 3, (ed. J. D. Mackenzie), Butterworth, London, pp. 1–28.
Pask, J. A. (1971) *Proc. A. I. Andrews Memorial Lecture, Ohio, USA.*
Pask, J. A. and Tomsia, A. P. (eds) (1981) *Surfaces on Interfaces in Ceramic and Ceramic–Metal Systems*, Plenum Press, New York and London, pp. 411–9.
Paul, A. (1982) *Chemistry of Glasses*, Chapman and Hall, New York.
Perley, G. A. (1941) *Anal. Chem.*, **21**, 391–559.
Pfaudler (1981) This is not a scientific reference; more information regarding the Pfaudler pH electrode is available from Pfaudler-Werk, A. G. Poorfach 1780, D-6830, Schwetzinger.
Salmeron, M. *et al.* (1981) *Surface Sci.*, 112–207.
Shannon, R. D. *et al.* (1982) *Inorg. Chem.*, **21**, 3372.
Shults, M. M. *et al.* (1981) *Sov. J. Glass Phys. Chem.*, **1**, 292–7.
Smets, B. M. and Lommen, T. P. A. (1983) *Phys. Chem. Glasses*, **24**, 35–6.
Smith, M. D., Genshaw, M. A. and Greyson, J. (1973) *Anal. Chem.*, **45**, 1972–84.
Subbarao, E. C. (ed.) (1980) *Solid Electrolytes*, Plenum Press, New York and London, p. 62.
Tomozawa, M. (1977) *Treatise on Material Science and Technology*, Vol. 12, *Glass Interaction with Electromagnetic Radiation*, Academic Press, New York.
Tomsia, A. P. and Pask, J. A. (1981) *J. Am. Ceram. Soc.*, **64**, 523–8.
Waser, J. and McClanahan, E. D. (1951) *J. Chem. Phys.*, **19**, 413–6.
Weyl, W. A. (1951) *Coloured Glasses*, Society of Glass Technology, Sheffield, England.
Wikby, A. (1974) *Phys. Chem. Glasses*, **15**, 37–41.
Wikby, A. and Karlberg, B. (1974) *Electrochem. Acta*, **19**, 323–8.
Yates, D. E., Levine, S. and Healy, T. W. (1974) *J. Chem. Soc. Faraday Trans. 1*, **70**, 1807.

10

Fibre reinforced glasses and glass-ceramics

Karl M. Prewo

10.1 INTRODUCTION

Historically it can be said that the creation of new types of materials has opened major new avenues to achieving advanced engineering systems and designs. Over the past two decades we have witnessed revolutionary changes in a broad spectrum of applications due to the development of fibre reinforced composites. From sporting goods to industrial and aerospace applications, the availability of these materials has freed the designer from the constraints of metals technology and permitted the development of higher performance systems. This progress has not taken place without difficulty. It has required a whole new way of designing to take into account the tailorability and anisotropy of material properties as well as a whole new way of thinking about material toughness and reliability. These advanced composites, such as carbon fibre reinforced epoxy, exhibit tensile failure strains of less than 1–2% yet they are crack growth resistant and extremely reliable if used properly. It is this successful experience of the past 20 years which has opened up the current real possibility for the success of fibre reinforced ceramics.

In this chapter it will be shown that fibre reinforced glasses and glass-ceramics can make an important contribution to this field of fibre toughened ceramics. The current status of these materials will be described and important aspects of their development traced. It will be clear that, while the potential for the use of these materials is now high, their development has many historical roots.

10.2 COMPOSITE SYSTEMS

Fibre reinforced composites of many types have been developed. Using resin, metal and ceramic matrices, broad classes of materials have been

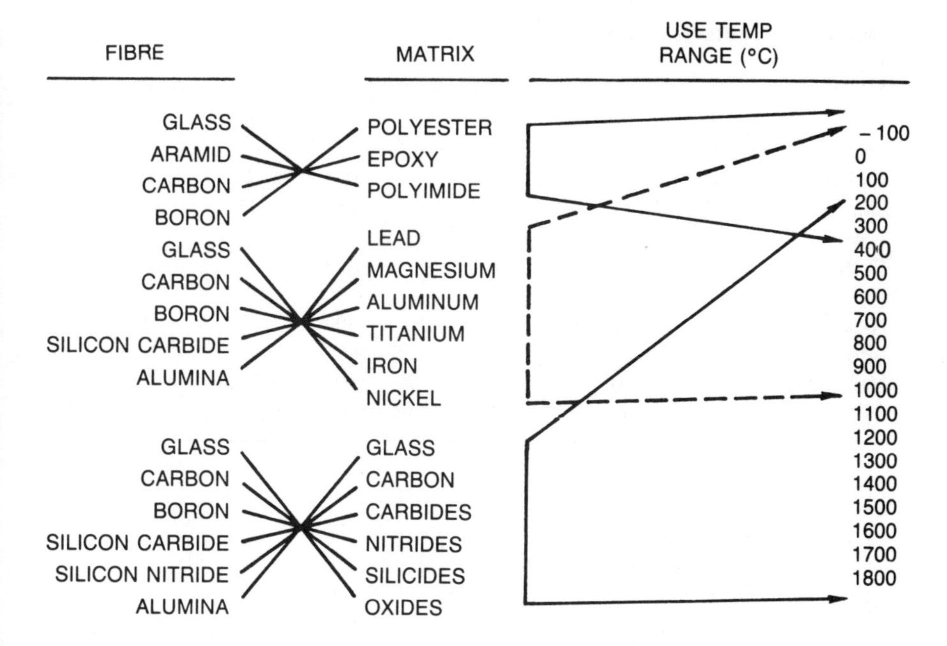

Fig. 10.1 Fibre reinforced composite systems.

created that can be conveniently divided on the basis of their matrix types and their potential use temperatures (Fig. 10.1). Of these systems, the glass and ceramic matrix composites offer the greatest range of utility both on a temperature basis as well as on other environmental considerations such as oxidation resistance, erosion, and chemical attack from acids etc.

Because of the successes achieved in the development of resin and metal matrix composites, a wide range of fibres for use in the reinforcement of glasses is now available. The fibres listed in Table 10.1 are those most prominent in current composite development. None of those listed was developed specifically for use in glass or ceramic matrices; however, development programmes currently under way around the world are aimed at this need, i.e. fibres tailored to high temperature glass and ceramic composite applications.

While a much wider variety of glasses and glass-ceramics is also available, the characteristics of several of the more important composition types are listed in Table 10.2. The glass-ceramics designated LAS I, II and III are compositions that were created to be compatible with Nicalon SiC type fibre. They provide examples of the unique capability to tailor compositions of glass ceramics to meet the requirements set down by fibre composition.

Table 10.1 Reinforcements used for glass and glass-ceramic matrices

Fibre	*Form*[5]	*Diameter* (μm)	*Density* (g cm^{-1})	E[6] (GPa)	*UTS*[7] (GPa)	*CTE*[8] ($10^{-6}\,°C^{-1}$)
Boron	M	100 to 200	2.5	400	2.75	4.7
Silicon carbide	M	140	3.3	425	3.45	4.4
Carbon	Y	7 to 10	1.7–2.0	200 to 700	1.4 to 5.5	−0.4 to −1.8
Nicalon[1]	Y	10 to 15	2.55	190	2.4	3.1
FP alumina[2]	Y	20	3.9	380	1.4	5.7
Nextel 312[3]	Y	10	2.5	150	1.7	–
VLSI-SiC[4]	W	6	3.3	580	8.4	–

[1] Nippon Carbon Company.
[2] Dupont Company.
[3] 3M Company.
[4] Los Alamos National Laboratory.
[5] M = monofilament; Y = yarn; W = whisker.
[6] E = elastic coefficient.
[7] *UTS* = ultimate tensile strength.
[8] *CTE* = coefficient of thermal expansion.

10.3 COMPOSITE FABRICATION

Glass matrix composites, as compared to other ceramic candidates, probably offer the greatest commercial potential due to their ease of densification, low cost and also achievement of high performance. The following attributes are important in this sense.

1. Glasses can be created with a broad range of chemistries to control fibre–matrix chemical interaction;
2. Glasses can be created with a wide range of thermal expansion coefficients to tailor them to nearly match those of reinforcing fibres;
3. The low elastic modulus of glasses (50–90 GPa) permits high modulus fibres to provide true reinforcement;
4. The ability to control the viscosity of glasses and to flow them easily under pressure permits the physical densification of fibre reinforced composites without mechanical damage to the fibres. It will be shown that relatively high fibre contents can be achieved by several techniques;
5. The composite densification process can be rapid since glass matrix flow is all that is required.

10.3.1 Hot pressing of tapes

Because of the fact that glass can be treated as a thermoplastic material, many of the processes developed for fibre reinforced glasses were made to

Table 10.2 Glass and glass-ceramic matrices of interest

Matrix type	*Major constituents*	*Minor constituents*	*Major crystalline phase*	*Maximum use temp.* (°C) *(in composite form)*
Glasses*				
7740 Borosilicate	B_2O_3,SiO_2	Na_2O,Al_2O_3	–	600
1723 Aluminosilicate	Al_2O_3,MgO,CaO,SiO_2	B_2O_3,BaO	–	700
7930 High silica	SiO_2	B_2O_3	–	1150
Glass-ceramics				
LAS-I	Li_2O,Al_2O_3,MgO,SiO_2	ZnO,ZrO_2,BaO	β-Spodumene	1000
LAS-II	Li_2O,Al_2O_3,MgO,SiO_2,Nb_2O_5	ZnO,ZrO_2,BaO	β-Spodumene	1100
LAS-III	Li_2O,Al_2O_3,MgO,SiO_2,Nb_2O_5	ZrO_2	β-Spodumene	1200
MAS	MgO,Al_2O_3,SiO_2	BaO	Cordierite	1200
BMAS	BaO,MgO,Al_2O_3,SiO_2	–	Barium osumilite	1250
Ternary mullite	BaO,Al_2O_3,SiO_2	–	Mullite	~1500
Hexacelsian	BaO,Al_2O_3,SiO_2	–	Hexacelsian	~1700

* 7740, 1723 and 7930 are Corning Glass Works designations.

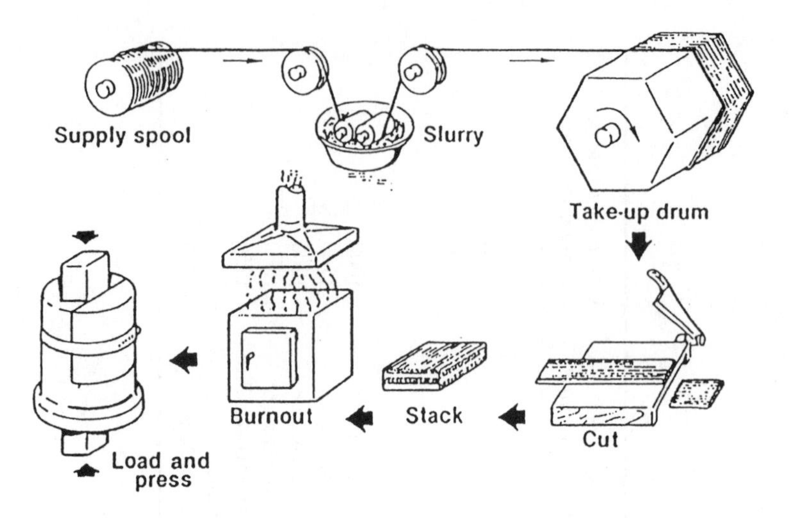

Fig. 10.2 Steps in tape lay-up of glass-matrix composites.

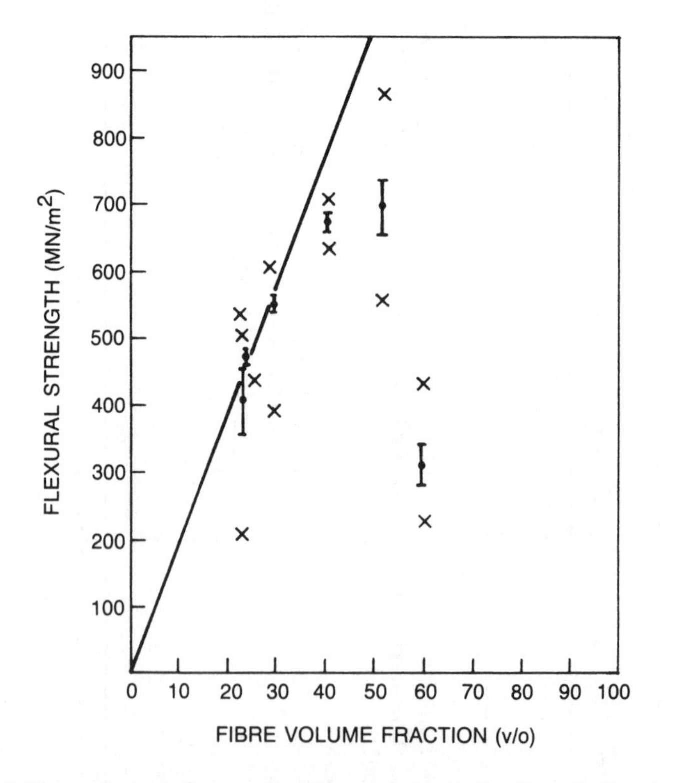

Fig. 10.3 Flexural strength as a function of volume fraction of fibre in aligned continuous carbon fibre reinforced glass. After Sambell *et al.* (1974).

emulate those previously used for polymer matrix systems. Sambell *et al.* (1974) described in detail the development of a procedure for the fabrication of carbon fibre reinforced glasses using a process closely resembling that for polymer matrix systems. Collimated fibres were wound on mandrels after having been infiltrated with a slurry of glass powder (Fig. 10.2). The resultant tapes could then be cut into plys, any organic binder burned out

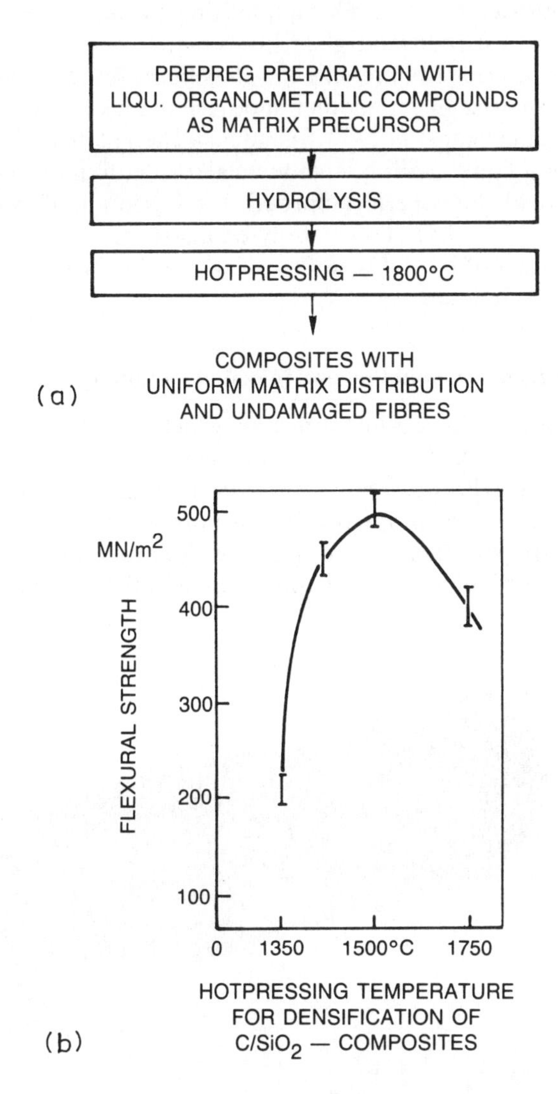

Fig. 10.4 Preparation of composites using metalalkoxides. After Fitzer (1978), Sahebkar *et al.* (1978) and Schlichting (1980). The following starting matrix materials were used: for SiO_2 tetramethoxy silane; for TiO_2 tetraethyl orthotitanate; for Al_2O_3 Al-tri-sec butylate.

and densified under pressure and high temperature to achieve nearly full density microstructures. Resultant composite flexural strengths were found to depend on bonding conditions and also on the volume percentage of fibre reinforcement (Fig. 10.3). While strength increased with fibre content in a manner expected from the rule of mixtures, no further increases, and in fact decreases in strength were noted at 40% fibre and above.

Another approach to precursor tape making has been reported (Fitzer, 1978) where both carbon and glass fibres were co-wound to provide a precursor tape. During hot pressing the glass fibres fused together to form the composite matrix. In yet another approach to making precursor tapes, Fitzer (1978), Sahebkar *et al.*, (1978) and Schlichting (1980) demonstrated that the use of a liquid hydrolysable metalalkoxide, to infiltrate precursor fibre tows, provided a superior approach leading to a uniform composite microstructure (Fig. 10.4). The hydrolysed matrix precursor plus fibres is then hot press densified. SiO_2, SiO_2–TiO_2 and Al_2O_3 matrices were all created by this technique.

10.3.2 Matrix transfer moulding and injection moulding

Other approaches to composite fabrication which do not require the use of precursor tapes and which simulate polymer matrix composite practice have also been reported (Prewo *et al.*, 1986). Matrix transfer moulding, i.e. the injection of hot matrix into a prealigned fibre array, has permitted the fabrication of shapes otherwise not possible. The tube shown in Fig. 10.5 contains fibres reinforcing it in both the circumferential and radial directions

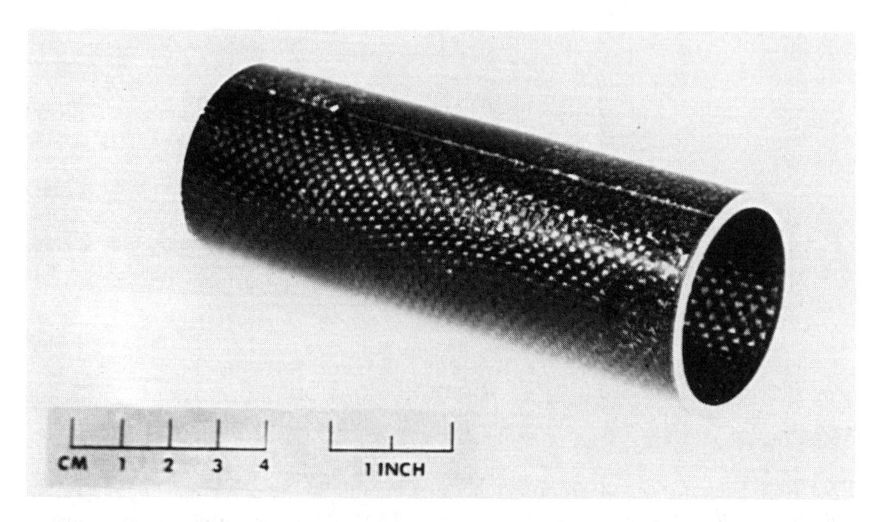

Fig. 10.5 Matrix transfer moulded woven graphite fibre reinforced borosilicate glass matrix tube.

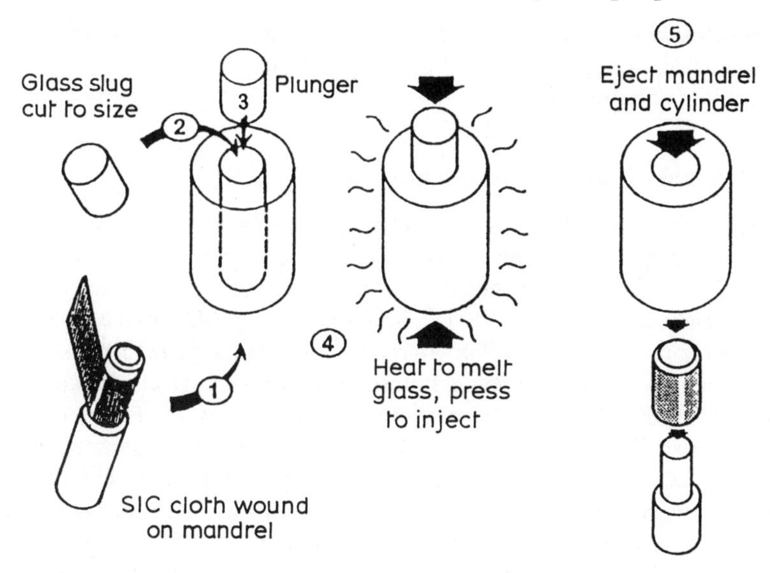

Fig. 10.6 Steps in matrix transfer moulding of a thin wall cylinder.

and could not be made by hot pressing. The process for making this tube is shown schematically in Fig. 10.6. Similarly tubes and other complex shapes can be made using chopped fibre moulding compounds (Prewo *et al.*, 1986) that contain both fibre and matrix and can be injected or compression moulded into shaped dies at high temperatures.

While the above processes were originally demonstrated with glasses, they can also be readily practised using glass-ceramics. The key here is that composite densification takes place while the matrix is in the substantially glassy and viscous form. After full densification the glass can be crystallized under controlled heat treatment conditions to achieve a matrix with superior toughness and high temperature strength.

As a final note, it should be pointed out that the exceptional processability of glasses as matrices has led to some rather ingenious concepts for the low cost continuous processing of composites (Siefert, 1971).

10.4 COMPOSITE PROPERTIES

While metal fibres had for many years formed the basis for the reinforcement of glasses and ceramics (Donald and McMillan, 1976, 1977; Lucas *et al.*, 1980) their relatively high density and low thermal stability prevented any major applications. In contrast the development and availability of advanced high performance fibres, Table 10.1, has led to composites with exceptional performance potential. It is the characteristics of these latter systems which will be used to illustrate the important aspects of this class of materials.

10.4.1 Carbon fibre reinforced composites

Carbon fibres offer the highest structural performance potential of any of the reinforcements. Available in a wide variety of elastic moduli and at relatively low cost, they offer the opportunity to create economically viable systems right now. Early evidence obtained using a SiO_2 matrix resulted in composites possessing excellent toughness and crack growth resistance as evidenced by non-catastrophic failure in bending (Crivelli-Visconti and Cooper, 1969) (Fig. 10.7). Despite being porous, these composites exhibited excellent strength of 350 MPa at both room room temperature and 800°C as well as being unaffected by a water quench from 1200°C. Further evidence for the ability to achieve improved composite toughness for carbon fibre reinforced borosilicate glass, glass-ceramic, MgO and Al_2O_3 was demonstrated for discontinuous fibres. However, only in the cases where fibres were carefully aligned or where they were continuous could composite strength exceed that of the matrix (Sambell *et al.*, 1972a, 1972b; Phillips *et al.*, 1972). The improved toughness was, in major part, attributed to low fibre–matrix interface strength which prevented matrix cracks from propagating from the matrix into the carbon fibres. As a consequence,

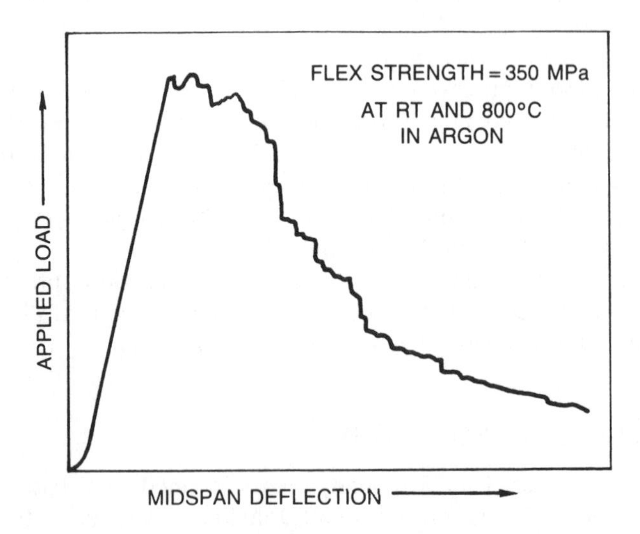

Fig. 10.7 Applied load against deflection for carbon fibre reinforced SiO_2 at room temperature. Water quenching from 1200°C has no effect on strength. After Crivelli-Visconti and Cooper (1969).

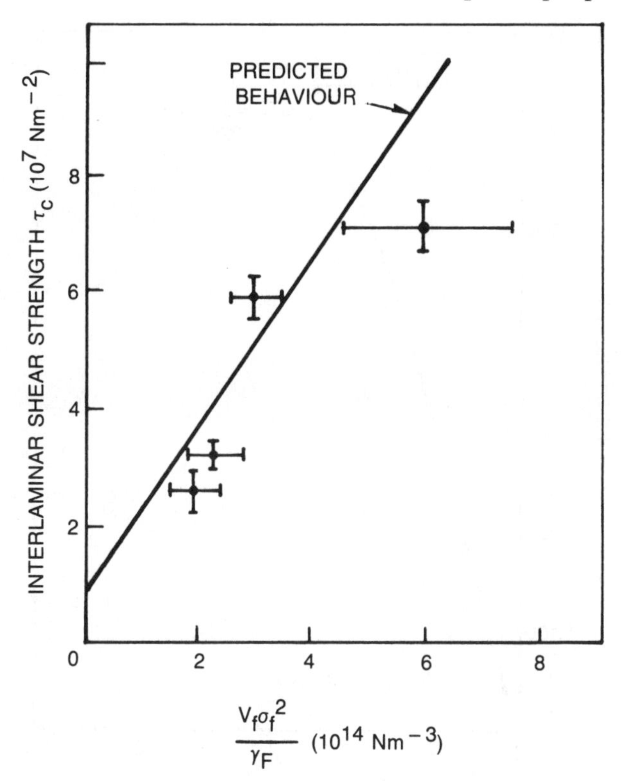

Fig. 10.8 Composite fracture energy (γ_f) due to fibre pull-out as it relates to composite shear strength. After Phillips (1974) (σ_f = fibre strength, V_f = fibre content).

however, the maintenance of a low fibre–matrix interfacial strength causes a low composite interlaminar shear strength (Phillips, 1974) (Fig. 10.8). This is a significant point in that it causes one to chose between composite 'off axis' strength and composite toughness, both of which are considered important performance parameters.

The ability to fabricate a broad range of carbon fibre reinforced glasses and glass-ceramics has been demonstrated by several investigators and resultant composite mechanical properties have been characterized for numerous types of test conditions (Levitt, 1973; Prewo and Bacon, 1978; Prewo and Minford, 1985; Prewo and Thompson, 1981; Prewo *et al.*, 1979a, 1979b). Through the use of very high elastic modulus pitch-based carbon fibre, composites with exceptionally high elastic moduli were obtained (Prewo and Thompson, 1981; Prewo and Minford, 1985) (Fig. 10.9). In contrast, using discontinuous carbon fibres of sufficient length permitted

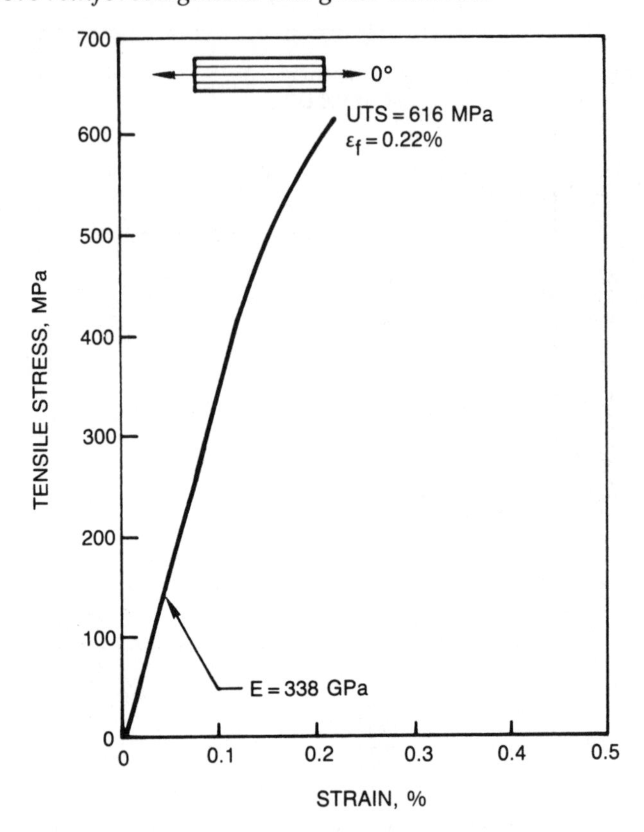

Fig. 10.9 Tensile stress–strain curve for unidirectional 54 v/o P-100 carbon fibre reinforced borosilicate glass matrix composite. After Prewo and Thompson (1981).

the development of composites whose elastic modulus is less than that of the glass matrix, whose failure strain can exceed 0.5%, and whose tensile stress–strain behaviour can be contrasted with that of resin matrix composites (Prewo, 1982) (Fig. 10.10). The very non-linear stress–strain curve shape for this latter glass matrix system was attributed to the microcracking of the matrix at its failure strain of 0.1–0.2% and hence a decrease in composite stiffness with increasing strain. This is unlike the epoxy matrix composite where the matrix failure strain exceeds that of the fibres. The increased compliance of the glass matrix system with increasing strain was also shown to be extremely beneficial to the glass matrix composite during bend testing where, unlike pure tension, the glass matrix composite flexural beam load-carrying capacity exceeded that for the epoxy matrix composite by nearly 50% (Table 10.3).

Further comparison with polymer matrix systems, over a wide test

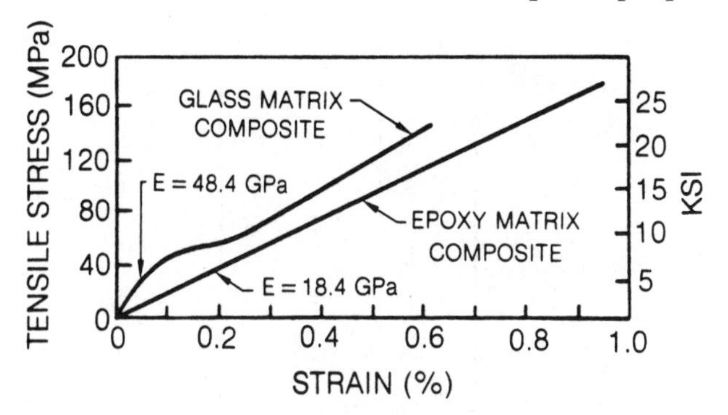

Fig. 10.10 Tensile stress–strain curves for discontinuous carbon fibre reinforced borosilicate glass and epoxy. The fibres are in a 2-D random array. The epoxy matrix composite contains 20 v/o fibres and the borosilicate glass matrix contains 30 v/o fibres. After Prewo (1982).

temperature regime (Prewo and Bacon, 1978) (Fig. 10.11), served to show that continuous carbon fibre reinforced glass matrix composites can exhibit strengths equivalent to their resin matrix counterparts but at much higher temperatures. The fact that the apparent 600°C strength of the glass matrix composite is equivalent to that of the resin matrix material at room temperature is due to the ability of the glass matrix system to redistribute loads more effectively at the higher temperatures. At temperatures above

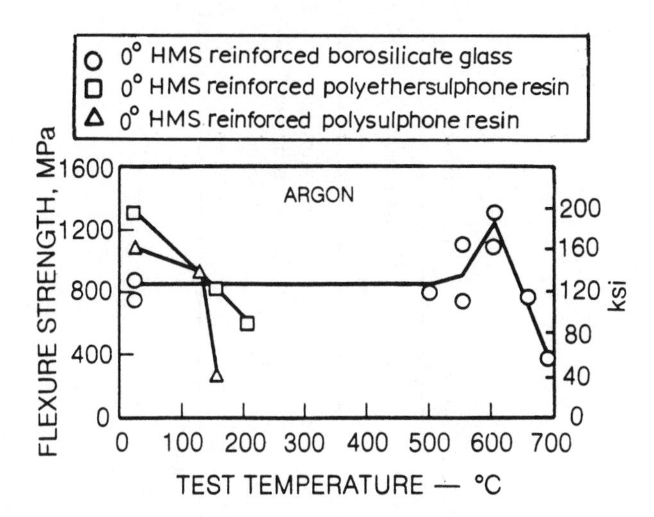

Fig. 10.11 Unidirectional carbon fibre reinforced borosilicate matrix composite flexural strength comparison for tests performed at temperature in inert argon atmosphere. After Prewo and Bacon (1978).

Table 10.3 Composite strength comparison

Fibre	*Vol %*	*Orientation*	*Matrix*	*Tensile strength* (MPa)	*Failure strain* (%)	*Flexural strength* (MPa)
Carbon	20	2D discontinuous	Epoxy	180	0.96	268
Carbon	30	2D discontinuous	Borosilicate	150	0.63	400
Nicalon	50	0° continuous	Epoxy	875	0.84	1240
Nicalon	44	0° continuous	LAS*	670	0.90	1380

* Unceramed.

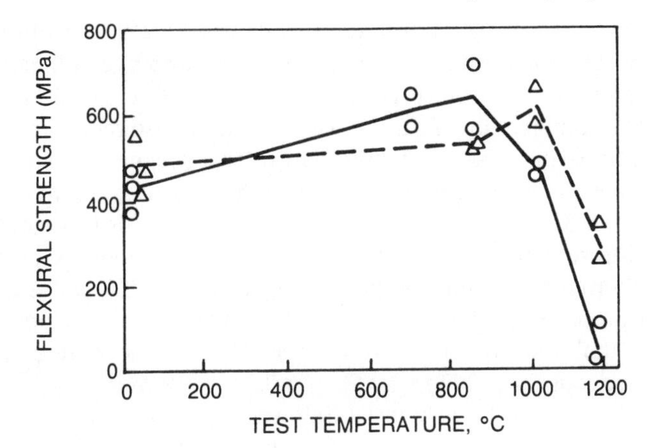

Fig. 10.12 Flexural strength against temperature in argon of unidirectional HM carbon fibre reinforced 96% SiO matrix composites. After Prewo and Thompson (1981).

600°C the glass matrix specimens just deformed and did not fracture at all. Glass-ceramic and silica matrix composites extend this strength retention region to over 1000°C (Sambell *et al.*, 1974; Fitzer, 1978; Prewo and Thompson, 1981; Prewo, 1982); however, only in non-oxidizing environments where fibre stability was not limiting (Fig. 10.12).

While most of the emphasis over the years has been on the development of carbon reinforced glass for its mechanical properties, it should not be forgotten that it is a system that can also be extremely useful for other

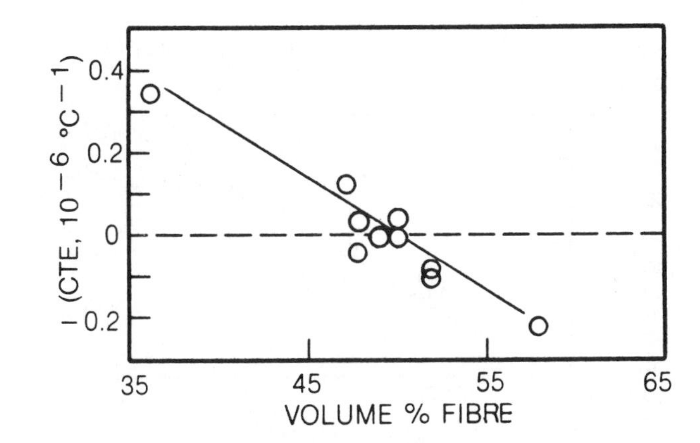

Fig. 10.13 In plane coefficient of thermal expansion (over −20°C to +80°C) as a function of fibre content for 0/90 cross ply HM carbon fibre reinforced borosilicate glass. After Prewo and Minford (1984).

reasons. As in the case of carbon reinforced resins, the carbon fibres impart lubricity to the composite surface and the glass matrix can impart higher harnesss and wear resistance (Khanna *et al.*, 1983; Minford and Prewo, 1985b). The combination of glass and carbon fibres also results in a material with exceptional dimensional stability equivalent or superior to even the most dimensionally stable glasses (Prewo, 1979; Prewo *et al.*, 1979a; Prewo and Thompson, 1981; Prewo and Minford, 1984, 1985). Also, because of the range of different glasses, carbon fibres and fibre distributions available, a range of desired coefficients of thermal expansion (CTE) can be achieved. The data presented in Fig. 10.13 were obtained in the development of a composite tailored to have an average CTE of near zero over the temperature range of −20 to +80°C.

All too frequently the non-structural aspects of composite performance are neglected when instead they could become the very first reasons for composite application.

10.4.2 Oxide fibre reinforced composites

Both Al_2O_3 and SiO_2 type fibres have been used to reinforce glasses in the hope of achieving systems with excellent high temperature oxidative stability. Several different types of alumina fibres were used to reinforce high silica glass matrices with the result that modest levels of strength were achieved (Bacon *et al.*, 1978) (Fig. 10.14). These levels were maintained up to 1000°C. Through qualitative observations of composite fracture surfaces, however, it was found that composite fracture toughness was much less than that of carbon fibre reinforced composites and this difference was associated with the formation of a much stronger fibre–matrix interfacial bond in these

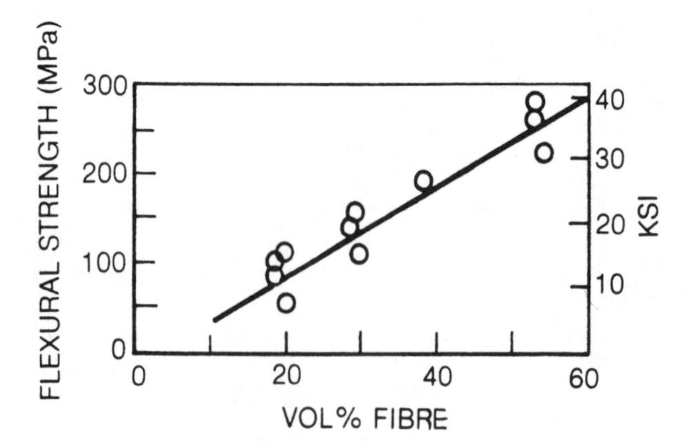

Fig. 10.14 FP alumina fibre reinforced silica composite strength as a function of fibre content. After Bacon *et al.* (1978).

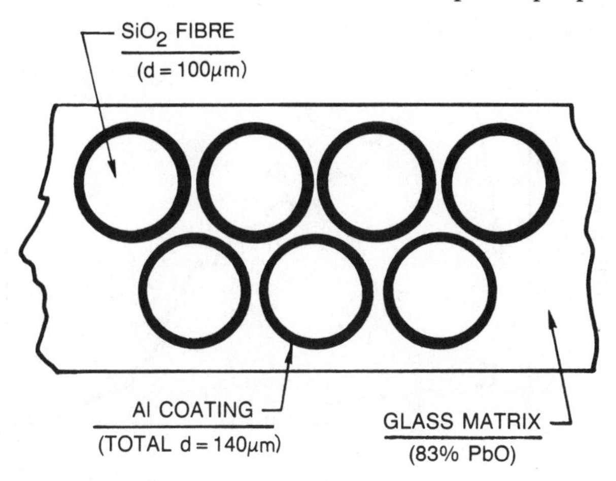

Fig. 10.15 Metal coated fibres used to create impact resistant inorganic composites. After Siefert (1972, 1975).

systems. Fracture surfaces exhibited only very short lengths of fibre pull out and cracks propagated through these composites much more readily.

A quantitative analysis of the dependence of composite fracture toughness on fibre reinforcement content for SiO_2 fibre/SiO_2 and Al_2O_3 fibre/Al_2O_3 composites showed that overall system toughness could be two to three times that of the unreinforced matrix (Fitzer, 1978). To achieve much higher levels of performance it was shown that an interfacial region could be artificially created between fibre and matrix (Siefert, 1972, 1975). Through the use of an aluminium metal coating on 100 μm diameter SiO_2 fibres, it was possible to incorporate these fibres into a low temperature glass matrix and achieve a notch insensitive impact resistant material (Fig. 10.15).

10.4.3 Silicon carbide and boron large diameter fibre reinforced composites

The availability of high strength fibres of boron and silicon carbide produced by chemical vapour deposition has also been actively pursued as an approach to achieving high performance composites. These filaments have been available for nearly as long as carbon fibres but because of their greater cost and their somewhat less convenient and less flexible composite fabrication possibilities they have not received quite as much attention. Boron fibre reinforced glass (Siefert, 1971), was shown to provide composites of exceptionally high strength, stiffness and toughness. A borosilicate glass matrix composite reinforced with 30% by volume boron fibres exhibited flexural strengths of 1120 MPa and 980 MPa at room

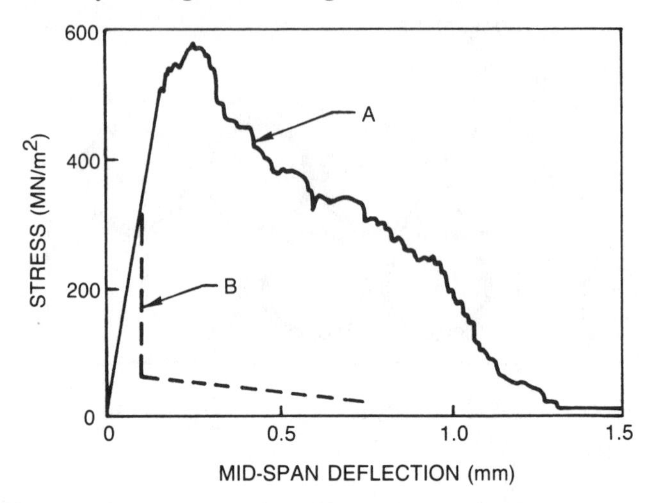

Fig. 10.16 Flexural stress against mid-span deflection for SiC large diameter fibre reinforced cordierite. After Aveston (1971). A, tough composite matrix CTE < fibre CTE; B, brittle composite with matrix CTE > fibre CTE.

temperature and 540°C respectively. Also, no loss in strength was experienced after 100 h exposure at 540°C. SiC monofilaments, however, were used to create composites with considerably higher use temperature possibilities (Aveston, 1971; Prewo and Brennan, 1980). The use of a cordierite matrix was found to introduce the complexity of achieving the proper heat treatment condition. Post fabrication composite heat treatment principally controlled matrix coefficient of thermal expansion (CTE) and was capable of drastically altering both composite strength and fracture mode. For processing conditions which resulted in composite matrices having a CTE greater than that of the SiC fibres, the composite fracture mode was brittle and the strength low, Fig. 10.16. This was attributed to a strong mechanical bond between fibre and matrix since, on cooling from the fabrication conditions, the matrix contracted enough to load the fibre–matrix interface in compression. However, by heat treating to create a matrix with CTE much less than that of the fibre, on cooling the fibre–matrix interface is loaded in tension so that, even if a chemical bond is formed during composite fabrication, it is fractured and disrupted during cooling after densification. The resultant composites were nearly 300% stronger than those with the high CTE matrix and the failure mode was extremely fibrous and tough (Aveston, 1971).

Using a newer form of large diameter SiC fibre that was deposited on a carbon core, borosilicate glass matrix composites were fabricated with high strength and toughness up to test temperatures at which the matrix began to

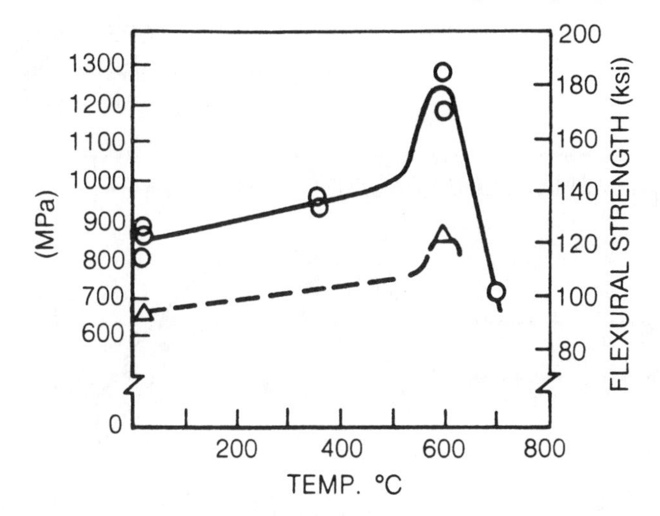

Fig. 10.17 Flexural strength against temperature in air for unidirectional SiC monofilament reinforced borosilicate glass (○, 65 vol % fibre; △, 35 vol % fibre). After Prewo and Brennan (1980).

soften (Prewo and Brennan, 1980). In this case the matrix CTE was less than that of the fibre and the as-produced fibre surface was carbon rich. Both of these factors combined in producing a low fibre–matrix interfacial bond strength and tough, strong composites (Fig. 10.17). In a manner similar to that noted in Fig. 10.11 for carbon fibre reinforced glass composites, the load carry capabilities of these flexure specimens increased significantly with increasing test temperature and peaked at a temperature (600°C) at which the matrix could deform significantly to redistribute stresses. At higher temperatures the matrix was too compliant and the specimens only deformed and did not fracture at all.

10.4.4 Silicon carbide yarn reinforced composites

A major increase of interest in the development of fibre reinforced ceramics and glasses can be attributed to the development of a high performance silicon carbide type yarn by Professor Yajima and his co-workers. Available under the name Nicalon with an average tensile strength and elastic modulus of 2060 MPa and 193 GPa respectively, this fibre has a unique non-stoichiometric chemistry that makes it particularly suited to the development of high strength glass and glass-ceramic matrix composites (Prewo and Brennan, 1980, 1982; Brennan and Prewo, 1982; Brennan, 1985). Initially, composites with excellent strength were achieved using glass matrices (Prewo and Brennan, 1980, 1982) (Fig. 10.18) and again an

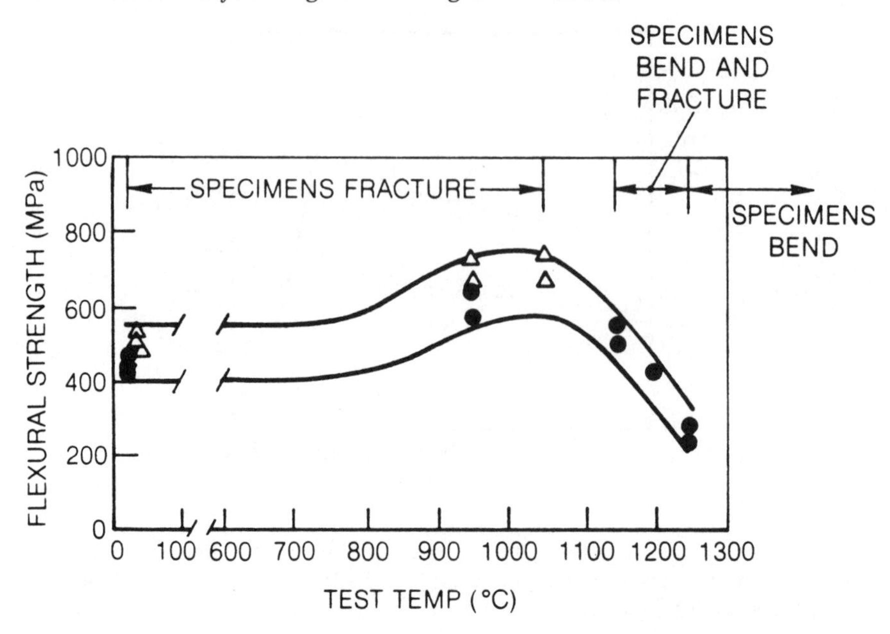

Fig. 10.18 Flexural strength against temperature in argon for unidirectional Nicalon reinforced 96% SiO_2 matrix composites. After Prewo and Brennan (1982).

increase in specimen strength was noted at temperatures at which the matrix permitted stress redistribution without excessive deformation. To achieve the highest levels of strength, however, the use of lithium aluminosilicate (LAS) matrices proved most advantageous (Brennan and Prewo, 1982). By densifying the composites while the matrix was in a glassy state and then crystallizing (ceraming) the matrix afterwards it was possible to fabricate easily yet end up with a very refractory composite. Resultant composite strength and toughness were high and, as in the case of fibre reinforced polymers, achievable in both multiaxially and uniaxially aligned fibre speciments (Figs 10.19 and 10.20). The fracture morphologies of these composites were exceptionally fibrous (Fig. 10.21) and have been related to the presence of a low strength carbon rich fibre–matrix interfacial region created during composite fabrication and attributable to fibre and matrix chemistry (Brennan, 1985). The interface chemistry of both strong (fibrous) and weak (brittle) composites was compared and it was shown that all strong composites possessed carbon rich interfacial regions while weak composites did not. The comparison between two different Nicalon fibre reinforced 96% silica glass matrix composites is made in Figs 10.22 and 10.23. In these examples, the chemistry of the Nicalon fibre surfaces was analysed after composite fabrication and fracture. The strong composite was associated

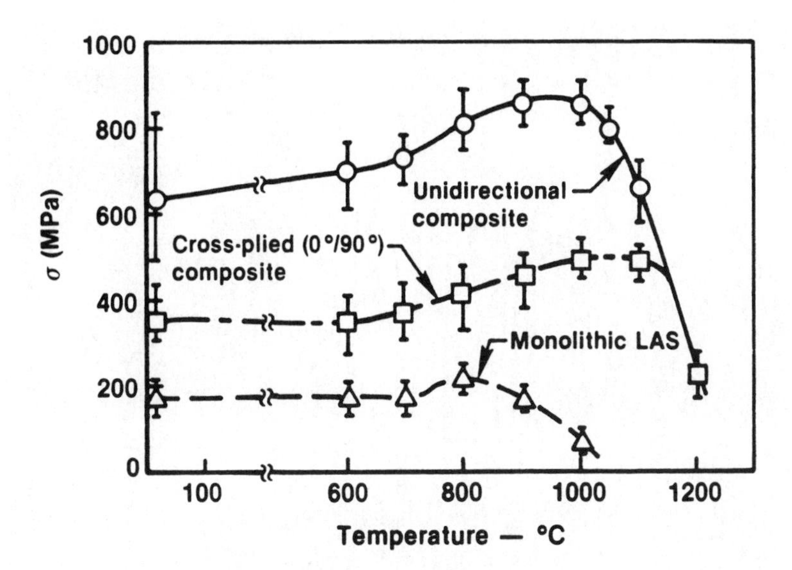

Fig. 10.19 Flexural strength in argon of Nicalon SiC fibre reinforced LAS-I glass-ceramic matrix composites. After Brennan and Prewo (1982).

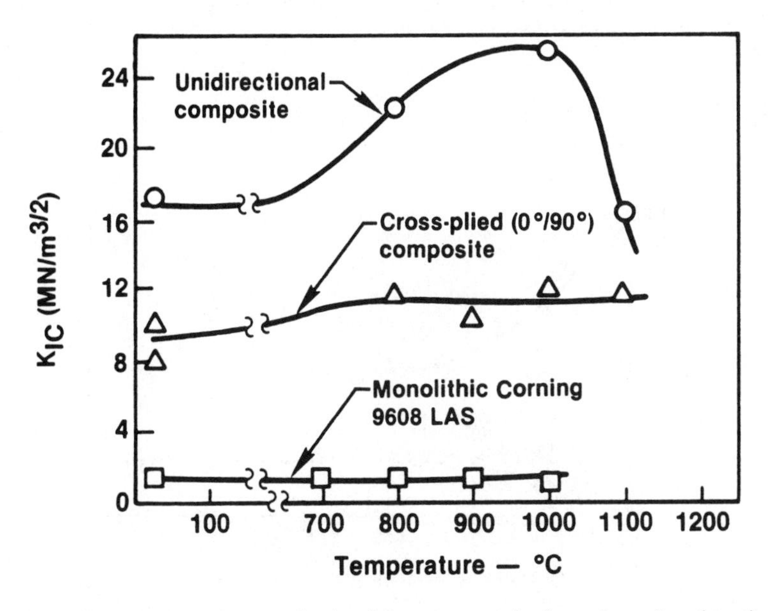

Fig. 10.20 Fracture toughness, obtained from prenotched specimen bend testing in argon, of Nicalon SiC fibre reinforced LAS-I glass-ceramic matrix composites. After Brennan and Prewo (1982).

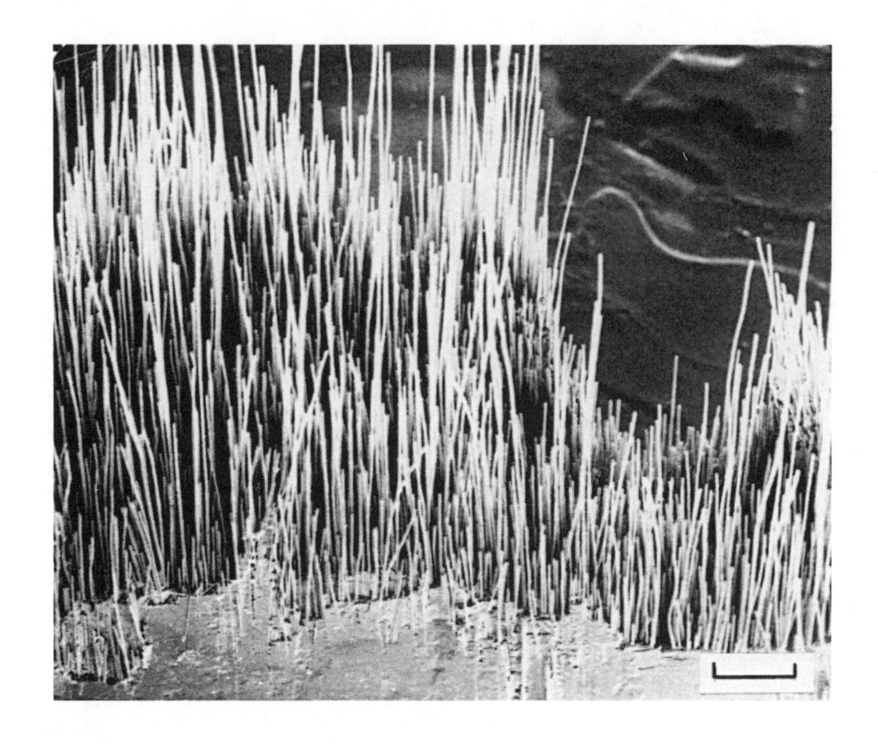

Fig. 10.21 Tensile fracture surface of unidirectional Nicalon fibre reinforced LAS-III glass-ceramic matrix composite tested at 22°C (bar = 500 μm). After Prewo (1986).

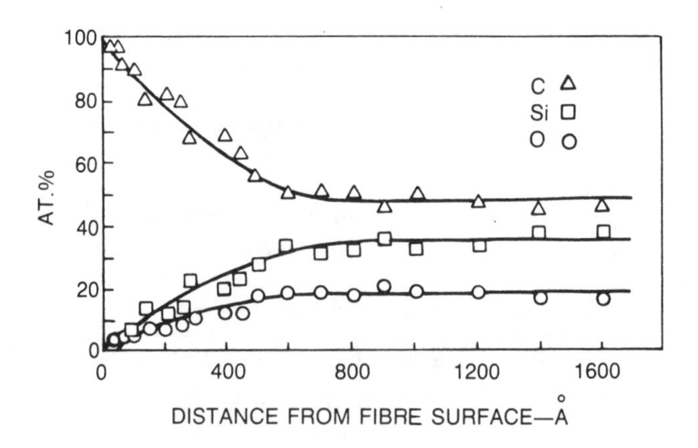

Fig. 10.22 Scanning Auger analysis of Nicalon SiC fibre from a strong 7930 glass matrix composite fracture surface. After Brennan (1985).

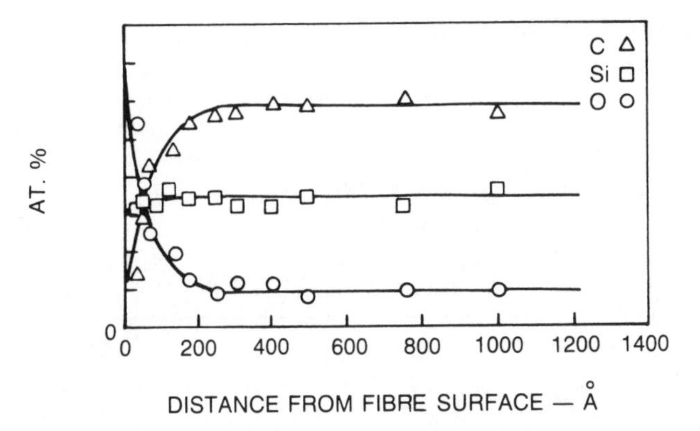

Fig. 10.23 Scanning Auger analysis of Nicalon SiC fibre from a weak 7930 glass matrix composite fracture surface. After Brennan (1985).

with fibres having a much higher carbon content than either the starting fibres themselves (Fig. 10.24) or much weaker composite counterparts (Brennan, 1985). As in the case of the carbon fibre reinforced glasses described above, the carbon rich interface provides an interfacial region of low strength which prevents cracks from propagating from matrix to fibre.

This low fibre–matrix interfacial strength, while desirable from a toughness point of view, is particularly deleterious to composite off axis strength. The data in Table 10.4 summarize some of the properties of the

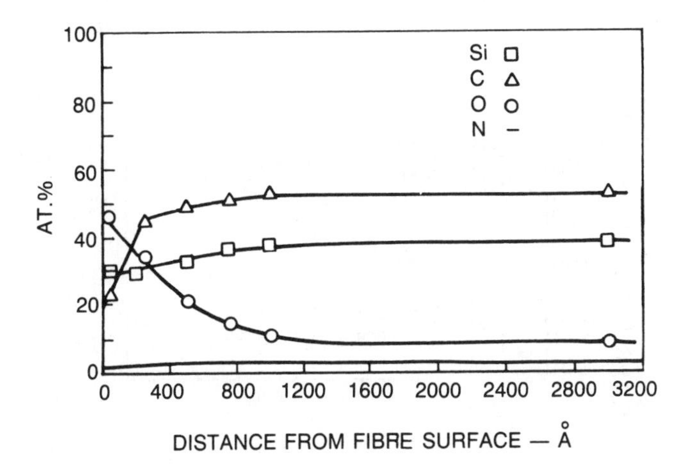

Fig. 10.24 Scanning Auger analysis of Nicalon SiC fibre after bunsen burner flame treatment to remove sizing and before composite fabrication. After Brennan (1985).

Table 10.4 Average properties of continuous Nicalon SiC reinforced glass-ceramic (ceramed LAS)

Unidirectional		*0/90 Cross ply*
ET = 130 GPa	E_{90}F = 45 GPa	E_0T = 76 GPa
σ_0T = 690 MPa	σ_{90}F = 25 MPa	σ_0T = 410 MPa
ε_{f_0}T = 0.9%	$\varepsilon_{f_{90}}$F = 0.06%	ε_{f_0}T = 0.9%
CTE$_0$ = 2.8×10^{-6} °C^{-1}	CTE$_{90}$ = 1.1×10^{-6} °C^{-1}	CTE = 2.3×10^{-6} °C^{-1}

Vol % fibre = 46 v/o.
$\varrho = 2.5\,\mathrm{g\,cm^{-3}}$.
T = tensile test; F = flexural test.

Nicalon SiC reinforced LAS composite system and show that, for unidirectionally reinforced composites, transverse (90°) strength is less than 5% of axial (0°) strength.

Further testing of these composites in tension has shown that their strength can, in part, be related to the *in-situ* Nicalon fibre tensile strength, and that, when fibre strength and failure strain are great enough to permit matrix failure to occur without overloading the fibres, a non-linear tensile stress–strain curve results (Prewo, 1986) (Fig. 10.25). Also shown in this figure is the tensile stress–strain curve for a Nicalon reinforced epoxy matrix composite (Strife and Prewo, 1982) which is perfectly linear to failure due to the fact that the epoxy matrix has a higher failure strain than the Nicalon fibres. Repeated mechanical tensile cycling of the LAS matrix composite at increasing values of strain, Fig. 10.26, results in the observation that composite elastic modulus decreases markedly with increasing strain. This is attributed to the progressive microcracking of the matrix and accompanying

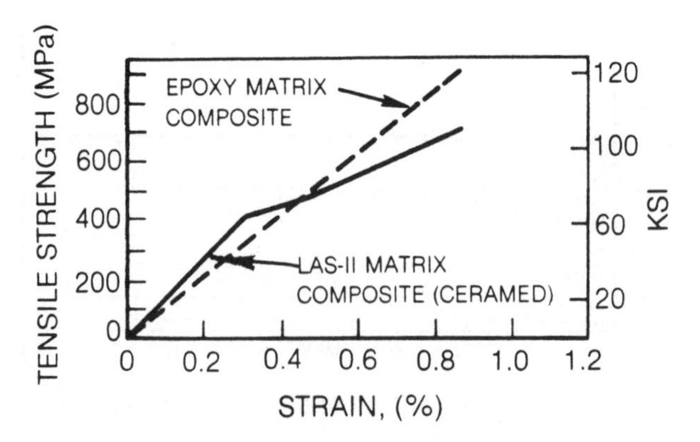

Fig. 10.25 Tensile stress–strain comparison for unidirectional Nicalon SiC fibre reinforced composites. After Prewo (1986).

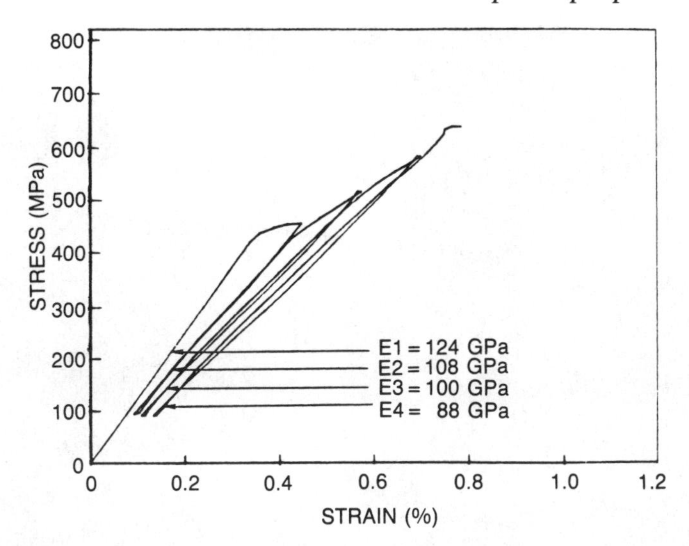

Fig. 10.26 Cycled tensile stress–strain curves for unidirectional Nicalon SiC fibre reinforced LAS-II (ceramed) at 22°C. After Prewo (1986).

decrease in its contribution to the composite stiffness. Eventually the final composite elastic modulus of 88 MPa can be attributed almost solely to the reinforcing fibres. Both the non-linear shape of this SiC/LAS composite's tensile stress–strain curve and also its decreasing stiffness with strain make this composite (just as in the case of the discontinuous carbon reinforced glass composite, Fig. 10.10) highly tolerant of stress concentrations. A summary of composite tensile and flexural strengths for both the discontinuous carbon and continuous Nicalon reinforced composites is presented in Table 10.3 where it can be seen that in the case of flexure, in which a non-uniform stress state is applied, the glass and LAS matrix composites are significantly stronger than their epoxy matrix counterparts.

The extent to which matrix cracking can occur in a non-uniform stress state is illustrated graphically in Fig. 10.27. The specimen shown in the figure consisted of 0/90 Nicalon reinforced LAS and was notched with a circular hole in the centre of its gauge length prior to testing. Subsequent tension–tension fatigue caused the cracking pattern shown to develop; however, ultimate composite failure occurred later after more fatigue cycles (Minford and Prewo, 1985a). In addition, specimens of this geometry tested in tension exhibited net section tensile strengths similar to those of unnotched specimens (Fig. 10.28). While not quantitative, these examples demonstrate the exceptional toughness of these materials.

While matrix microcracking is the factor which causes this unique stress–strain behaviour and accompanying stress gradient tolerance, it also has

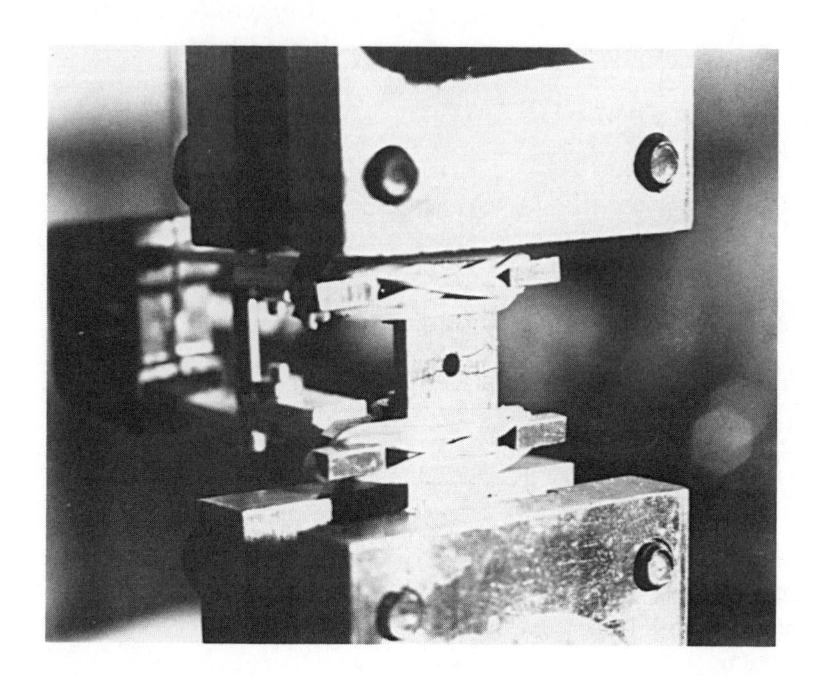

Fig. 10.27 Cracking pattern emanating from premachined hole in 0/90 Nicalon SiC fibre reinforced LAS-II composite during tension–tension fatigue. After Minford and Prewo (1985).

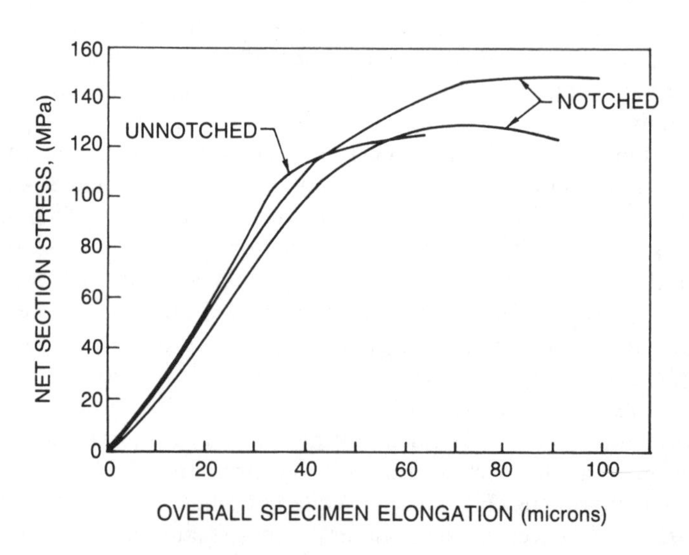

Fig. 10.28 Comparison of notched (circular 15 mm diameter hole) and unnotched tensile tests of 0/90 Nicalon SiC fibre reinforced LAS-III composites.

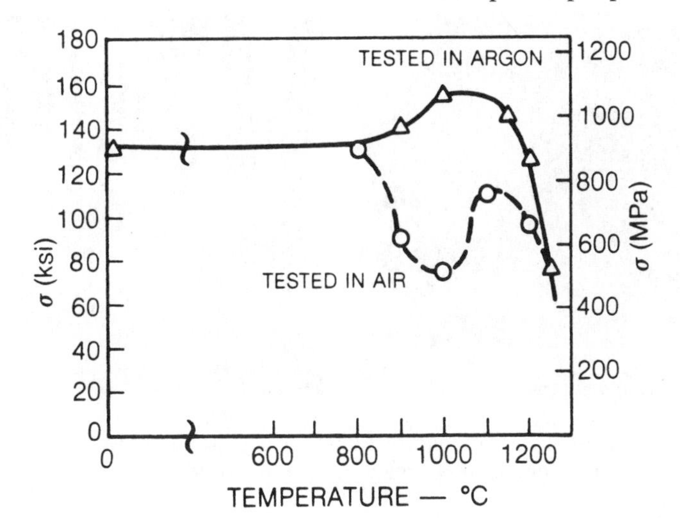

Fig. 10.29 Flexural strength against temperature in argon and air for unidirectional Nicalon SiC fibre reinforced LAS-III glass-ceramic.

implications for composite fatigue performance and environmental stability. Composite fatigue resistance is found to be quite good, however dependent on whether the applied stress exceeds the matrix cracking point (Minford and Prewo, 1985a). Similarly, composite environmental stability is related to this same point (Prewo, 1986). When applied stresses exceed those necessary to cause matrix cracking the surrounding test environment can attack the fibre–matrix interface and change composite fracture morphology from fibrous to relatively brittle. The data in Fig. 10.29 were obtained by flexure testing in both air and argon. At 700°C and above in air composite strength decreased significantly in comparison with the data for tests performed in argon. The testing in air at these temperatures has lowered composite strength to the stress level at which matrix cracking begins to occur (Prewo, 1986). The air then infiltrates the composite and attacks the formerly low strength carbon rich fibre–matrix interface in such a way as to cause an embrittlement of the fracture process (Brennan, 1985; Prewo, 1986). The fracture surface morphology shown in Fig. 10.30 was obtained by flexure testing in air and contrasts markedly from that of Fig. 10.21. The tensile side of the fracture surface, Fig. 10.30, is almost completely flat and exhibited comparatively little crack propagation resistance. The remaining fibrous portion of the specimen seen in the background of Fig. 10.30 was broken at room temperature to complete the separation of the two specimen halves to permit examination and is still very fibrous indicating that embrittlement occurred only during crack propagation in tension at temperature.

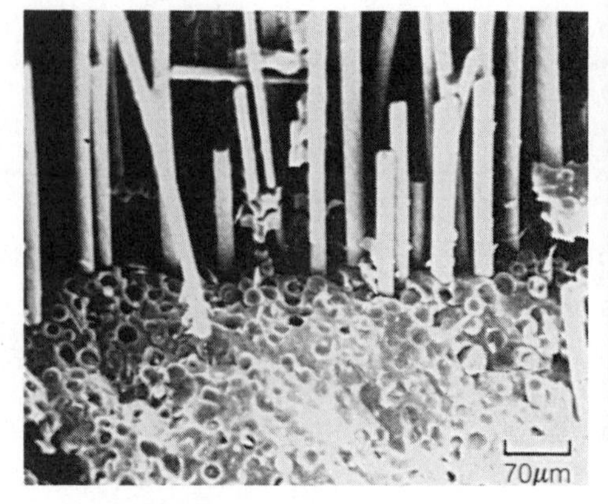

Fig. 10.30 Fracture surface of unidirectional Nicalon fibre reinforced LAS-III glass-ceramic tested in flexure at 900°C in air. After Prewo (1986).

10.4.5 Silicon carbide whisker reinforced composites

While the major emphasis in composite development has been through the use of continuous fibres, whiskers provide an alternative approach. Whiskers can be produced, in some cases, inexpensively and more important, they can be produced with high purity and high strength and stiffness. While a wide variety of whisker compositions have been explored, recent composite fabrication has emphasized the use of SiC derived either from the VLS (vapour–liquid–solid) process (Petrovic *et al.*, 1985) or from

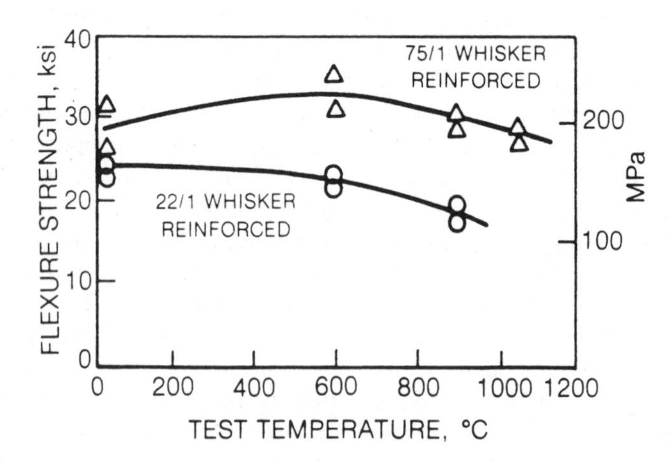

Fig. 10.31 Flexural strength against temperature in air for SiC whisker reinforced LAS-III glass-ceramic matrix composites (30 vol % whiskers). After Layden and Prewo (1985).

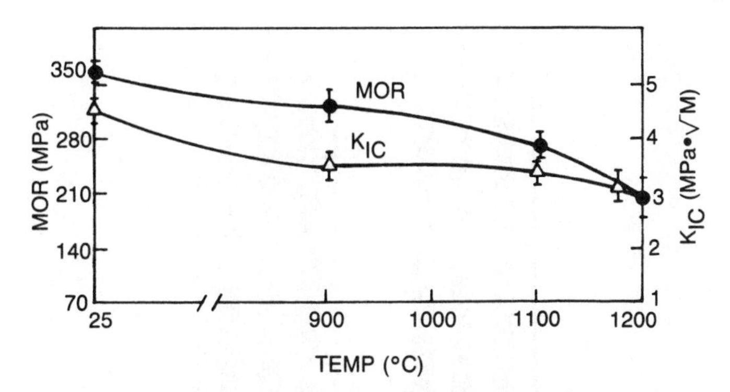

Fig. 10.32 Variation in properties of SiC whisker reinforced Ba-stuffed cordierite matrix composites with temperature. After Gadkaree and Chyung (1986).

other larger scale chemical means such as the conversion of rice hulls to SiC. In either case, the incorporation of these whiskers has succeeded in raising composite strength (Layden and Prewo, 1985; Gadkaree and Chyung, 1986). This strength increase was found to be substantial and could be maintained to very high temperatures using LAS matrices similar to those used with Nicalon yarn (Fig. 10.31) or with even higher temperature matrices (Fig. 10.32). In these cases, however, composite toughness was greater than that of the parent matrix but not nearly as great as that achievable with the Nicalon fibre. This less notable toughening is due to several factors. First, the chemistry of these whiskers did not lead to the formation of carbon rich, low strength, fibre–matrix interfaces during composite fabrication. This lack of a carbon interface did lead to oxidative stability, however. Second, the smaller diameter and shorter length of these fibres was less effective in causing fibre pull out to absorb energy on fracture. Finally, whiskers, while easy to use, are more difficult to distribute uniformly to avoid matrix rich low toughness regions prone to easy crack growth. Despite these drawbacks, whisker reinforced glasses and glass-ceramics offer a group of materials easily fabricated and more performant than most monolithic glass-ceramics and ceramics.

10.5 SUMMARY

From the above review of glass matrix composite development it can be seen that a broad range of material combinations has already been explored. Numerous suitable reinforcing fibres are available, matrix compositions have been identified, and fabrication processes have been demonstrated. Also, in all cases it has been found that these composites must be treated as

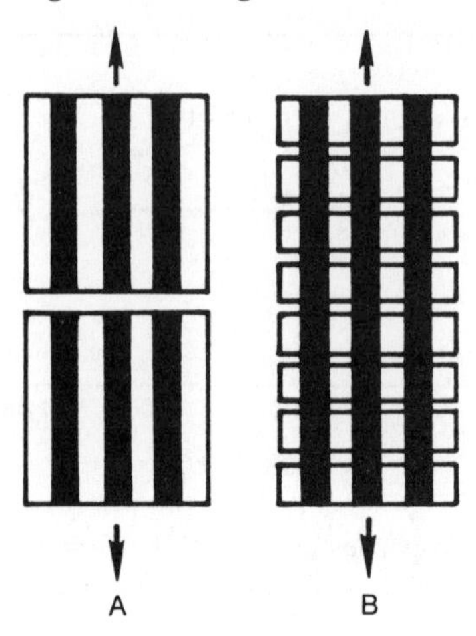

Fig. 10.33 Fracture types for fibre reinforced composites. (A), matrix and fibre fail simultaneously in a strongly bonded composite; (B), matrix failure occurs first in a composite with low fibre–matrix interfacial strength.

three component systems, i.e. fibre, matrix and fibre–matrix interfacial region. It is this last region of transition which appears to control the fracture process in these composites and hence their relative toughness. The two different types of potential composite fracture sequences are shown in Fig. 10.33 for the case where the *in-situ* matrix failure strain ε_m is less than that of the reinforcing fibres ε_f. In case A, Fig. 10.33 total composite failure has occurred when the matrix has failed. A strong fibre–matrix interface has permitted the first cracks, formed in the matrix, to propagate straight through the entire composite. This would typically occur at the matrix failure strain, ε_m, and composite failure would be given by the following expression where σ_f is the fibre strength at a strain equal to that at matrix failure

$$\sigma_c = \sigma_m V_m + \sigma_f' V_f$$
$$\sigma_c = \varepsilon_m(E_m V_m + E_f V_f).$$

A plot of composite tensile strength as a function of fibre volume fraction, V_f, for this type of failure mechanism is shown as line A in Fig. 10.34 for the case where the fibre elastic modulus E_f is greater than the matrix modulus E_m.

If the fibre–matrix interfacial strength is weak enough to prevent cracks in the matrix from propagating into the fibre, it is possible to achieve significantly stronger composites above a critical fibre content V^*. In this case B, Fig. 10.33, matrix cracking can occur throughout the composite without causing immediate failure. Instead composite failure will occur at

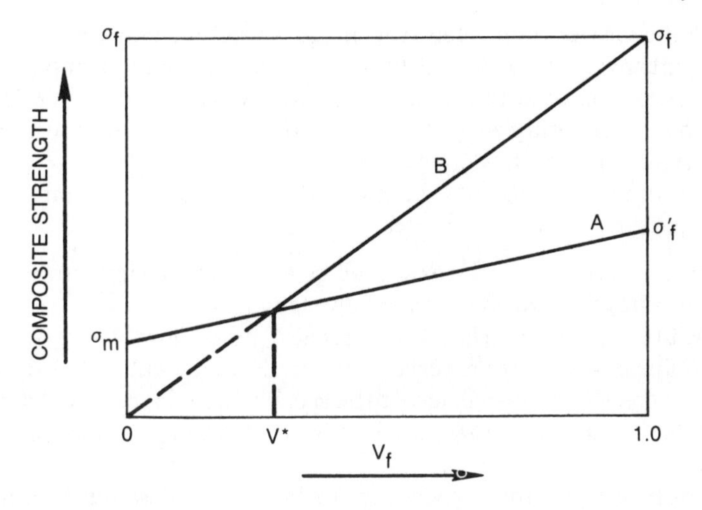

Fig. 10.34 Dependence of glass matrix composite strength on fibre content for cases A and B where matrix failure occurs first; V_f = volume fraction fibre.

the *in-situ* fibre failure strain and composite ultimate axial strength is given by

$$\sigma_c = \sigma_f V_f$$

In the simplest case there is no longer any effective contribution of matrix to this expression because of the general deterioration of the matrix at its failure strain and σ_f is some complex measure of *in-situ* fibre strength that takes strength population and an effective gauge length into account.

It should be noted that composite elastic modulus is affected by this particular sequence of events. For case A the composite elastic modulus can be given by

$$E_0 = E_m V_m + E_f V_f$$

For case B this same formula holds, however, only until matrix failure has taken place. Subsequent to matrix failure the composite effective elastic modulus is decreased progressively as the matrix deteriorates

$$E_0 = E_f V_f$$

Thus, for a high strength B-type composite the total stress–strain curve must be non-linear as shown in Figs 10.10 and 10.25.

Major questions still exist as to the significance of this non-linear stress–strain behaviour and whether the added increment of composite strength denoted by the region between curves A and B of Fig. 10.34 can be used in actual engineering applications. Clearly, in the case where the surrounding environment can attack the reinforcing fibre or the important interfacial region, the answer is no. However, in cases where the environment is not very aggressive it may indeed by possible to take full advantage of the strengths associated with a type B fracture. As has been shown, in some cases these composites may even prove to out-perform their resin matrix counterparts (Table 10.3).

The following points may be concluded as being important for fibre reinforced glass matrix composites:

1. Fibre, matrix and interface must all be tailored to achieve strong, tough composites that are environmentally stable;
2. Low fibre–matrix interfacial strength also implies low off-axis strength;
3. Non-linear stress–strain curves and an elastic modulus that decreases with increasing strain will have to be accounted for in successful designs if composites are to see strain levels in use above ε_m, the matrix failure strain;
4. Flexure tests by themselves are inadequate in describing composite stress–strain behaviour or strength;
5. Environmental stability and composite structural life prediction will be the key issues for successful composite implementation;
6. Non-structural composite performance such as tribology and dimensional stability may provide the keys to first composite usage at minimum risk of failure.

It is anticipated that the potential for fibre reinforced glass matrix composites is very significant and that, at the time of writing, the precise compositions of those materials to be most important in the long run are yet to be developed.

REFERENCES

Aveston, J. (1971) Strength and toughness in fibre reinforced ceramics. *Proc. Conf. on Properties of Fibre Composites*, IPC Science and Technology Press.

Bacon, J., Prewo, K. and Veltri, R. (1978) Glass matrix composites II: alumina reinforced glass. *Proc. Int. Conf. on Composite Materials, Toronto, Canada, 1978*, AIME.

Brennan, J. J. (1985) Interfacial characterization of glass and glass-ceramic matrix-Nicalon SiC fibre composites. *Proc. 21st University Conference on Ceramic Science, Pennsylvania State University, July 1985.*

Brennan, J. J. and Prewo, K. M. (1982) *J. Mater. Sci.*, **17**, 2371.

Crivelli-Visconti, I. and Cooper, G. A. (1969) *Nature*, **221**, 754–5.

Donald, I. W. and McMillan, P. W. (1976) *J. Mater. Sci.*, **11**, 949.

Donald, I. W. and McMillan, P. W. (1977) *J. Mater. Sci.*, **12**, 290.

Fitzer, E. (1978) Fiber reinforced ceramics, *Proc. Int. Symp. Factors in Densification and Sintering of Oxide and Non-Oxide Ceramics, 1978, Japan*, pp. 618–73.

Gadkaree, K. P. and Chyung, K. (1986) *Am. Ceram. Soc. Bull.*, **65** (z), 370–6.

Khanna, V. D. *et al.* (1983) Friction and wear of glass matrix-graphite fiber composites, *Proc. Mechanical Behavior of Metal Matrix Composities* (ed. J. Hock), AIME.

Layden, G. K. and Prewo, K. M. (1985) *Study of SiC Whisker Reinforced Glass and Glass-Ceramic Matrix Composities*, Office of Naval Research Contract N00014-84-C-0386, Final Report, June.
Linger, K. R. and Pratchett, A. G. (1977) *Composites*, July, p. 139.
Lucas, J. P., Toth, L. E. and Gerberich, W. W. (1980) *J. Am. Ceram. Soc.*, **63**, 280.
Minford, E. J. and Prewo, K. M. (1985a) Fatigue behavior of SiC fiber reinforced LAS glass ceramic. *Proc. 21st University Conference on Ceramic Science, Pennsylvania State University, July 1985.*
Minford, E. and Prewo, K. (1985b) Friction and wear of graphite fibre reinforced glass matrix composites. *Wear*, to be published.
Petrovic, J. J., Milewski, J. V., Rohr, D. L. and Gac, F. D. (1985) *J. Mater. Sci.*, **20**, 1167–70.
Prewo, K. M. (1979) Development of a new dimensionally and thermally stable composite. *Proc. Special Topics in Advanced Composites Mtg, El. Segundo, Calif., 1979.*
Prewo, K. M. (1982) *J. Mater. Sci.*, **17**, 3549.
Prewo, K. M. (1986) Tension and flexural strength of SiC fibre reinforced glass ceramics, *J. Mater. Sci.*, to be published.
Prewo, K. M. and Bacon, J. F. (1978) Glass matrix composites – I, Graphite fiber reinforced glass. *Proc. 2nd Int. Conf. on Composites* (ed. B. Noton), AIME.
Prewo, K. M., Bacon, J. F. and Dicus, D. L. (1979a) *SAMPE Q.*, **10**, 42.
Prewo, K. M., Bacon, J. F. and Thompson, E. R. (1979b) Graphite fibre reinforced glass. *Proc. Conf. Advanced Fibers and Composites for Elevated Temperatures* (eds I. Ahmad and B. Noton), AIME.
Prewo, K. M. and Brennan, J. J. (1980) *J. Mater. Sci.*, **15**, 463.
Prewo, K. M. and Brennan, J. J. (1982) *J. Mater. Sci.*, **17**, 1201.
Prewo, K. M., Brennan, J. J. and Layden, G. K. (1986) *Am. Ceram. Soc. Bull.*, **65**, 305–22.
Prewo, K. M. and Minford, E. J. (1984) *Proc. SPIE Int. Soc. Opt. Eng.*, **505**, Aug.
Prewo, K. M. and Minford, E. J. (1985) *SAMPE J.*, **21–2**, March.
Prewo, K. M. and Thompson, E. R. (1981) *Research on Graphite Reinforced Glass Matrix Composites*, NASA Contract Report 165711, May.
Levitt, S. R. (1973) *J. Mater. Sci.*, **8**, 793.
Phillips, D. C. (1974) *J. Mater. Sci.*, **9**, 1847.
Phillips, D. C., Sambell, R. A. and Bowen, D. H. (1954) *J. Mater. Sci.*, **7**, 1454.
Sahebkar, M., Schlichting, J. and Schubert, P. (1978) *Ber. Dtsch. Keram. Ges.*, **55**, 265–8.
Sambell, R. A., Bowen, D. and Phillips, D. C. (1972a) *J. Mater. Sci.*, **7**, 663.
Sambell, R. A., Briggs, A., Phillips, D. C. and Bowen, D. H. (1972b) *J. Mater. Sci.*, **7**, 676.
Sambell, R. A., Phillips, D. C. and Bowen, D. H. (1974) *The Technology of Carbon Reinforced Glasses and Ceramics*, Harwell Report AERE-R-7612, Feb.
Schlichting, J. (1980) Ceramics from metalloxides. Science of ceramics, *Proc. 10th Int. Conference, German Ceramic Society, 1980.*
Siefert, A. C. (1971a) *Fiber Ceramic Composites and Method of Producing Same*, US Patent 3 575 789 April 20 (applied Dec. 1966).
Siefert, A. C. (1971b) *Fiber Reinforced Ceramics*, US patent 3 607 608, Sept. 21, (applied Jan. 1966).
Siefert, A. C. (1972) *Method of Making Impact Resistant Inorganic Composites*, US Patent 3 702 240, Nov. 7 (applied April 1969).
Siefert, A. C. (1975) *Impact Resistant Inorganic Composites*, US Patent 3 869 335, March 4 (applied April 1969).
Strife, J. R. and Prewo, K. M. (1982) *J. Mater. Sci.*, **17**, 65

Thompson, E. R. and Prewo, K. M. (1980) Glass reinforced by graphite, silicon carbide and alumina fibres, AIAA Paper 80-0756-CP, *Proc. 21st Structures, Structural Dynamics and Materials Conference, May 1980*.

Yajima, S., Okamura, K., Hayashi, J. and Omori, M. (1976) *J. Am. Ceram. Soc.*, **58**, 324.

Glass systems index

Al–Si–O–N 108
AlF_3 156
Al_2O_3–SiO_2 15
As_2S_3 181

Ba–Si–Al–O–N 108
BaO–Al_2O_3–SiO_2 339
BaO–MgO–Al_2O_3–SiO_2 339
BaO–SiO_2 73, 75, 93
Be–Si–Al–O–N 108
BeF_2 156, 163, 181

Ca–Si–Al–N 108, 134
Ca–Si–O–N 108
CaO–SiO_2 75, 204
CdF_2 160
CdF_2–$CdCl_2$–$BaCl_2$ 174
Cs_2O–B_2O_3–SiO_2 208
Cs_2O–SiO_2 12, 25

FeF_3 160

GeO_2 181

HfF_4 158
HfF_4–BaF_2–ThF_4 169

K–P–O–N 109
KI–$BiCl_2$ 160
K_2O–B_2O_3–SiO_2 208
K_2O–SiO_2 12

La–Si–Al–O–N 108, 145
Li–P–O–N 109
Li–Si–Al–O–N 108, 132
Li_2O–Al_2O_3–SiO_2 89
Li_2O–B_2O_3–SiO_2 208
Li_2O–BaO–SiO_2 81
Li_2O–SiO_2 12, 23, 97
Ln–Si–Al–O–N (Ln = Nd, La, Sc) 108, 145

Mg–Si–Al–O–N 107, 134, 142
Mg–Si–O–N 108, 138
MgO–Al_2O_3–SiO_2 229, 237, 250
MnF_2 160

Na–Ca–Si–O–N 108, 127, 132
Na–P–O–N 109, 125, 132
Na–Si–B–O–N 109
Na–Si–O–N 127
Na–(Ca, Sr, Ba)–P–O–N 109
Na_2O–B_2O_3–SiO_2 208
Na_2O–BaO–SiO_2 81
Na_2O–CaO–SiO_2 73, 82, 89
Na_2O–SiO_2 12, 24, 74, 204
Nd–Si–Al–O–N 108, 145

P_2O_5–SiO_2 15
PbO–SiO_2 14, 25

R_2O–B_2O_3–SiO_2 (R = Li, Na, K, Rb, Cs) 208
R_2O–SiO_2 12
Rb_2O–B_2O_3–SiO_2 208
Rb_2O–SiO_2 12

Sc–Si–Al–O–N 108, 145
SiO 35
SiO_2 9

$ThCl_4$ 160
ThF_4 158, 160

UF_4 160

YbF_4 160

$ZnCl_2$ 158, 163, 181
$ZnCl_2$–KBr–$PbBr_2$ 176
ZnF_2 160
ZnF_2–BaF_2–ThF_4 170
ZnF_2–BaF_2–YbF_3–ThF_4 171

ZrF_4–BaF_2 156, 158, 162
ZrF_4–BaF_2–AlF_3 158
ZrF_4–BaF_2–AlF_3–NaF 158
ZrF_4–BaF_2–$BaCl_2$ 170
ZrF_4–BaF_2–CaF_2 158
ZrF_4–BaF_2–LaF_3–AlF_3–NaF 158, 181
ZrF_4–BaF_2–ThF_4 174

Glass-ceramic phases and other compounds

AlN 228
Al_2O_3 33, 228
$Al_6Si_2O_{13}$ 141, 237

$BaSi_2Al_2O_8$ 146
$BaSi_2Al_2O_8(OH)$ 148
$BaSi_2O_5$ 73
$Ba_3Si_5O_{13}$ 75
$BaTiO_3$ 273
$Ba_2TiGe_2O_8$ 279, 283
$Ba_2TiSi_2O_8$ 279, 281
$BaZr_2F_{10}$ 158
BeO 228

$CaAl_2O_4$ 33
$CaAl_2Si_2O_8$ 78
$CaSiO_3$ 75
$Ca_2(Si, Al)_3(O, N)_7$ 146

$LaAlO_3$ 146
$LaAl_{11}O_{18}$ 146
$La_2Si_2O_7$ 146
$La_4Si_3Al_3(O, N)$ 146
$La_4Si_4Al_3(O, N)$ 146
$Li_2B_4O_7$ 279, 282
Li_2GeO_3 279
$Li_2Ge_2O_5$ 279
Li_3PO_4 279
Li_2SiO_3 279
$Li_2Si_2O_5$ 89, 274, 279, 283
$LiSiAlO_4$ 146
$LiTaO_3$ 274
$LnAlO_3$ (Ln = Nd, La) 146
$LnAl_{11}O_{18}$ 146
$Ln_2Si_2O_7$ 146
$Ln_4Si_3Al_3(O, N)$ 146
$Ln_4Si_4Al_3(O, N)$ 146

$MgAl_2O_4$ 34, 237
$Mg_2Al_4Si_5O_{18}$ 237
$MgSiO_3$ 237

$NaNbO_3$ 274
Na_2SiO_3 279
$Na_2Si_2O_5$ 78, 279
$NdAlO_3$ 146
$NdAl_{11}O_{18}$ 146
$Nd_2Si_2O_7$ 146
$Nd_4Si_3Al_3(O, N)$ 146
$Nd_4Si_4Al_3(O, N)$ 146

PbB_4O_7 279
$Pb_5Ge_3O_{11}$ 274
$PbTiO_3$ 274

$Si_{12}Al_{18}O_{33}N_9$ 138, 142
$Si_{3-x}Al_xO_xN_{4-x}$ (β'Sialon) 32, 115
$Si_2Mg_xAl_{4-x}O_{4+x}$ (β''Sialon) 32, 138, 145
Si_3N_4 150
Si_2N_2O 32
SiO_2 181, 237, 351
SrB_4O_7 279
$Sr_2TiSi_2O_8$ 279, 282

$Y_3Al_5O_{12}$ 141
Y_2SiAlO_5N 141
$YSiO_2N$ 129, 141

ZnP_2O_5 279
Zn_2SiO_4 234
ZrO_2 237

Subject index

Abbé number 188
Acid leaching 208
Acoustic impedance 301
Activation
  enthalpy 67
  free energy 62
Adams catalyst 327
Adsorption isotherms 212
Alamosite 25
Alkali
  phosphosilicates 17
  silicates 12, 23
Alkaline-earth silicates 16
Alumina
  fibres 350
  gels 34
  β-alumina 146
Aluminium
  oxides 33
  phosphates 33
Aluminosilicates 16, 339
Amorphous phase separation 91
Antireflection coatings 216
Asymmetry parameter 1
Auger analysis 356

Bayerite 33
Ba-stuffed cordierite 363
BET theory 212
Binary distribution 8
Binary glasses 12
Binodal phase separation 93, 97, 206
Boehmite 33
Bond angle 9, 10
  distribution 25
Boron fibre reinforced glass 351
Borosilicate glass 228, 339
Bose–Einstein function 176
Bridging oxygens 6, 8
Brillouin scattering 172, 178

Calcium aluminosilicates 17
Carbon fibre
  reinforced composites 344
  reinforced glass 340
Carbon rich interface 357
Catalyst supports 222
Cements 33
Chalcogenide glasses 41, 43, 55, 167
Chemical
  bonding 321
  durability 134, 207
  properties 134
  shift 1, 5, 6, 20
Classical nucleation theory 60
Coatings 194
Colloidal dispersion 332
Composite
  cross-plied composite 355
  fabrication 338
  properties 359
    elastic modulus 365
    environmental stability 361
    fatigue 359
    strength 348, 350
    tensile strength 364
Conditional glass formers 157
Conduction mechanisms 324
Conductor tracks 253
Congruent melting point 75
Contact angle 60, 84, 323, 325
Cordierite 22, 250, 352

Corning
  015 glass 321
  7740 Vycor glass 217
  7930 high silica glass 339
  7930 Vycor glass 210, 331
Covalent glasses 55
Cristobalite 234, 237
Critical nucleus 63
Cross linking 129
Cross-polarization 11
Crystal nucleation 59
Crystallization 59, 142, 158, 226, 230
  of glasses 280
  products 138
  temperature 285
Crystallographic point groups 277
Cymrite 148

Density 131, 192
Desalination 210
Devitrification 22, 59
Dialysis 223
Die pressing 247
Dielectric
  breakdown strength 267
  constant 134, 227, 232, 235, 240, 276, 290, 292, 293, 301
  losses 134, 233, 236, 241
  properties 232, 235, 240, 288
Differential scanning calorimetry, DSC 158
Diffusion coefficient, D 61
Dipolar
  broadening 3
  interaction 1
Dispersion 187, 190
Dissipation factor 290, 292, 293
Domain structure 276
Durability 193

Elastic coefficients 276
Electric field gradient 2,7
Electrical conductivity 134, 207
Electrically reversible phases 319
Electromechanical
  coupling coefficients 298
  properties 299
Electron
  diffraction 145
  micrograph 262
  microscopy 64
  probe micro-analysis, EPMA 262
  scattering 43
Electronegativity 126
Electrophoretic deposition 258
Enzyme immobilization 222
β-eucryptite 146
Eutectic temperatures 113
Excitons 177
Extended X-ray absorption fine structure, EXAFS 42
Extrinsic losses 182
Extrinsic scattering 186
E-glass 210

Fabrication 195
Far infra-red, FIR 167
Fatigue 359
Ferroelectric glass-ceramics 273
Fibre
  pull-out 363
  stability 349
  volume fraction 364
Fibre–matrix interface 361
Fibre-reinforced composites 336
Fibre-reinforced glasses and glass-ceramics 336
Fictive temperature, T_f 178
Firing 249, 258
Flexural strength 151, 351, 355
Fluoroaluminates 157
Fluoroberrylate glass 157
Fluorozirconates 21
  fibres 182
Forsterite 237
Fourier transform techniques 49
Fracture
  energy 345
  surfaces 351, 356
  toughness 133, 195, 350, 355
Free energy 71, 80
Free energy–composition curve 205
Fresnoite 284
Fulcher parameters 91
Fused silica 228

Gas adsorption 211
Gel layer 328, 331
Gibbs energies of formation 109
Gibbs free energy 80
Glass and glass-ceramic matrices 339
Glass
  formation 156
  melting 279
    temperature 285
  preparation 229
  structure 7, 50, 156, 207
  transition temperature, T_g 78, 133
Glass-ceramics 59, 226, 272, 337
  coatings 258
    properties 267
  composites 308
  properties 232, 235, 241
  substrates 245
Glass-ceramic-coated metal substrates 253
Glass-forming systems 278
Glass–metal interface 260, 321
Glass–solution interface 328
Gradient–index surface 217
Grain boundary glass 107
Grain boundary glass crystallization 150
Grain size 239

Halide fibres 195
Halide glasses 21, 156
Heat treatment 227, 230, 284
Heavy metal fluoride glasses, HMF 157
Heterogenous nucleation 62, 83
Hexacelsian 148
High performance fibres 343
High temperature insulation 222
Hollow glass fibre 213
Homogenuous nucleation 60
Hot pressing 338
Hydrolysis 160, 329
Hydrophones 301
Hydrostatic voltage coefficient 299
Hydroxyl groups 89
Hydroxyl ions 184

Immiscibility boundary 92
Incongruent melting point 75
Induction time 67
Inelastic scattering 48
Infra-red
  absorption 166, 167, 174
  absorption edge 165
  optical fibres 21
  reflectivity 168
  spectra 126
  transmission 164
Injection moulding 342
Interface chemistry 354
Interfacial free energy 60
Intermediate phases 141
Intermediates 52
Intrinsic scattering 177
Ion exchange 328
Ion-selective electrodes 316
Isostatic pressing 247
Isothermal crystallization 287
Isotopes 7, 36, 37
Isotropic shift 1

Janecke prism 112, 139

Landau–Placzek ratio 178
Liquid phase sintering 107
Lithium disilicate 64

Magic angle spinning, MAS 4
  nuclear magnetic resonance, NMR 1, 124
Magnesium aluminoborates 23
Magnesium aluminosilicates 22
Matrix microcracking 359
Matrix transfer moulding 342
Mechanical
  properties 241
  quality factor 298
  strength 281
Melting techniques 150
Metalalkoxides 341, 342
Metallic glasses 52
Metallic nucleating agents 83
Metal-coated fibres 351
Metastable β'' phase 145
Metastable crystallization 145

Metastable immiscibility 203
Microhardness 131, 151
Microporous glasses 203, 207
Microstructure 141, 146, 207, 239, 261, 284, 288
Mineral glasses 20
Minimum loss V-curves 180
Modifier ion 129
Modifier oxide 116
Modulus of rupture 257
Molecular absorptions 184
Mullite 113, 141, 237
Multilayer substrates 253
Multiphonon adsorption 173

Nernst equation 319, 331
Network formers 50
Network modifying ions 157
Network stabilizing fluorides 157
Network structure 157
Nicalon 337, 353, 355
  monofilaments 352
Non-bridging oxygen 8
Non-destructive testing 303
Non-metallic nucleating agents 87
Nuclear magnetic resonance, NMR 1
Nuclear waste disposal 221
Nucleating agents 63, 148, 232, 239
Nucleation 59, 226, 230, 284
  kinetics 59
  rate 64

Optical
  fibres 164
  microscopy 64
  properties 156
  transmission 164
  waveguides 189, 216, 218
Ortho and pyrophosphates 19
Osmotic pressure 210
Oxidation 135, 151
Oxidation–reduction reactions 110, 324
Oxide fibre reinforced composites 350
Oxynitride
  glass
    crystallization 138
    preparation 116
    structure 123, 128
    systems 108
  glasses 22, 106
    properties 129, 151
  glass-ceramics 137
Particle size analyser 248
Phase separation 8, 28, 114, 159, 203
Phosphorous oxynitride glasses 118
Photochemical machining 83
Piezoelectric
  coefficients 294, 306
  constants 276
  devices 272
  noise 312
  properties 297
Polar glass-ceramics 272, 274, 275
Polar orientation 277, 307
Polytypoids 31
Pore characterization 211
Pore size distribution 211
Powder techniques 246
Preferred orientation 275, 286, 288
Pyrex glass 207
Pyroelectric
  coefficients 276, 281, 293, 294, 295, 306
  detectors 293
  devices 272
  figures of merit 293, 296, 297
  noise 312
  properties 292

Quadrupolar nuclei 1

Radiation resistance 185
Raman
  scattering 178
    stimulated, SRS 172
  spectra 168, 171
Random network 51
Rayleigh scattering 165, 178
Reactive atmosphere processing, RAP 161
Redox reactions 324
Refractive index 131, 187
Refractory foams 221
Reinforced borosilicate glasses 347
Relaxation time 29

Resistance thermometers 219
Reverse osmosis 210, 213

Scanning electron microscopy 64, 145, 147, 148
Scattering
  loss 165
    coefficient 178
  Raman 178
    stimulated 172
  Rayleigh 165, 178
Screen printing 244, 247, 253, 258
Sellmeier curve 190
Sessile drop experiments 323, 324
Sialon
  ceramics 150
  β'-sialon 113
Silicon
  carbide 33
    whisker reinforced composites 362
    yarn reinforced composites 353
  dioxide 9
  oxynitrides 31, 113
Sintering 249
  additives 111
Site binding model 331
Small angle X-ray scattering, SAXS 92
Sodium aluminosilicates 17
Sol-gel 34, 52, 89, 118, 258, 267
Specific heat 293
Spherulites 65, 287
Spinels 34, 237
Spinning sidebands 4, 5
Spinodal
  curve 93, 97
  phase separation 206
Spin-lattice relaxation 7
Stable immiscibility 203
Statistical distribution 8
Steady state nucleation 69
Step index fibre 218
Stishovite 20
Stokes–Einstein relation 61
Stress–strain curve 346
Structure determination 1
Substrate materials 226
  properties 228
Superconducting materials 219
Supercooled liquid 60
Suprasil 11
Surface acoustic wave (SAW) devices 303
Surface
  charge 332
  crystallization 247
  nucleation 64, 159, 286, 306
  treatment 214
Synchrotron radiation 55
Szigeti equation 165

Tape casting 227, 247, 253
Telecommunication fibres 164
Temperature coefficient of resonance 298
Temperature gradient crystallization 286
Tensile stress–strain curve 358
Tensor properties 276
Ternary glasses 16
Tetrahedral groups 124
Theory of EXAFS 45
Thermal
  conductivity 227, 253
  diffusivity 293
  dissipation 267
  expansion 132, 192, 228, 230, 235, 237, 243, 244, 251, 254, 349
  properties 243
Thick film
  circuitry 244
  inks 253, 268
  pH electrodes 317
Thin film circuitry 246
Total intrinsic losses 165
Transient nucleation 62
Transition metal and rare earth absorption 182
Transmission electron microscopy 65, 145
Transverse rupture strengths 133
Turnbull parameter 77, 85

Ultraviolet
  absorption 176
    edge 165

Unidirectional composite 355
Urbach
  edge 165
  rule 177

van der Waals bonding 321
Viscosity 69,83, 133, 192, 207
Vitreous enamels 254
Vycor 7930 glass 209
Vycor process 208

Willemite 234
N-wollastonite 141

X-phase 113, 142
X-ray
  absorption 41
    coefficients 45
    near edge structure, XANES 41
    spectroscopy 41
  diffraction 280
  photoelectron spectra, XPS 127

Young's modulus 131
Yttrium aluminium garnet, YAG 113, 141

Zachariesen's rules 156
Zeta potential 332
Zirconia 237

PL 71732
H1B5.
£40.00